Measurement and Control Basics

Fourth Edition

Measurement and Control Basics

Fourth Edition

Thomas A. Hughes

Notice

The information presented in this publication is for the general education of the reader. Because neither the author nor the publisher have any control over the use of the information by the reader, both the author and the publisher disclaim any and all liability of any kind arising out of such use. The reader is expected to exercise sound professional judgment in using any of the information presented in a particular application.

Additionally, neither the author nor the publisher have investigated or considered the effect of any patents on the ability of the reader to use any of the information in a particular application. The reader is responsible for reviewing any possible patents that may affect any particular use of the information presented.

Any references to commercial products in the work are cited as examples only. Neither the author nor the publisher endorses any referenced commercial product. Any trademarks or tradenames referenced belong to the respective owner of the mark or name. Neither the author nor the publisher makes any representation regarding the availability of any referenced commercial product at any time. The manufacturer's instructions on use of any commercial product must be followed at all times, even if in conflict with the information in this publication.

Printed in the United States of America.
10 9 8 7 6 5 4

ISBN-10: 1-55617-916-2
ISBN-13: 978-1-55617-916-7

ISA
67 Alexander Drive
P.O. Box 12277
Research Triangle Park, NC 27709

Library of Congress Cataloging-in-Publication Data

Hughes, Thomas A.
Measurement and control basics / Thomas A. Hughes. -- 4th ed.
p. cm.
Includes index.
ISBN 1-55617-764-X
1. Process control--Instruments. 2. Measuring instruments. I. Title.
TS156.8.H78 2007
670.42'7--dc22

2007029105

ISA Resources for Measurement and Control Series

- *Measurement and Control Basics, 4th Edition* (2007)
- *Industrial Flow Measurement, 3rd Edition* (2005)
- *Programmable Controllers, 4th Edition* (2005)
- *Control Systems Documentation: Applying Symbols and Identification, 2nd Edition* (2005)
- *Industrial Data Communications, 3rd Edition* (2002)
- *Control System Safety Evaluation and Reliability, 2nd Edition* (1998)

THIS BOOK IS DEDICATED TO

my wife Ellen, my daughter Audrey, and my grandson Ian
for their love

Contents

About the Author

Thomas A. Hughes, a Senior Member of ISA, has 30 years of experience in the design and installation of instrumentation and control systems, including 20 years in the management of instrumentation and control projects for the process and nuclear industries. He is the author of two books: *Measurement and Control Basics, 4th Edition, (2007)* and *Programmable Controllers, 4th Edition, (2005),* both published by ISA.

Mr. Hughes received a B. S. in engineering physics from the University of Colorado, and an M.S. in control systems engineering from Colorado State University. He holds professional engineering licenses in the states of Colorado and Alaska, and has held engineering and management positions with Dow Chemical, Rockwell International, EG&G Rocky Flats, Topro Systems Integration, and the International Atomic Energy Agency.

Mr. Hughes has taught numerous courses in electronics, mathematics, and instrumentation systems at the college level and in industry. He lives in Arvada, Colorado.

Preface

The fourth edition of *Measurement and Control Basics* is a thorough and comprehensive treatment of the basic principles of process control and measurement. It is designed for engineers, technicians, management, and sales personnel who are new to process control and measurement. It is also valuable as a concise and easy-to-read reference source on the subject.

This new edition provides expanded coverage of pressure, level, flow, temperature, analytical measurement, and process control computers. Material on the proper tuning of control loops was added to Chapter 1, and expanded coverage of control loops was added to Chapter 2. Chapter 3 includes a more complete discussion of electrical and electronic fundamentals needed in process control and instrumentation.

The discussion of the basic principles underlying pressure measurement has been expanded to include a discussion of sensor characteristics and potentiometric-type pressure sensors. Extensive coverage was added on typical pressure transmitter applications. The discussion on level measurement has been increased with the addition of several common level instruments and switches such as displacers, tape floats, microwave, and radar. The chapter on temperature measurement has been improved by adding new illustrations and a section on radiation pyrometers. Coverage of analytical measurement and control in Chapter 8 was increased by the addition of a section on the principles of electromagnetic radiation and its application to analytical measurement. Three sections were also added to Chapter 8 on photoconductive sensors, photomultiplier tubes, and turbidity analyzers.

Chapter 9 on flow measurement contains new coverage on Reynolds Number and fluid flow profiles. The discussion of the basic principles of fluid flow has been expanded and improved in Chapter 9. A discussion on types of control valves and control valve actuators was added to Chapter 10 and the section on control valve sizing was expanded and improved.

All of the chapters have been supplemented with new or improved example problems and exercises. Most of the illustrations in the book have been revised and improved.

1 Introduction to Process Control

Introduction

To study the subject of industrial process control effectively, you must first gain a general understanding of its basic principles. To present these control principles clearly and concisely, an intuitive approach to process control is used. First, however, some basic definitions and concepts of process control are presented.

Definition of Process Control

The operations that are associated with process control have always existed in nature. Such "natural" process control can be defined as any operation that regulates some internal physical characteristic that is important to a living organism. Examples of natural regulation in humans include body temperature, blood pressure, and heart rate.

Early humans found it necessary to regulate some of their external environmental parameters to maintain life. This regulation could be defined as "artificial process control" or more simply as "process control," as we will refer to it in this book. This type of process control is accomplished by observing a parameter, comparing it to some desired value, and initiating a control action to bring the parameter as close as possible to the desired value. One of the first examples of such control was early man's use of fire to maintain the temperature of his environment.

The term *automatic process control* came into wide use when people learned to adapt automatic regulatory procedures to manufacture products or process material more efficiently. Such procedures are called automatic because no human (manual) intervention is required to regulate them.

All process systems consist of three main factors or terms: the manipulated variables, disturbances, and the controlled variables (Figure 1-1). Typical manipulated variables are valve position, motor speed, damper position, or blade pitch. The controlled variables are those conditions—such as temperature, level, position, pressure, pH, density, moisture content, weight, and speed—that must be maintained at some desired value. For each controlled variable there is an associated manipulated variable. The control system must adjust the manipulated variables so the desired value or "set point" of the controlled variable is maintained despite any disturbances.

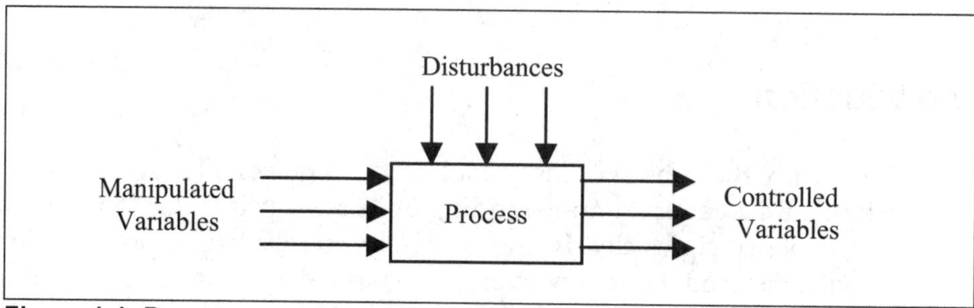

Figure 1-1. Process control variables

Disturbances enter or affect the process and tend to drive the controlled variables away from their desired value or set point condition. Typical disturbances include changes in ambient temperature, in demand for product, or in the supply of feed material. The control system must adjust the manipulated variable so the set point value of the controlled variable is maintained despite the disturbances. If the set point is changed, the manipulated quantity must be changed to adjust the controlled variable to its new desired value.

For each controlled variable the control system operator selects a manipulated variable that can be paired with the controlled variable. Often the choice is obvious, such as manipulating the flow of fuel to a home furnace to control the temperature of the house. Sometimes the choice is not so obvious and can only be determined by someone who understands the process under control. The pairing of manipulated and controlled variables is performed as part of the process design.

Elements of a Process Control System

Figure 1-2 illustrates the essential elements of a process control system. In the system shown, a level transmitter (LT), a level controller (LC), and a control valve (LV) are used to control the liquid level in a process tank. The purpose of this control system is to maintain the liquid level at some prescribed height (H) above the bottom of the tank. It is assumed that the rate of flow into the tank is random. The level transmitter is a device that *measures* the fluid level in the tank and converts it into a useful measurement signal, which is sent to a level controller. The level controller *evaluates* the measurement, compares it with a desired set point (SP), and produces a series of corrective actions that are sent to the control valve. The valve *controls* the flow of fluid in the outlet pipe to maintain a level in the tank.

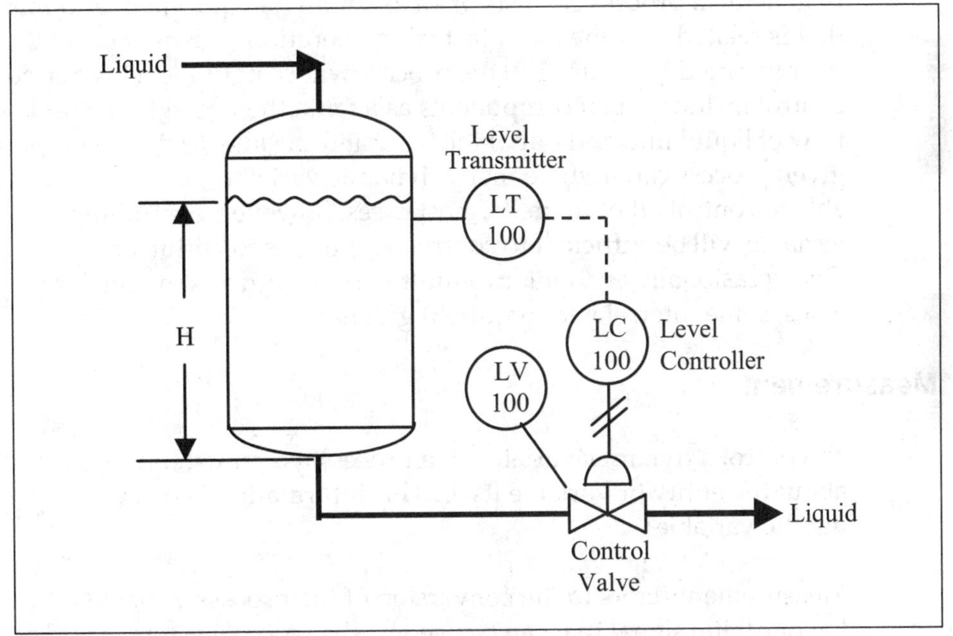

Figure 1-2. Process level control: Example

Thus, a process control system consists of four essential elements: *process, measurement, evaluation,* and *control*. A block diagram of these elements is shown in Figure 1-3. The diagram also shows the disturbances that enter or affect the process. If there were no upsets to a process, there would be no need for the control system. Figure 1-3 also shows the input and output of the process and the set point used for control.

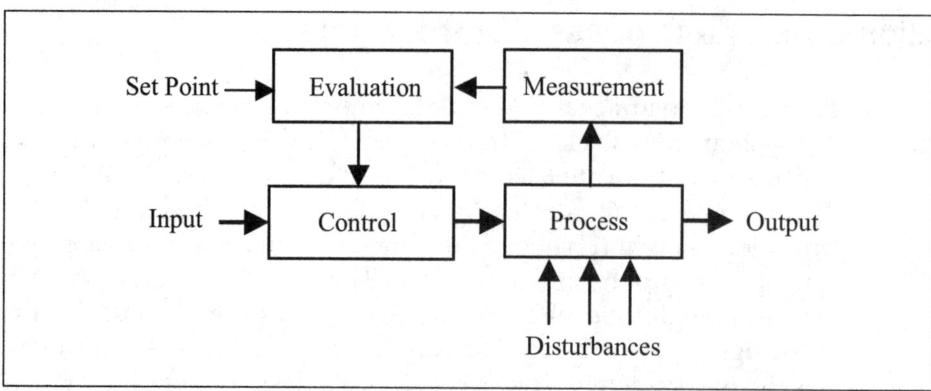

Figure 1-3. Four elements of a control system

Process

In general, a process consists of an assembly of equipment and material that is related to some manufacturing operation or sequence. In the example presented in Figure 1-2, the process whose liquid level is placed under control includes such components as a tank, the liquid in the tank, the flow of liquid into and out of the tank, and the inlet and outlet piping. Any given process can involve many dynamic variables, and it may be desirable to control all of them. In most cases, however, controlling only one variable will be sufficient to control the process to within acceptable limits. One occasionally encounters a multivariable process in which many variables, some interrelated, require regulation.

Measurement

To control a dynamic variable in a process, you must have information about the entity or variable itself. This information is obtained by measuring the variable.

Measurement refers to the conversion of the process variable into an analog or digital signal that can be used by the control system. The device that performs the initial measurement is called a *sensor* or *instrument*. Typical measurements are pressure, level, temperature, flow, position, and speed. The result of any measurement is the conversion of a dynamic variable into some proportional information that is required by the other elements in the process control loop or sequence.

Evaluation

In the evaluation step of the process control sequence, the measurement value is examined, compared with the desired value or set point, and the amount of corrective action needed to maintain proper control is determined. A device called a *controller* performs this evaluation. The controller can be a pneumatic, electronic, or mechanical device mounted in a control panel or on the process equipment. It can also be part of a computer control system, in which case the control function is performed by software.

Control

The control element in a control loop is the device that exerts a direct influence on the process or manufacturing sequence. This final control element accepts an input from the controller and transforms it into some proportional operation that is performed on the process. In most cases, this final control element will be a control valve that adjusts the flow of fluid in a process. Devices such as electrical motors, pumps, and dampers are also used as control elements.

Process and Instrumentation Drawings

In the measurement and control field, a standard set of symbols is used to prepare drawings of control systems and processes. The symbols used in these drawings are based on the standard ISA-5.1-1984 (R1992) Instrumentation Symbols and Identification, which was developed by ISA. A typical application for this standard is process and instrumentation diagrams (P&IDs), which show the interconnection of the process equipment and the instrumentation used to control the process. A portion of a typical P&ID is shown in Figure 1-4.

In standard P&IDs, the process flow lines, such as process fluid and steam, are indicated with heavier solid lines than the lines that are used to represent the instrument. The instrument signal lines use special markings to indicate whether the signal is pneumatic, electric, hydraulic, and so on. Table A-1 in Appendix A lists the instrument line symbols that are used on P&IDs and other instrumentation and control drawings. In Figure 1-4, two types of instrument signals are used: double cross-hatched lines denote the pneumatic signals to the steam control valve and the process outlet flow control valve, and a dashed line is used for the electrical control lines between various instruments. In process control applications, pneumatic signals are almost always 3 to 15 psi (i.e., pounds per square inch, gauge pressure), and the electric signals are normally 4 to 20 mA (milliamperes) DC (direct current).

A balloon symbol with an enclosed letter and number code is used to represent the instrumentation associated with the process control loop. This letter and number combination is called an instrument identification or instrument *tag number*.

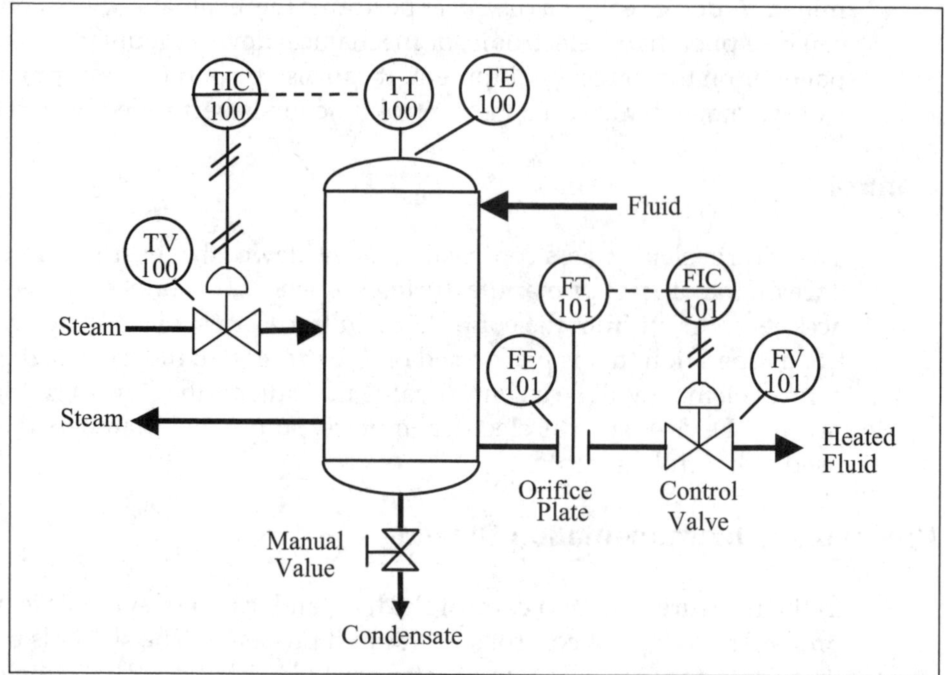

Figure 1-4. P&ID: Example

The first letter of the tag number is normally chosen so that it indicates the measured variable of the control loop. In the sample P&ID shown in Figure 1-4, T is the first letter in the tag number that is used for the instruments in the temperature control loop. The succeeding letters are used to represent a readout or passive function or an output function, or the letter can be used as a modifier. For example, the balloon in Figure 1-4 marked TE represents a temperature element and that marked TIC is a temperature-indicating controller. The line across the center of the TIC balloon symbol indicates that the controller is mounted on the front of a main control panel. No line indicates a field-mounted instrument, and two lines mean that the instrument is mounted in a local or field-mounted panel. Dashed lines indicate that the instrument is mounted inside the panel.

Normally, sequences of three- or four-digit numbers are used to identify each loop. In our process example (Figure 1-4), we used loop numbers 100 and 101. Smaller processes use three-digit loop numbers; larger processes or complex manufacturing plants may require four or more digits to identify all the control loops.

Special marks or graphics are used to represent process equipment and instruments. For example, in our P&ID example in Figure 1-4, two parallel lines represent the orifice plate that is used to detect the discharge flow from the process heater. The two control valves in the figure also use a special symbol. See Appendix A for a more detailed discussion of the instrumentation and process symbols that are used on P&IDs.

General Requirements of a Control System

The primary requirement of a control system is that it be reasonably stable. In other words, its speed of response must be fairly fast, and this response must show reasonable damping. A control system must also be able to reduce the system error to zero or to a value near zero.

System Error

The system error is the difference between the value of the controlled variable set point and the value of the process variable maintained by the system. The system error is expressed in equation form by the following:

$$e(t) = PV(t) - SP(t) \tag{1-1}$$

where

$e(t)$ = system error as a function of time (t)
$PV(t)$ = the process variable as a function of time
$SP(t)$ = the set point as a function of time

System Response

The main purpose of a control loop is to maintain some dynamic process variable (pressure, flow, temperature, level, etc.) at a prescribed operating point or set point. System response is the ability of a control loop to recover from a disturbance that causes a change in the controlled process variable.

There are two general types of good response: underdamped (cyclic) and damped. Figure 1-5 shows an underdamped or cyclic response of a system in which the process variable oscillates around the set point after a process disturbance. The wavy response line shown in the figure represents an acceptable response if the process disturbance or change in set point is large, but it will not be an acceptable response if the change from the set point is small.

Figure 1-6 shows a damped response where the control system is able to bring the process variable back to the operating point with no oscillations.

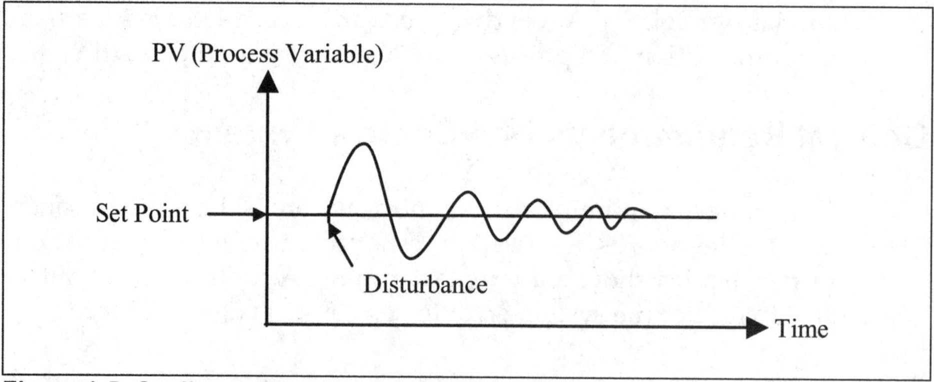

Figure 1-5. Cyclic response to process disturbance

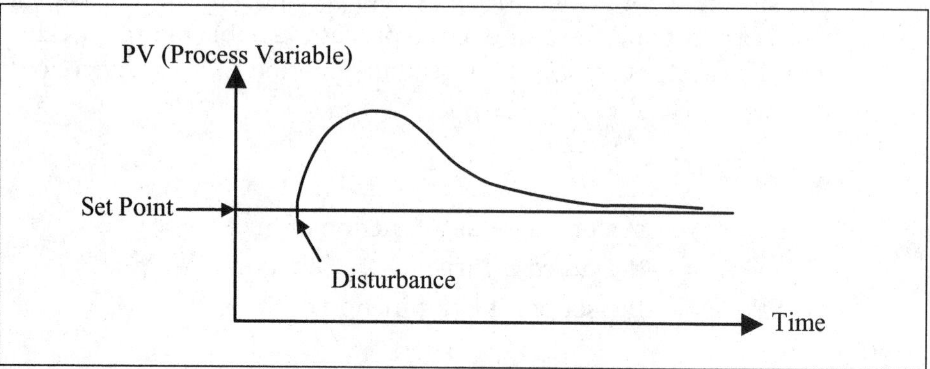

Figure 1-6. Damped response to process disturbance

Control Loop Design Criteria

Many criteria are employed to evaluate the process control's loop response to an input change. The most common of these include settling time, maximum error, offset error, and error area (Figure 1-7).

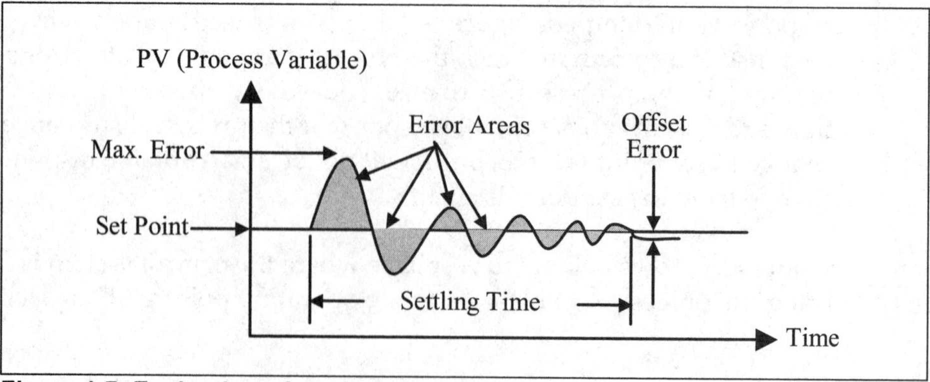

Figure 1-7. Evaluation of control loop response

When there is a process disturbance or a change in set point, the *settling time* is defined as the time the process control loop needs to bring the process variable back to within an allowable error. The *maximum error* is simply the maximum allowable deviation of the dynamic variable. Most control loops have certain inherent linear and nonlinear qualities that prevent the system from returning the process variable to the set point after a system change. This condition is generally called *offset error* and will be discussed later in this chapter. The *error area* is defined as the area between the response curve and the set point line as shown by the shaded area in Figure 1-7.

These four evaluation criteria are general measures of control loop behavior that are used to determine the adequacy of the loop's ability to perform some desired function. However, perhaps the best way to gain a clear understanding of process control is to take an intuitive approach.

Intuitive Approach to Process Control Concepts

The practice of process control arose long before the theory or analytical methods underlying it were developed. Processes and controllers were designed using empirical methods that were based on intuition ("feel") and extensive process experience. Most of the reasoning involved was nonmathematical. This approach was unscientific trial and error, but it was a successful control method.

Consider, for example, an operator looking into an early metal processing furnace to determine whether the product was finished. He or she used flame color, amount of smoke, and process time to make this judgment. From equally direct early methods evolved most of the control concepts and hardware used today. Only later did theories and mathematical techniques emerge to explain how and why the systems responded as they did.

In this section, we will approach the study of control fundamentals in much the same way that control knowledge developed—that is, through a step-by-step procedure starting from manual control and moving to ever-increasing automatic control.

Suppose we have a process like that shown in Figure 1-8. A source of feed liquid flows into a tank at a varying rate from somewhere else in a process plant. This liquid must be heated so that it emerges at a desired temperature, T_d, as a hot liquid. To accomplish this, hot water, which is available from another part of the plant, flows through heat exchanger coils in the tank. By controlling the flow of hot water, we can obtain the desired temperature, T_d. A further process requirement is that the level of the tank must neither overflow nor fall so low that it exposes the heater coils.

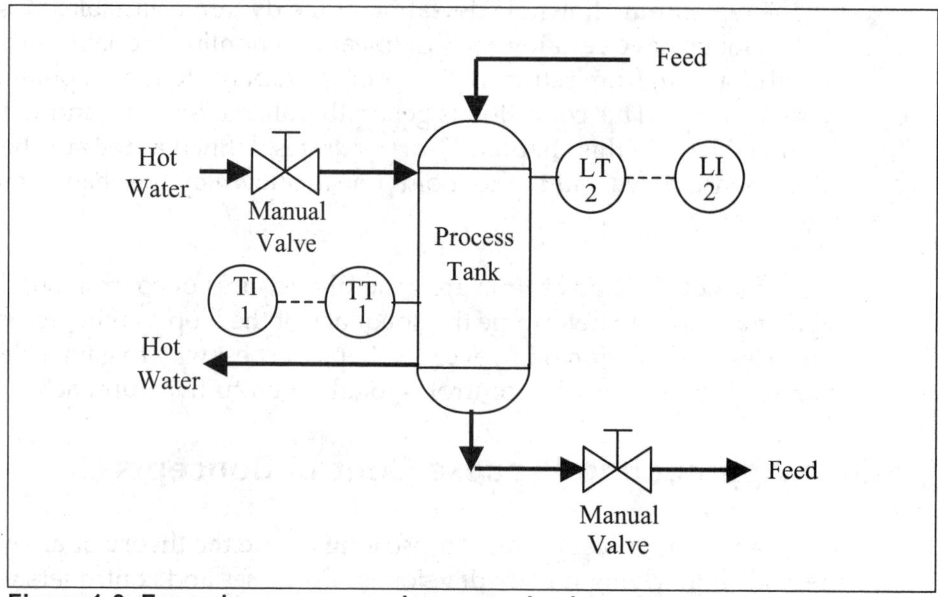

Figure 1-8. Example process – using manual valves

The temperature is measured in the tank, and a temperature transmitter (TT-1) converts the signal into a 4-20 mA direct current (DC) signal to drive a temperature indicator (TI-1) mounted near the hot water inlet valve. Similarly, a level indicator (LI-2) is mounted within the operator's view of the hot feed outlet valve (HV-2).

Suppose a process operator has the task of holding the temperature, T, near the desired temperature, T_d, while making sure the tank doesn't overflow or the level get too low. The question is how the operator would cope with this task over a period of time. He or she would manually adjust the hot water inlet valve (HV-1) to maintain the temperature and occasionally adjust the outlet valve (HV-2) to maintain the correct level in the tank.

The operator would face several problems, however. Both indicators would have to be within the operator's view, and the manual valves would have to be close to the operator and easy to adjust.

On/Off Control

To make the operator's work easier, suppose we install electrically operated solenoid valves in place of the manual valves, as shown in Figure 1-9. We can also install two hand switches (HS-1 and HS-2) so the solenoid valves can be operated from a common location. The valves can assume two states, either fully open (on) or fully closed (off). This type of control is called two-position or *on/off control.*

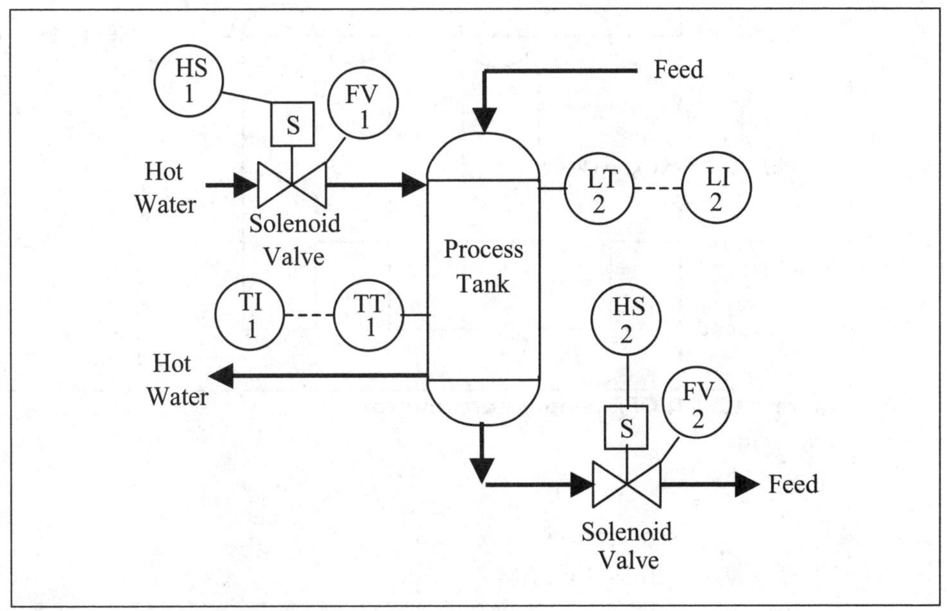

Figure 1-9. Sample process: Solenoid valves

Assume for the moment that the level is holding steady and that the main concern is controlling temperature. The operator has been told to keep the temperature of the fluid in the tank at 100°F. He or she compares the reading of the temperature indicator with the selected set point of 100°F. The operator closes the hot water valve when the temperature of the fluid in the tank rises above the set point (Figure 1-10). Because of process dead time and lags the temperature will continue to rise before reversing and moving toward the set point. When the temperature falls below 100°F, the operator opens the hot water valve. Again, dead time and lags in the process create a delay before the temperature begins to rise. As it crosses the set point, the operator again shuts off the hot water, and the cycle repeats.

This cycling is normal for a control system that uses on/off control. This limitation exists because it's impossible for the operator to control the process exactly with only two options.

This on/off type of control can be expressed mathematically as follows:

$$e = PV - SP \tag{1-2}$$

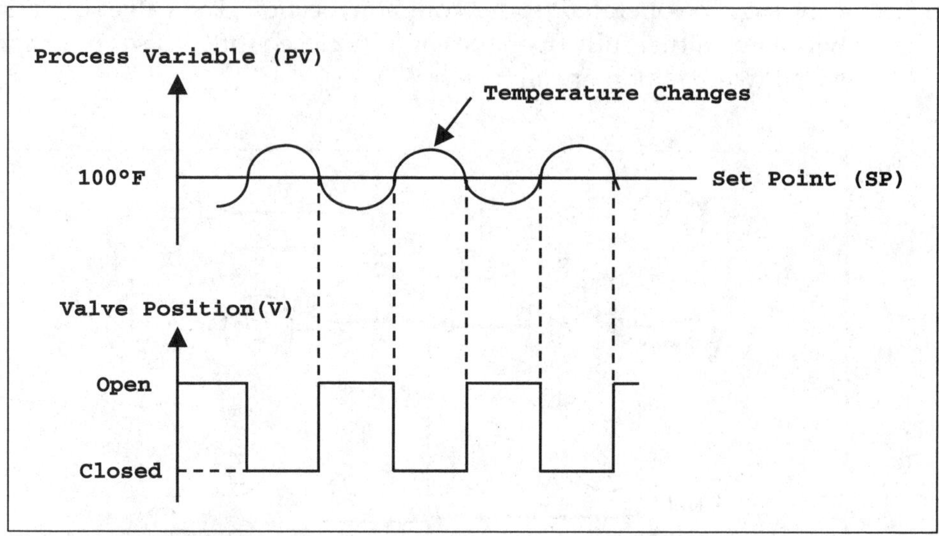

Figure 1-10. ON/OFF temperature control

where

e	=	the error
SP	=	the set point
PV	=	the process variable

In the on/off control mode, the valve is opened when the error (e) is positive (+), and the valve is closed when e is negative (–).

Proportional Control

When we view the process as a balance between energy in and energy out, it is clear that smoother control would result if a steady flow of hot water were maintained rather than the sudden changes between ON and OFF. The problem is finding the correct value for the steady flow required for proper control. Obviously, for each rate of feed flow in and out of the tank, some ideal amount of inlet water flow exists that will hold the outlet temperature, T, at 100°F.

This suggests that we should make two modifications to our *control mode* or strategy. The first is to establish some steady-flow value for the hot water that, at average operating conditions, tends to hold the process variable (temperature) at the desired value or set point (100°F). Once that average flow value has been established for the hot water, increases or decreases of error ($e = SP - PV$) must be allowed to cause corresponding increases and decreases in water flow from this normal value. This illustrates the concept of *proportional control* (i.e., initiating a corrective action to a value that is in some proportion to the change in error or deviation of the process variable from set point).

Before proportional control can be implemented on our sample process, we must change the solenoid valves to adjustable control valves. Such valves can be positioned to any degree of opening—from fully closed to fully opened—depending on the type of valve actuator mechanism you choose (generally either an electrically or pneumatically operated diaphragm actuator). Our sample process now looks like Figure 1-11, which now shows the use of pneumatically operated control valves (TV-1 and LV-2) and process controllers (TIC-1 and LIC-2). Control valves and controllers in the system make it possible to achieve better control of the process.

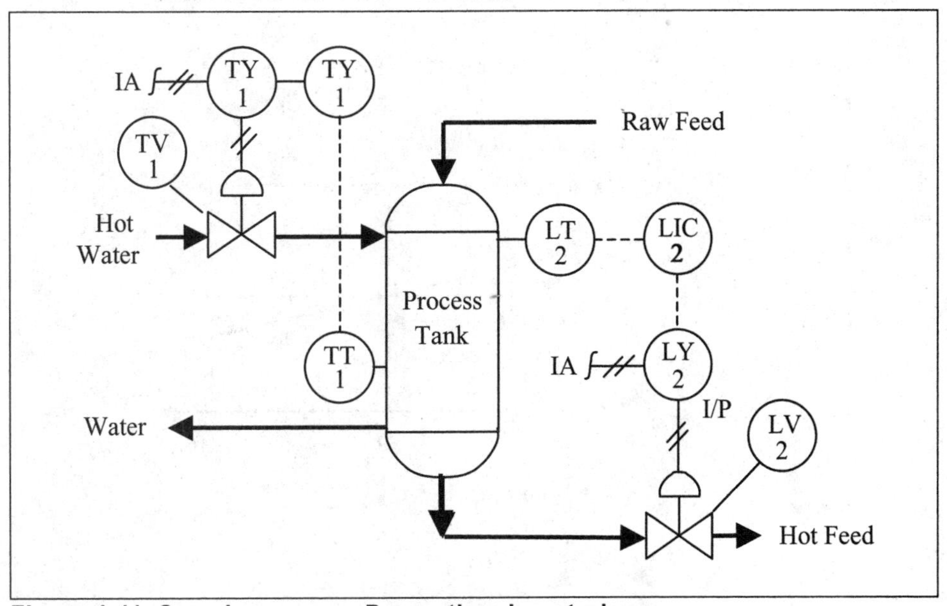

Figure 1-11. Sample process: Proportional control

Proportional control can be described mathematically as follows:

$$V = K_c e + m \quad (1\text{-}3)$$

where

V = the control valve position

K_c = the adjustable proportional gain of a typical process controller

m = a constant, which is the position of the control valve when the system error (e) is zero.

Proportional control can be illustrated by using the three graphs in Figure 1-12 and setting the proportional constant to three different values (i.e., $K_c = 1$, $K_c < 1$, and $K_c > 1$).

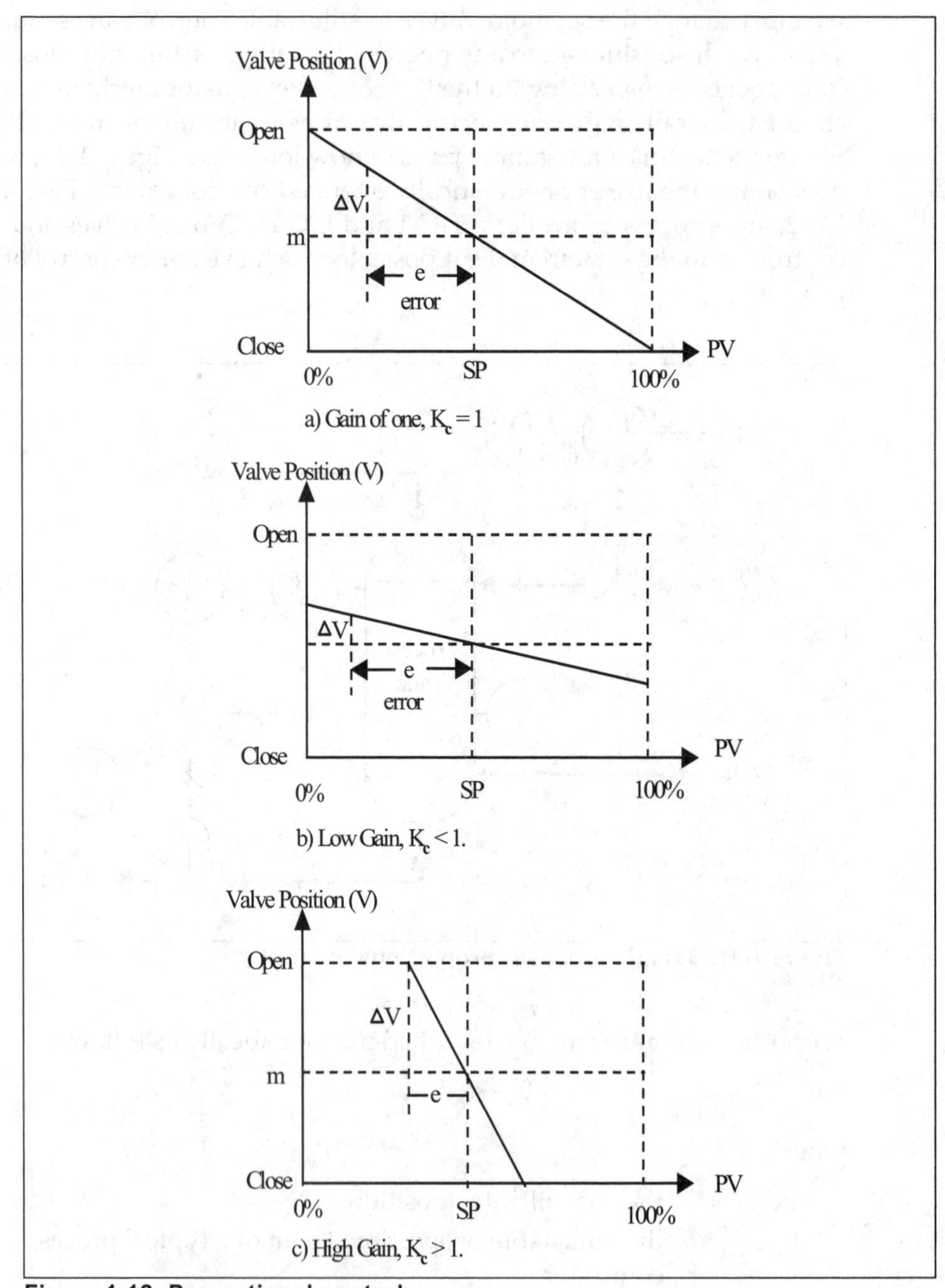

Figure 1-12. Proportional control

As these graphs show, the amount of valve change (ΔV) for a given error can vary substantially. A one-to-one relationship is shown in Figure 1-12(a). In that example, the control valve would move 1 percent of its full travel for a corresponding 1 percent change in error or in a one-to-one ratio. In Figure 1-12(b), where a low gain (K_C <1) is selected, a large change in error is required before the control valve would be fully opened or closed. Finally, Figure 1-12(c) shows the case of high gain (K_C >1), where a very small error would cause a large change in the control valve position.

The term *proportional gain*, or simply *gain*, arose as a result of the use of analytical methods in process control. Historically, the proportionality between error and valve action was called proportional band (PB). Proportional band is the expression that states the percentage of change in error that is required to move the valve full scale. Again, this had intuitive plausibility because it gave an operator a feel for how small of an error caused full corrective action. Thus, a 10 percent proportional band meant that a 10 percent error between *SP* and *PV* would cause the output to go full scale.

This definition can be related to proportional gain K_c by noting the following equation:

$$PB = \frac{1}{K_C} \text{ x } 100 \tag{1-4}$$

An example will help you understand the relationship between proportional band and gain.

EXAMPLE 1-1

Problem: For a proportional process controller:

What proportional band corresponds to a gain of 0.4?

What gain corresponds to a PB of 400?

Solution: a) $PB = \frac{1}{K_C} \times 100 = \frac{100}{0.4} = 250\%$

b) $K_C = \frac{1}{PB} \times 100 = \frac{100}{400} = 0.25$

The modern way of considering proportional control is to think in terms of *gain* (K_C). The *m* term, as Equation 1-3 shows, has to be that control valve position that supplies just the right amount of hot water to make the temperature 100°F, that is, *PV* = *SP*. The position, *m*, indicated in Figures 1-12(a), (b), and (c), is often called the *manual (m)* reset because it is a manual controller adjustment.

When a controller is designed to provide this mode of control, it must contain at least two adjustments: one for the K_C term and one for the *m* term. Control has become more complicated because it is now necessary to know where to set K_C and *m* for best control.

It would not take too long for the operator of our sample process to discover a serious problem with proportional control. Proportional control rarely ever keeps the process variable at the set point if there are frequent disturbances to the process. For example, suppose the flow to the tank suddenly increases. If the temperature of the tank is to be maintained at 100°F at this new rate of feed flow, more hot water must be supplied. This calls for a change in valve position. According to Equation 1-3, the only way that the valve position (*V*) can be changed is for the error (*e*) to change. Remember that *m* is a constant. Thus, an error will occur, and the temperature will drop below 100°F until an equilibrium is reached between the hot water flow and new feed flow. How much this drop will be depends on the value of K_C that was set in the controller as well as on the characteristics of the process. The larger K_C is, the smaller this offset will be in a given system.

However, it can be shown that K_C cannot be increased indefinitely because the control loop will become unstable. So, some error is inevitable if the feed rate changes. These points are illustrated in Figure 1-13, which shows a plot of hot feed temperature versus hot water flow rate (valve position) for both low raw feed flow and high raw feed flow.

For the hot water valve in position 1 and the raw feed coming into the process tank at the low flow rate, the process would heat the fluid and produce hot feed fluid at temperature T_2. If suddenly the feed went to the high flow rate and the valve position was not changed, the temperature would drop to T_1. At this new high flow rate, the hot water valve must be moved to position 2 if the original temperature T_2 is to be restored. Figure 1-14 shows the extent to which proportional control of the temperature valve can achieve this restoration.

One way to cope with the offset problem is by manually adjusting the *m* term. When we adjust the *m* term (usually through a knob on a process controller), we are moving the valve to a new position that allows *PV* to

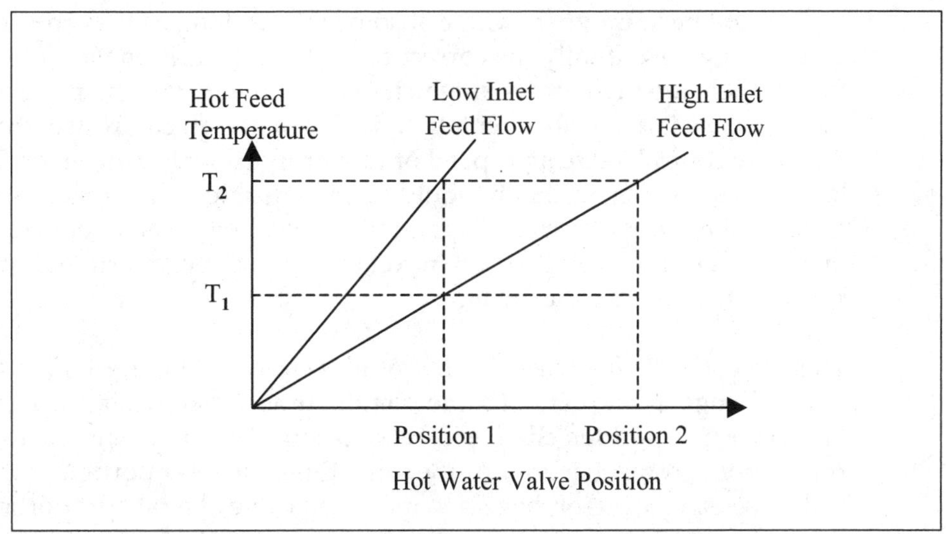

Figure 1-13. Sample process: Temperature vs. valve position

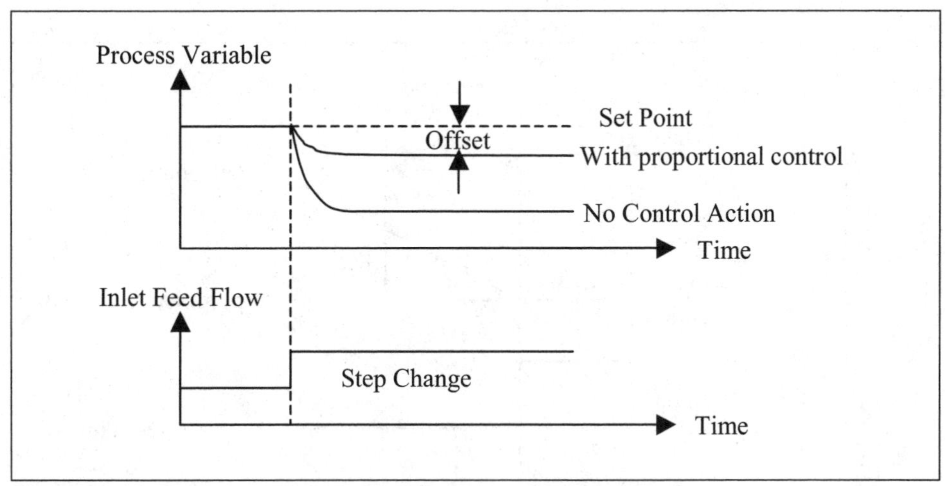

Figure 1-14. Process response with proportional control

equal *SP* under the new conditions of load. In this case, with an increase in feed flow, Equation 1-3 clearly shows that the only way to obtain a new value for *V*, if *e* is to be zero, is by changing the *m* term. If process changes are frequent or large it may become necessary to adjust *m* frequently. It is apparent that some different type of control mode is needed.

Proportional-Plus-Integral Control

Suppose that the controller rather than the operator manually adjusts the proportional controller described in the previous section. This would eliminate the offset error caused by process changes. The question then is, on what basis should the manual reset be automated? One innovative concept would be to move the valve at some rate, as long as the error is not zero. Though eventually the correct control valve position would be found, there are many rates at which to move the valve. The most common practice in the instrumentation field is to design controllers that move the control valve at a speed or rate proportional to the error. This has some logic to it, in that it would seem plausible to move the valve faster as the error got larger. This added control mode is called *reset* or *integral action*. It is usually used in conjunction with proportional control because it eliminates the offset.

This *proportional-plus-integral (PI) control* is shown in Figure 1-15. Assume a step change in set point at some point in time, as shown in the figure. First, there is a sudden change in valve position equal to K_ce due to the proportional control action. At the same time, the reset portion of the controller, sensing an error, begins to move the valve at a rate proportional to the error over time. Since the example in Figure 1-15 had a constant error, the correction rate was constant.

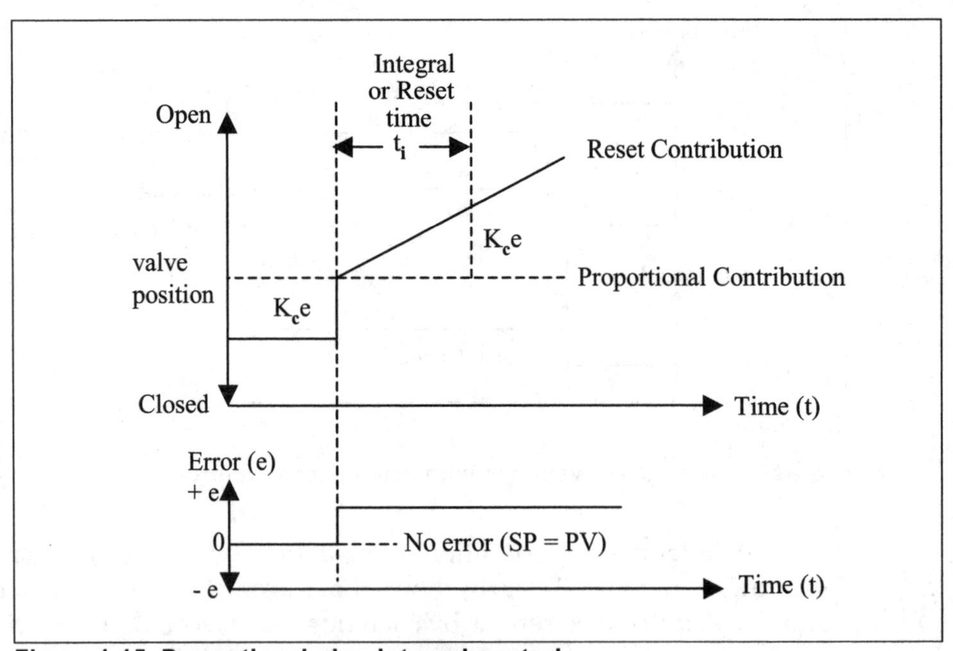

Figure 1-15. Proportional-plus-integral control

When time is used to express integral or reset action, it is called the *reset time*. Quite commonly, its reciprocal is used, in which case it is called *reset rate* in "repeats per minute." This term refers to the number of times per minute that the reset action is repeating the valve change produced by proportional control alone. Process control systems personnel refer to reset time as the *integral time* and denote it as t_i.

The improvement in control that is caused by adding the integral or reset function is illustrated in Figure 1-16. The same process change is used that was previously assumed under proportional-only control. Now, however, after the initial upset the reset action returns the error to zero and there is no offset.

Recognizing that the reset action moves the control valve at a rate proportional to error, this control mode is described mathematically as follows:

$$\frac{dV}{dt} = K_i e \tag{1-5}$$

where

dV/dt = the derivative of the valve position with respect to time (t)
K_i = an adjustable constant

We can find the position of the valve at any time by integrating this differential equation (Equation 1-5). If we integrate from time 0 to time, t, we obtain:

$$V = K_i \int_0^t e\,dt \tag{1-6}$$

This equation shows that the control valve position is proportional to the integral of the error. This fact leads to the "integral control" label. Finally, combining proportional and integral control gives the total expression of a two-mode proportional-plus-integral (PI) controller:

$$V = K_c e + K_i \int_0^t e\,dt \tag{1-7}$$

If we let $K_i = K_c/t_i$, we obtain an alternate form of the PI control equation in terms of the proportional constant, K_c, and the integral time, t_i, as follows:

$$V = K_c e + \frac{K_c}{t_i} \int_0^t e\,dt \tag{1-8}$$

One problem with PI control bears mentioning. If a control loop is using PI control, the possibility exists with the integral (reset) mode that the controller will continue to integrate and change the output even outside the operating range of the controller. This condition is called "reset windup."

For example, the heat exchanger shown in Figure 1-17 can be designed and built to heat 50 gal/min of process fluid from 70°F to 140°F. If the

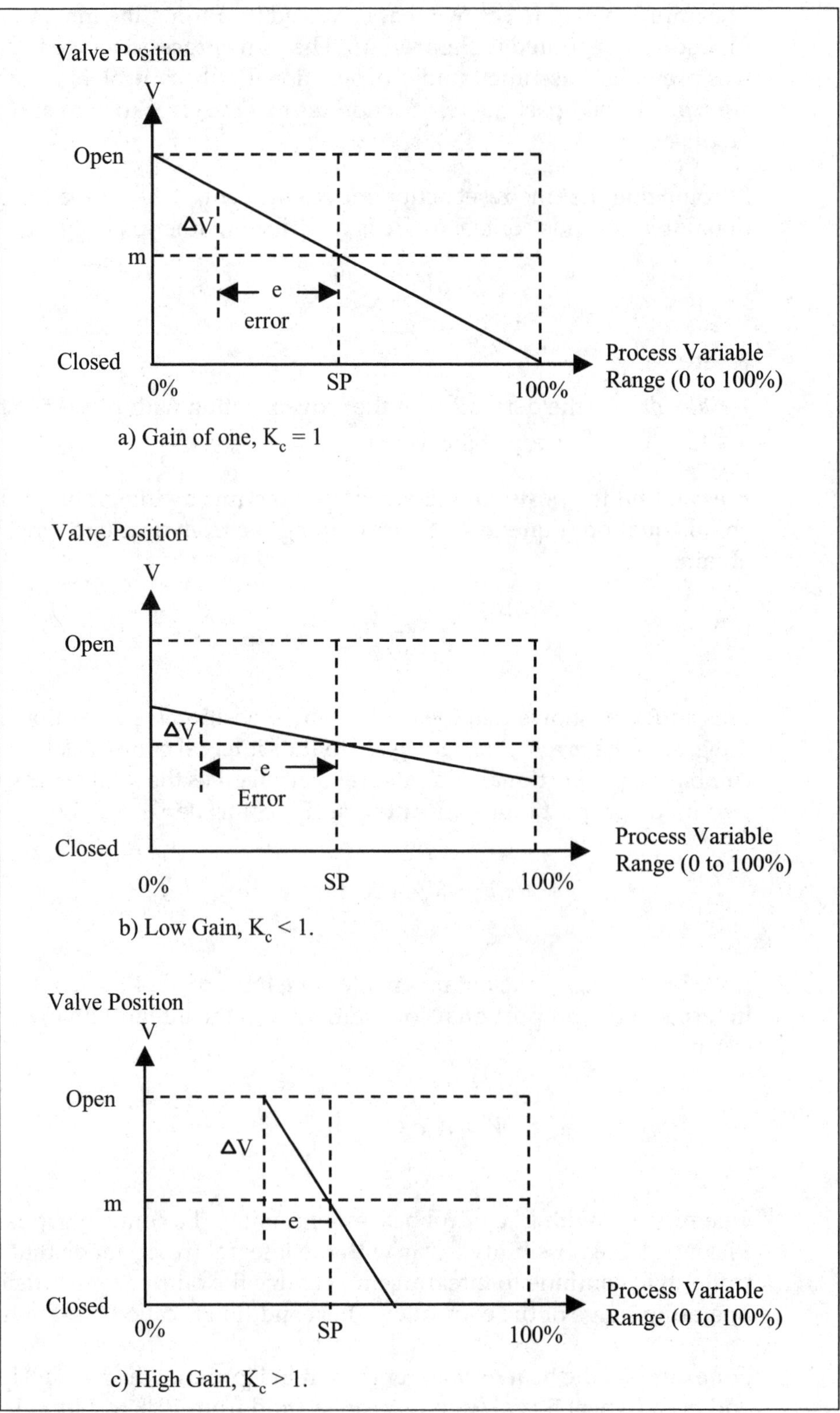

Figure 1-16. Process response with PI control

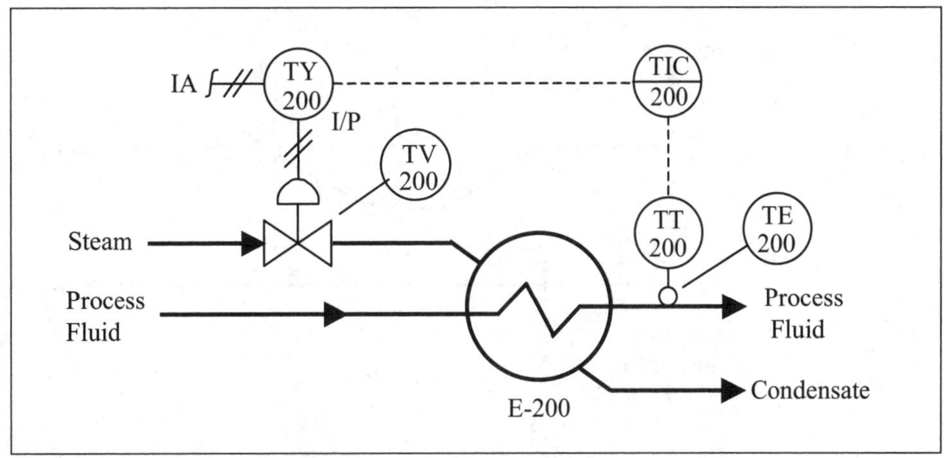

Figure 1-17. Heat exchanger temperature control

process flow should suddenly increase to 100 gal/min, it may be impossible to supply sufficient steam to maintain the process fluid temperature at 140°F even when the control valve is wide open (100%), as shown in Figure 1-18. In this case, the reset mode, having opened the valve all the way (the controller output is perhaps 15 psi), would continue to integrate the error signal and increase the controller output all the way in order to supply pressure from the pneumatic system. Once past 15 psi, the valve will open no further, and the continued integration serves no purpose. The controller has "wound up" to a maximum output value.

Further, if the process flow should then drop to 50 gal/min (back to the operable range of the process), there would be a period of time during which the controlled temperature is above the set point while the valve remains wide open. It takes some time for the integral mode to integrate (reset) downward from this wound-up condition to 15 psi before the valve begins to close and control the process.

It is possible to prevent this problem of controller-reset windup by using a controller operational feature that limits the integration and the controller output. This feature is normally called *anti-reset windup* and is recommended for processes that may periodically operate outside their capacity.

Proportional-Plus-Derivative (PD) Control

We can now add to proportional control another control action called *derivative action*. This control function produces a corrective action that is proportional to the rate of change of error. Note that this additional correction exists only while the error is changing; it disappears when the error stops changing, even though there may still be a large error.

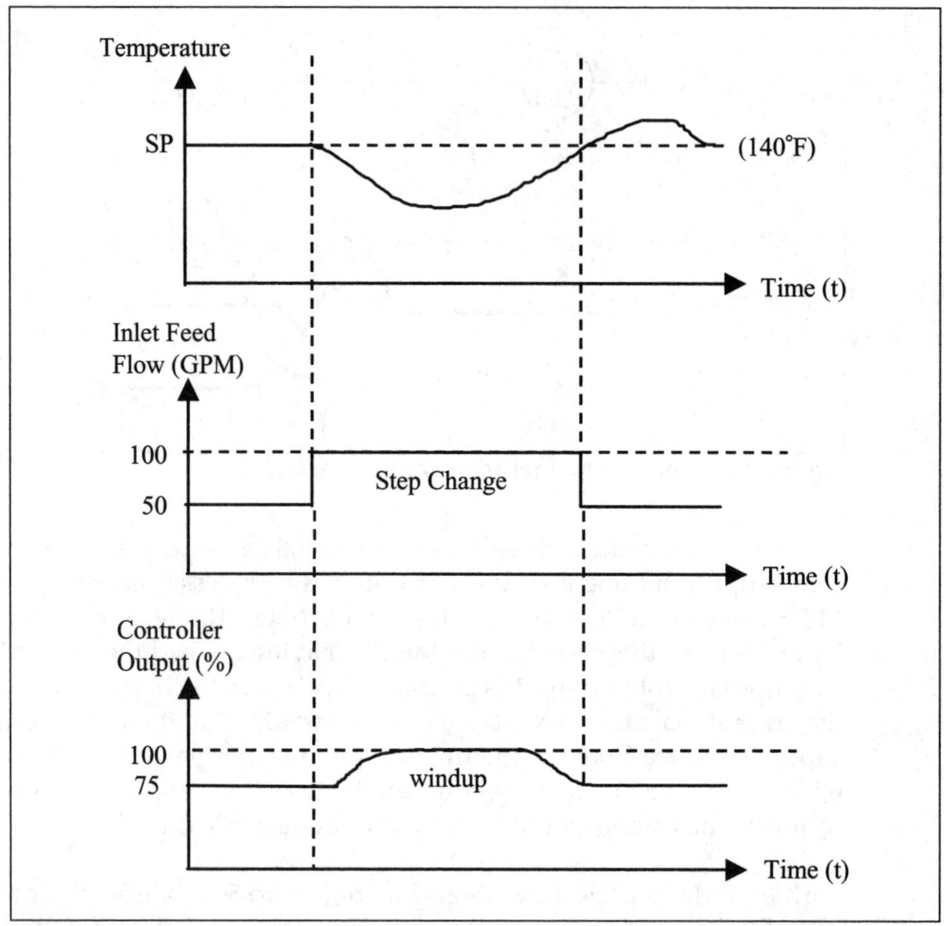

Figure 1-18. Reset windup control

Derivative control can be expressed mathematically as follows:

$$V = K_d \frac{de}{dt} \tag{1-9}$$

where

K_d = the derivative constant

de/dt = the derivative of the control system error with respect to time

The derivative constant K_d can be related to the proportional constant K_c by the following equation:

$$K_d = K_c t_d \tag{1-10}$$

where t_d is the derivative control constant. If we add derivative control to proportional control, we obtain

$$V = K_c e + K_c t_d \frac{de}{dt} \tag{1-11}$$

To illustrate the effects of PD control, let's assume that the error is changing at a constant rate. This can be obtained by changing the set point at a constant rate (i.e., $SP = Ct$), as shown in Figure 1-19.

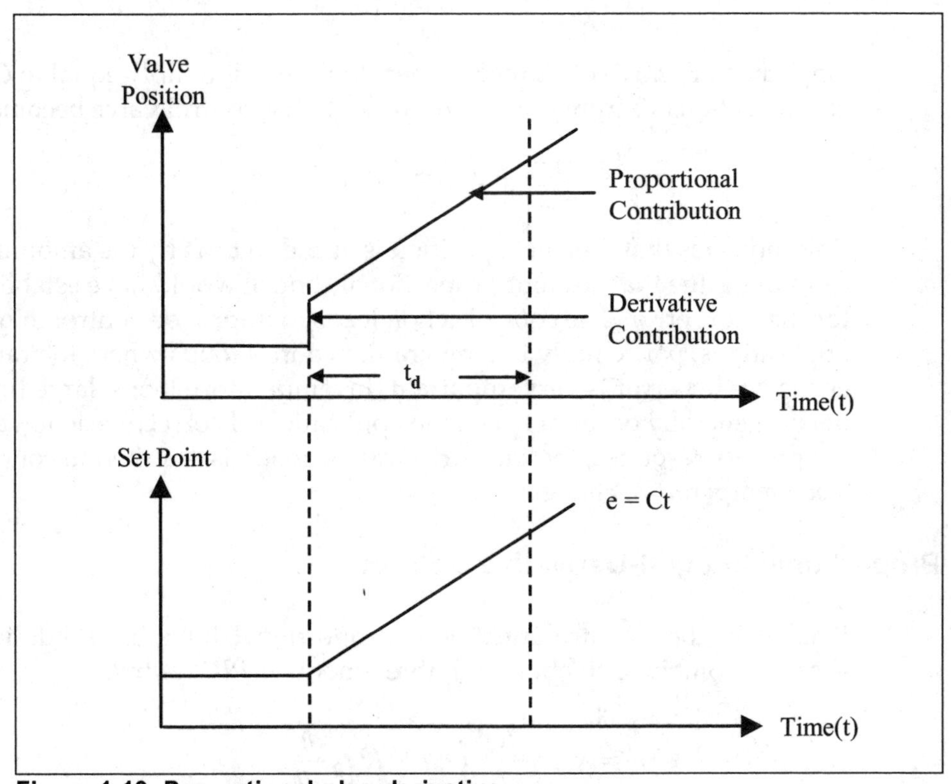

Figure 1-19. Proportional-plus-derivative

Derivative action contributes an immediate valve change that is proportional to the rate of change of the error. In Figure 1-19, it is equal to the slope of the set point line. As the error increases, the proportional action contributes additional control valve movement. Later, the contribution of the proportional action will have equaled the initial contribution of the rate action. The time it takes for this to happen is called the *derivative time,* t_d. The ramped error can be expressed mathematically as follows:

$$e = Ct \tag{1-12}$$

where

e	=	the control loop error
C	=	a constant (slope of set point change)
t	=	time

If we substitute this value ($e = Ct$) for the control loop error into the equation for a PD controller, we obtain

$$V = K_c Ct + K_c t_d \frac{d(Ct)}{dt} \tag{1-13}$$

Since the derivative of Ct with respect to time, t, is simply equal to C, the control action (V) from the PD controller to the control valve becomes

$$V = K_c C(t + t_d) \tag{1-14}$$

This indicates that the valve position is ahead in time by the amount t_d from the value that straight proportional control would have established for the same error. The control action leads to improved control in many applications, particularly in temperature control loops where the rate of change of the error is very important. In temperature loops, large time delays generally occur between the application of corrective action and the process response; therefore, derivative action is required to control steep temperature changes.

Proportional-Integral-Derivative Control

Finally, the three control functions—proportional, integral, and derivative—can be combined to obtain full three-mode or *PID control*:

$$V = K_c e + \frac{K_c}{t_i} \int_0^t e\,dt + K_c t_d \frac{de}{dt} \tag{1-15}$$

Deciding which control action (i.e., PD, PID, etc.) should be used in a control system will depend on the characteristics of the process being controlled. Three-mode control (PID) cannot be used on a noisy measurement process or on one that experiences stepwise changes because the derivative contribution is based on the measurement of rate of change. The derivative of a true step change is infinite, and the derivatives of a noisy measurement signal will be very large and lead to unstable control.

The PID controller is used on processes that respond slowly and have long periods. Temperature control is a common example of PID control because the heat rate may have to change rapidly when the temperature measurement begins to change. The derivative action shortens the response of the slow process to an upset.

In the next chapter we discuss such important characteristics of processes as time constants and dead time. By understanding these concepts you will be better able to select the proper control action type for effective control.

EXERCISES

1.1 What are the three main factors or terms found in all process control systems? List examples of each type.

1.2 List the four essential elements of a process control system.

1.3 What function is performed by a process controller in a control loop?

1.4 What type of instrument is identified by each of the following instrument tag numbers: (a) PIC-200, (b) FV-250, (c) LC-500, and (d) HS-100?

1.5 What is the primary requirement of any process control system?

1.6 Define the term *system error* with respect to a control system.

1.7 List the four most common design criteria used to evaluate a typical process control loop's response to an input change.

1.8 For a proportional controller, (a) what gain corresponds to a proportional band of 150 percent and (b) what proportional band corresponds to a gain of 0.2?

1.9 What is the main reason to use integral action with proportional control?

1.10 Explain the concept of "reset windup" encountered in proportional-plus-integral controllers.

1.11 What type of controller is used on the heat exchanger shown in Figure 1-17? Where is the controller located?

1.12 Discuss the type of process that can most benefit from the use of PID control.

2 Process Control Loops

Introduction

We discussed the general concepts of process control in Chapter 1. In this chapter, we will cover the basic principles of process control loops. *Single-loop* feedback control is the most common type of control used in industrial processes, so it will be discussed in the greatest detail. We will then discuss other types of control loops, such as *cascade, ratio,* and *feedforward*. Finally, we will examine several common methods used to tune control loops.

Single-loop Feedback Control

In a feedback control loop, the variable to be manipulated is measured. This measured process value (PV) is then compared with a set point (SP) to generate an error signal (e = PV - SP). If a difference or error exists between the actual value and the desired value of the process, a process controller will take the necessary corrective action to return the process to the desired value. A block diagram of a single-feedback control loop is shown in Figure 2-1.

The measured process variable is sensed or measured by the appropriate instrumentation, such as temperature, flow, level, or analytical sensors. This measured value is then compared with the set point. The controller uses this comparison to adjust the manipulated variable appropriately by generating an output signal. The output signal is based in turn on whichever control strategy or algorithm has been selected. Because in the process industries the manipulated variable is most often flow, the output of

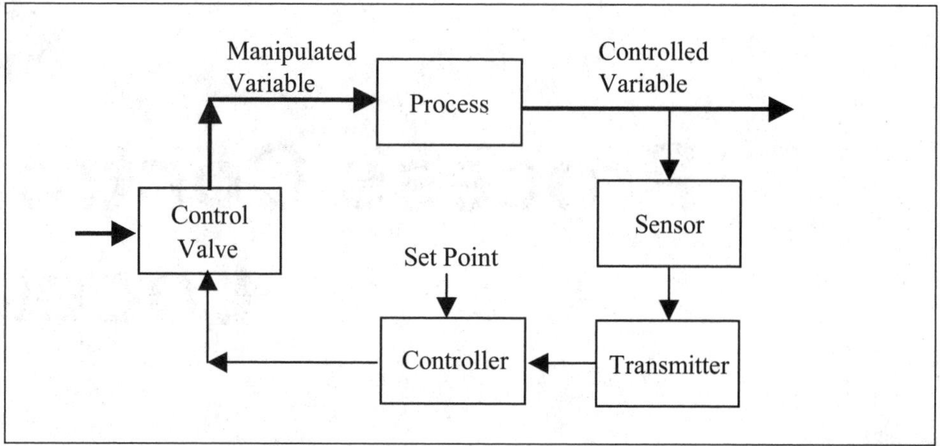

Figure 2-1. Feedback control loop

the controller is usually a signal to a flow control valve, as shown in Figure 2-1.

During the operation of the process, disturbances can enter the process and drive the process variable in one direction or another. The single manipulated variable is used to compensate for all such process changes produced by the disturbances. Furthermore, if changes occur in the set point, the manipulated variable is altered to produce the needed change in the process output.

Process Controllers

The most dynamic device in a feedback control loop is the process controller. There are three types of controllers—mechanical, pneumatic, and electronic—and they all serve the same function. They compare the process variable with the set point and generate an output signal that manipulates the process to make the process variable equal to its set point. Figure 2-2 shows a block diagram of a feedback control loop with an expanded view of its common functions. In this diagram the measurement transducer has been expanded into its two components: the sensor and the transmitter. The sensor measures the process variable, and then the transmitter converts the measurement into a standard signal such as 4 to 20 mA DC or 3 to 15 psi.

The controller consists of an input transmission system, a comparator with a set point input, controller functions, and an output transmission system. The comparator block measures the difference between the set point and the process variable. For this comparison to be useful, the set point and the process variable must have the same units of measure. For example, if the set point has the units of 0 to 10 mV, then the signal from the sensor must

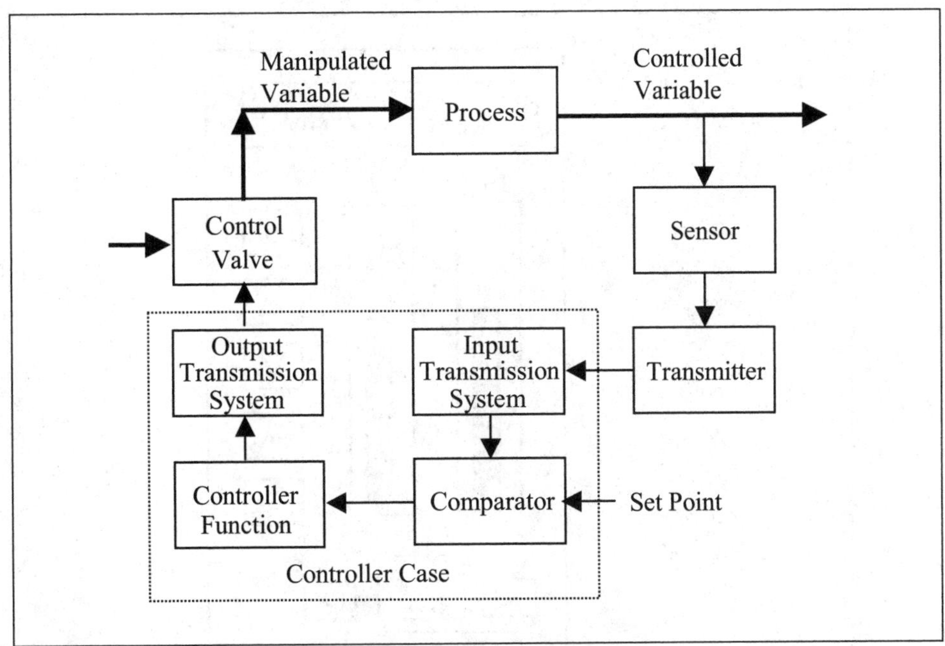

Figure 2-2. Functional block diagram of feedback loop

be converted into the same units. The purpose of the input transmission system is to convert the sensor signal into the correct units. For example, if the input signal is 4 to 20 mA DC the input circuit in the controller will convert the signal to 0 to 10 mV. The function of the output transmission system is to convert the signal from the controller circuit into the form required by the final control device. The four common controller functions are proportional, proportional plus integral (PI), proportional plus derivative (PD), and proportional plus integral plus derivative (PID).

A front-panel view of a typical electronic process controller is shown in Figure 2-3. The controller has two vertical bar displays to give the operator a pictorial view of the process variable and the set point. It also has two short horizontal digital displays just above the vertical bars to give the operator a direct digital readout of the process variable and the set point. The operator uses dual push buttons with indicating arrows to adjust the set point and the manual output functions. The operator must depress the manual ("M") push button to activate the manual output function.

During normal operation, the operator will select automatic ("A") mode. Manual is generally used only during system startup or during a major upset condition when the operator must take control to stabilize the process. The controller shown in Figure 2-3 has both a horizontal bar display and a digital indicator to provide the operator with the value of the output signal from the controller. The square indicator marked "RSP" is used to indicate that the controller is using a remote set point.

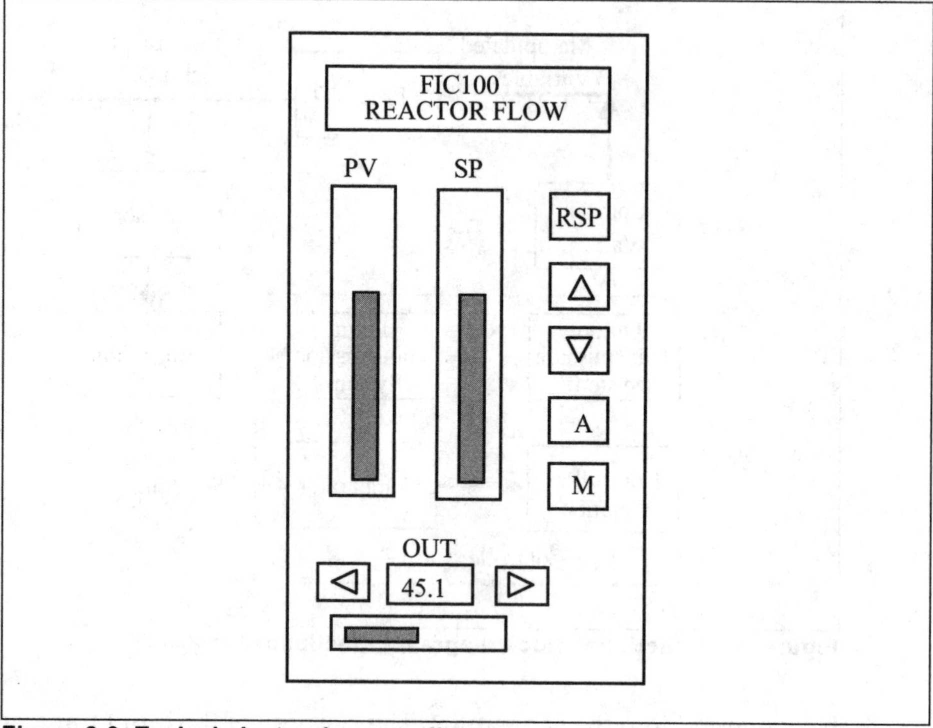

Figure 2-3. Typical electronic controller

Time Elements of a Feedback Loop

The various components of the feedback control loop shown in Figure 2-2 need time to sense an input change and transform this new condition into an output change. The time of response of the control loop is the combination of the responses of the sensor, the transmitter, the controller, the final control element, and the process.

An important objective in control system design is to correctly match the time response of the control system to that of the process. To reach this objective, it is necessary to understand the concept of time delays or "lags" in process control systems.

Time Lags

In process control, the term *lag* means any relationship in which some result happens after some cause. In a feedback control loop, lags act in series, the output of one being the input to another. For example, the lags around a simple temperature control loop would be the output of the electric controller to the input to a valve lag. The output of the valve lag is the input to a process heat lag. The output of process heat lag is the input to the measurement sensor lag. We will start our discussion of time response and time lag with sensor time response.

Sensor Time Response

In process sensors, the output lags behind the input process value that is being measured. Sensor output changes smoothly from the moment a change in measurement value occurs, even if the disturbance is sudden and discontinuous. It is interesting to note that the nature of the sensor time-response curve is the same for virtually all sensors, even though the sensors measure different physical variables.

A typical response curve for a process sensor is shown in Figure 2-4, where the input has been changed suddenly at time equal to zero.

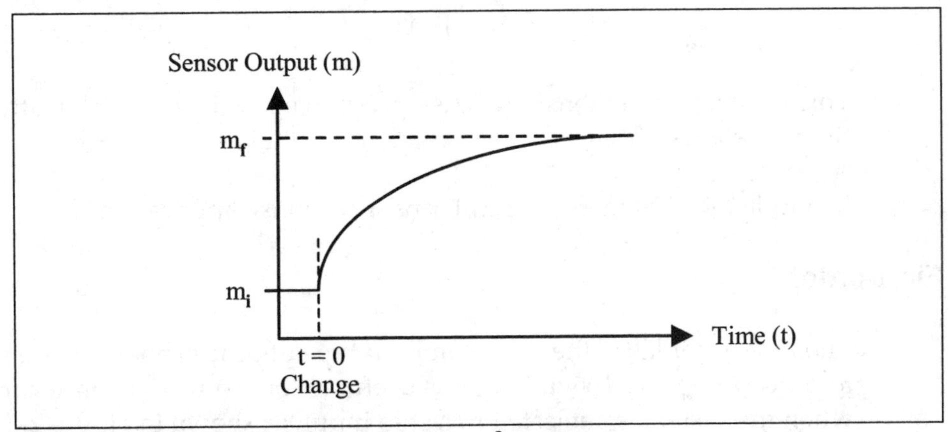

Figure 2-4. Exponential time response of a sensor

This curve is described by the following equation for the output measurement $m(t)$ as a function of time:

$$m(t) = m_i + (m_f - m_i)(1 - e^{-t/\tau}) \qquad (2\text{-}1)$$

where

m_i	=	the initial sensor output measurement
m_f	=	the final sensor output value
τ	=	the sensor time constant

Note that the sensor output is in error during the transition time of the output value from m_i to m_f. The actual process variable was changed instantaneously to a new value at $t = 0$. Equation 2-1 relates initial sensor output, final sensor output, and the time constant that is a characteristic of the sensor. The significance of the time constant τ can be found by looking at the equation for the case where the initial sensor output is zero. In this special case, the sensor output value is as follows:

$$m(t) = m_f(1 - e^{-t/\tau}) \tag{2-2}$$

If we wish to find the value of the output exactly τ seconds after a sudden change occurs, then

$$m(\tau) = m_f(1 - e^{-1}) \tag{2-3}$$

$$m(\tau) = 0.632m_f \tag{2-4}$$

Thus, we see that one time constant (1τ) represents the time at which the output value has changed by 63.2 percent of the total change. If we solve Equation 2-2 for time equal to 5τ, or five time constants, we find that

$$m(5\tau) = 0.993m_f \tag{2-5}$$

This means that the sensor reaches 99.3 percent of its final value after five time constants.

Example 2-1 illustrates a typical sensor response application.

First-order Lag

The first-order lag is the most common type of time element encountered in process control. To study it, it is useful to look at the response curves when the system is subjected to a step input, as shown in Figure 2-5. The advantage of using a step input as a forcing function is that the input is at steady state before the change and then is instantaneously switched to a new value. When the output curve (y) is studied, the transition of the system can be observed as it passes from one steady state to a new one. The output or response to the step input applied at time zero (t_o) is not a step output but an output that lags behind the input and gradually tries to reach some final value.

The equation for the system shown in Figure 2-5 is as follows:

$$\tau\frac{dy(t)}{dt} + y(t) = Kx(t)\frac{dy_c}{dt} \tag{2-6}$$

where

$y(t)$	=	the output y as a function of time
$x(t)$	=	the input x as a function of time
K	=	a constant
τ	=	the system time constant

EXAMPLE 2-1

Problem: A sensor measures temperature linearly with a transfer function (i.e., output/input = m_f/m_i) of 30 mV/°C and has a one-second time constant. Find the sensor output two seconds after the input changes rapidly from 25°C to 30°C. Also find the process temperature.

Solution: First, find the initial and final values of the sensor output:

$$m_i = (30 \text{ mV/°C})\,(25\text{°C})$$

$$m_i = 750 \text{ mV}$$

$$m_f = (30 \text{ mV/°C})(30\text{°C})$$

$$m_f = 900 \text{ mV}$$

Then, use Equation 2-1 to solve for the sensor output at $t = 2$ s. Note that $e = 2.718$.

$$m(t) = m_i + (m_f - m_i)(1 - e^{-t/\tau})$$

$$m(2) = 750 \text{ mV} + (900 - 750) \text{ mV}(1 - e^{-2})$$

$$m(2) = 879.7 \text{ mV}$$

This corresponds to a process temperature at $t = 2$ s of

$$T(2) = (879.7 \text{ mV})/(30 \text{ mV/°C})$$

$$T(2) = 29.32\text{°C}$$

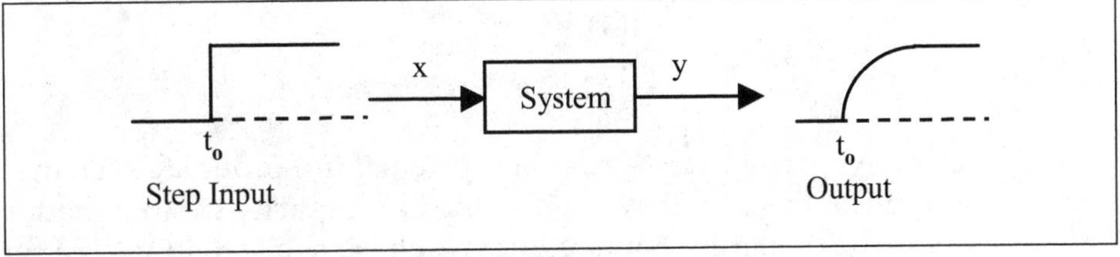

Figure 2-5. Response of a system to step

The system response is called a "first-order lag" because the output lags behind the input, and the differential equation for the system shown in Figure 2-5 is a linear *first-order* differential equation.

Differential equations are difficult to understand in some cases. If the system as a whole contains several components with their own differential equations, it is very difficult to understand or solve the entire system.

The French mathematician Pierre-Simon Laplace developed a method to transform differential equations into algebraic equations so as to simplify the calculations for systems governed by differential equations. We will avoid most of the rigorous math of the Laplace transform method and simply give the steps required to transform a normal differential equation into an algebraic equation:

1. Replace any derivative symbol, d/dt, in the differential equation with the transform symbol s.
2. Replace any integral symbol, $\int ...dt$, with the symbol $1/s$.
3. Replace the lowercase letters that represent variables with their corresponding uppercase letter in the transformed equation.

We can use the Laplace transform method to convert differential Equation 2-6 for our system into an algebraic equation. Since the system equation contains only a single derivative and no integral, it can be transformed using steps 1 and 3. When we transform the equation, it becomes

$$\tau sY(s) + Y(s) = KX(s) \tag{2-7}$$

The transfer function for our system is defined as follows:

$$\frac{Output}{Input} = \frac{Y(s)}{X(s)} \tag{2-7a}$$

Thus, solving Equation 2-7 results in the following equation:

$$\frac{Y(s)}{X(s)} = \frac{K}{\tau s + 1} \tag{2-7b}$$

This is the form of a first-order lag system. First-order lag systems in process applications are characterized by their capacity to store matter or energy. The dynamic shape of their response to a step input is a function of their time constant. This system time constant, designated by the Greek letter τ (tau), is meaningful both in a physical and a mathematical sense. Physically, it determines the shape of the response of a process or system to a step input. Mathematically, it predicts, at any instant, the future time period that is required to obtain 63.2 percent of the change remaining. The response curve in Figure 2-6 illustrates the concept of a time constant by showing the response of a simple first-order system to a step input. In this system, the output is always decreasing with time, that is, the rate of response is the maximum in the beginning and continuously decreases from that time on.

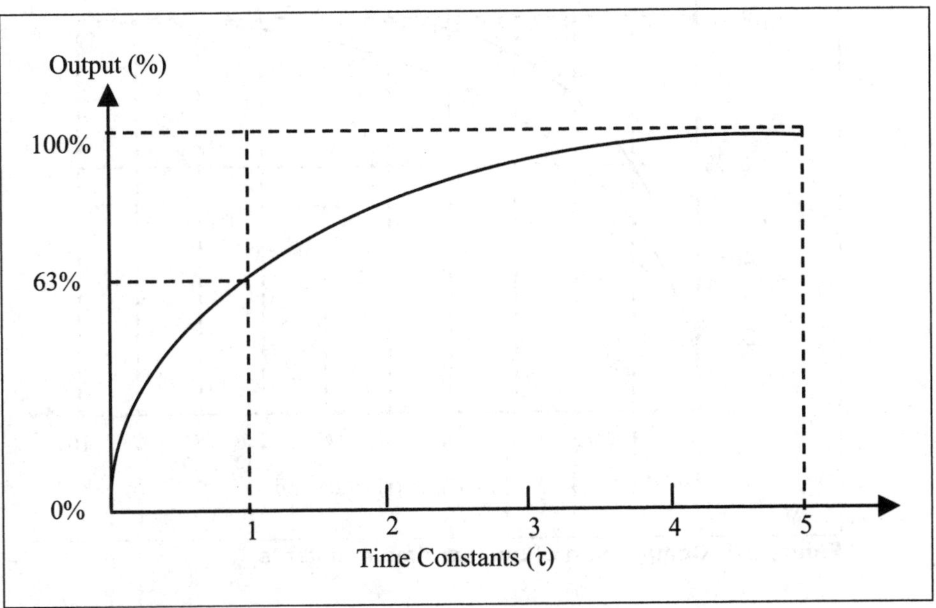

Figure 2-6. Response of a first-order system to step

The time constant also provides an insight into the physical response of a system. Figure 2-7 shows the output response curves for two different "first-order" systems where each system has a different time constant. It can be seen from the two different curves that as the time constant increased, the response of the system to a step change becomes slower. The first system has a time constant of one second, and it takes five seconds to reach 99 percent of full-scale output. The second system has a time constant of two seconds and takes ten seconds to 99 percent of full scale. So, the second system responds more slowly to a step change at the input.

Comparison of Basic Physical Systems

To understand the effect of system time constants on process behavior, we will now look at three common physical systems: electrical, liquid, and thermal. All three systems can be said to have resistance and capacitance. It will become clear that the resistance multiplied by the capacitance of a system produces the process time constant.

Electrical Systems

The characteristics of an electric circuit that has pure resistance and pure capacitance in series are analogous to the resistance and capacitance of

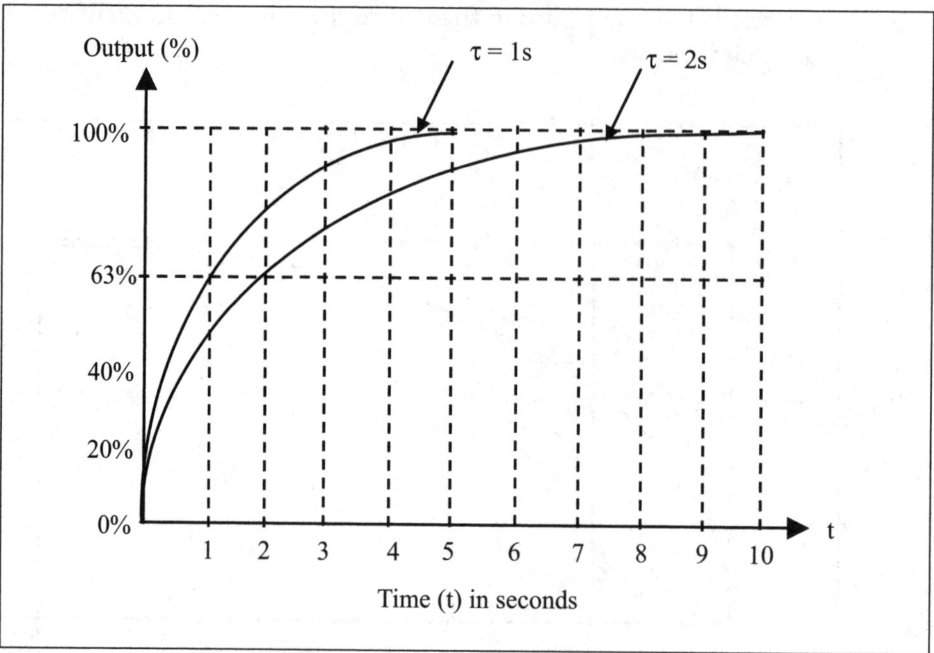

Figure 2-7. Comparison of system time constants

most liquid and thermal systems. Resistance in a purely resistive circuit is defined by Ohm's law, which states that the potential or voltage required to force an electric current through a resistor is equal to the current times the resistance. In equation form, Ohm's law is

$$V = IR \tag{2-8}$$

where

V	=	electrical potential in volts
I	=	current in amps or coulombs per second
R	=	resistance in ohms

The electrical resistance is

$$R = V/I = \text{Potential/Flow} \tag{2-9}$$

The relationship for capacitance states that the charge on a capacitor is equal to the capacitance times the voltage (potential) across the capacitor, or

$$q = CV \tag{2-10}$$

where

q	=	charge in coulombs
C	=	capacitance in farads
V	=	potential in volts

Thus, the capacitance is given by

$$C = \frac{q}{V} = \frac{\text{Charge quantity}}{\text{Potential}} \tag{2-11}$$

A series electrical RC network is shown in Figure 2-8.

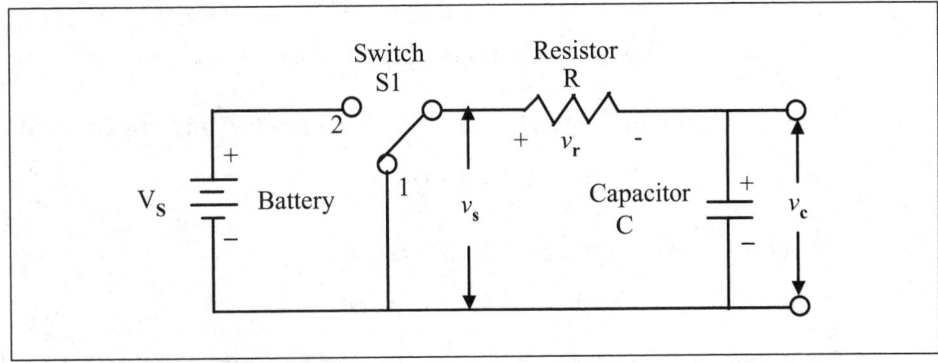

Figure 2-8. RC series circuit

A step input to the circuit is provided by moving switch 1 from position 1 to position 2. This applies the battery voltage to the series circuit and current flows to charge the capacitor (C) to the applied battery voltage V_S. The voltage across the capacitor, $v_c(t)$, is the output of the system and the voltage $v_s(t)$ is the input to the system. The RC series circuit is the signal processing system. When switch 1 is placed into position 2, current flows in the circuit and the capacitor charges up to the applied battery voltage V_S as shown in Figure 2-9. To understand the concept of time constant, we will next derive the system equation for the RC circuit.

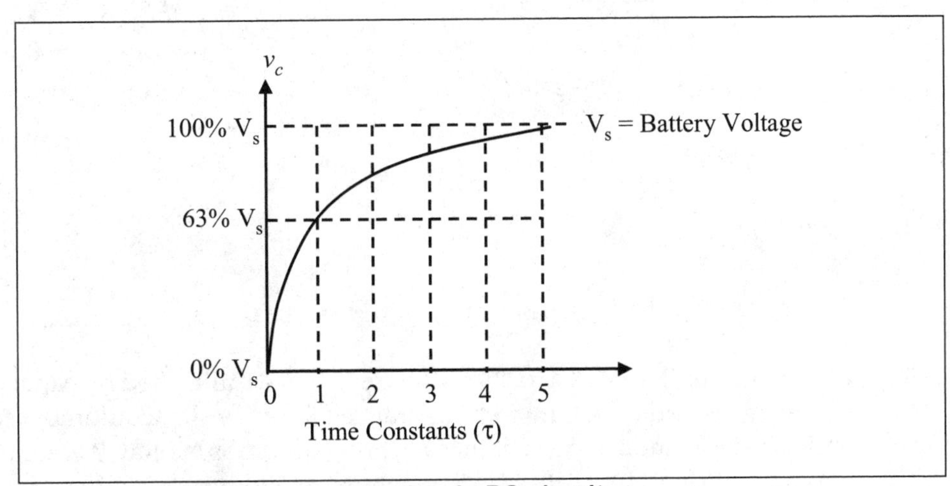

Figure 2-9. Voltage across capacitor in RC circuit

After switch 1 is placed into position 2, the sum of the voltages in the RC section of the circuit equals the applied voltage V_s as follows:

$$V_s(t) = V_r(t) + V_c(t) \quad (2\text{-}12)$$

where

V_s = the instantaneous voltage measured from the common terminal on switch 1 to the negative terminal of the battery
$V_r(t)$ = the instantaneous voltage drop across the resistor
$V_c(t)$ = the instantaneous voltage across the capacitor

According to Ohm's law, the voltage drop across the resistor is given by

$$V_r(t) = i(t)R \quad (2\text{-}13)$$

Therefore, system Equation 2-12 becomes

$$V_s(t) = i(t)R + V_c(t) \quad (2\text{-}14)$$

The equation for charge on a capacitor is given by Equation 2-10, where $q = CV$. To find the instantaneous rate of change of the charge on the capacitor as it charges, the derivative of Equation 2-10 must be taken as follows:

$$\frac{dq}{dt} = C\frac{dV_c}{dt} \quad (2\text{-}15)$$

$$i(t) = \frac{dq}{dt} \quad (2\text{-}16)$$

Substituting the current in Equation 2-15 into the circuit or system Equation 2-14, gives the following:

$$V_s = RC\frac{dV_c}{dt} + V_c \quad (2\text{-}17)$$

Therefore,

$$V_s = \tau\frac{dV_c}{dt} + V_c \quad (2\text{-}18)$$

where $\tau = RC$ is the time constant for the system.

We can use unit analysis to show that resistance multiplied by capacitance results in seconds. The units of resistance (R) are [volts/coulombs/seconds] and the units of capacitance (C) are [coulombs/volts]. If we multiply the units of R by the units of C, we obtain seconds.

The electric circuit *RC* time constant is the time required to charge the capacitor to 63.2 percent of its maximum value in the series circuit after the switch S1 is moved from position 1 to position 2. We can perform a Laplace transform on system Equation 2-18 to obtain

$$V_s = \tau s V_c + V_c \quad (2\text{-}19)$$

or

$$\frac{V_c}{V_s} = \frac{1}{\tau s + 1} \quad (2\text{-}20)$$

Note that the transformed system equation has the same form as the "first-order lag" systems we discussed earlier.

To investigate the concept of time constants in other physical systems, we will next discuss forms of resistance and capacitance in liquid and thermal systems.

Liquid Systems

Figure 2-10 shows liquid flow in a pipe, with a restricting device (a valve) providing a hydraulic resistance (R_h) to the flow. Note that the walls of the pipe will also provide a small amount of resistance to flow, depending on how rough they are.

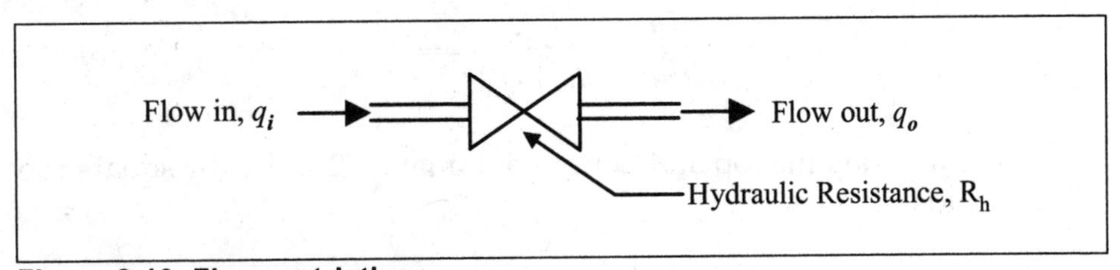

Figure 2-10. Flow restricting

Two main types of flow take place in liquid systems: *laminar* and *turbulent* flow. Laminar flow occurs when the fluid velocity is relatively low and the liquid flows in layers, so the flow is directly proportional to differential pressure or head on the liquid. Unfortunately, laminar flow is seldom encountered in actual practice and usually occurs only with very viscous fluids at low velocities.

Turbulent flow occurs when the fluid velocity is relatively high and the velocity of the liquid at any point varies irregularly. When turbulent flow occurs from a tank discharging under its own head or pressure, the flow is found by the following equation:

$$q = KA\sqrt{2gh} \tag{2-21}$$

Where q is the flow rate (ft^3/s), K is a flow coefficient, A is the area of the discharge orifice (ft^2), g is gravitation constant (ft/s^2), and h is pressure head of liquid (ft). We can define hydraulic resistance (R_h) to flow as follows:

$$R_h \equiv \frac{\text{Potential}}{\text{Flow}} = \frac{h}{q} \tag{2-22}$$

Therefore, the instantaneous rate of change of hydraulic resistance to flow is

$$R_h = \frac{dh}{dq} \tag{2-23}$$

Rearranging Equation 2-21, we arrive at

$$\sqrt{h} = \frac{q}{KA\sqrt{2g}} \tag{2-24}$$

and differentiating Equation 2-24 with respect to q gives us

$$\frac{dh}{dq} = \frac{2\sqrt{h}}{KA\sqrt{2g}} \tag{2-25}$$

Multiplying the top and bottom of Equation 2-25 by the square root of h yields

$$\frac{dh}{dq} = \frac{2h}{KA\sqrt{2gh}} \tag{2-26}$$

According to Equation 2-21, the denominator of Equation 2-26 is q, so substituting q into Equation 2-26 gives us

$$\frac{dh}{dq} = \frac{2h}{q} \tag{2-27}$$

Therefore, instantaneous hydraulic resistance is

$$R_h = \frac{2h}{q} \tag{2-28}$$

The hydraulic resistance is analogous to electrical resistance in that it is inversely proportional to flow q but directly proportional to two times the differential pressure, h, or the driving potential. The difference lies in the fact that turbulent flow involves the square root of the driving potential or head h.

However, we will now show that liquid capacitance is directly analogous to electrical capacitance. Figure 2-11 shows a tank being filled with a liquid. The equation for the volume (V) of the liquid in the tank is given by the following equation:

$$V(t) = Ah(t) \tag{2-29}$$

where

$V(t)$	=	the volume of liquid as a function of time
$h(t)$	=	height of liquid
A	=	the surface area of the liquid in the tank

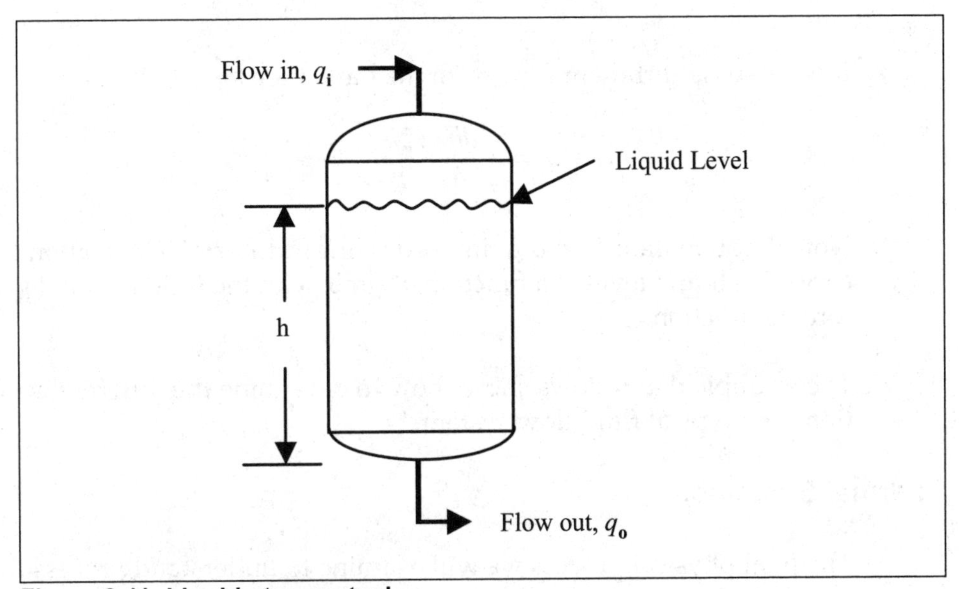

Figure 2-11. Liquid storage tank

Note that the volume V of the tank and the liquid height or head are a function of time. The flow of liquid into the tank, q_i, and the flow liquid out of the tank, q_o, vary with time.

If we solve Equation 2-29 for A, we obtain the following:

$$A = \frac{V(t)}{h(t)} = \frac{quantity}{potential} \tag{2-30}$$

Comparing this equation to the equation for electrical capacitance (i.e., $C = q/V$) clearly shows that liquid capacitance C_1 is simply the surface area of the liquid in the tank, or $C_1 = A$. Furthermore, taking the derivative of Equation 2-29 with respect to time yields

$$\frac{dV(t)}{dt} = A\frac{h(t)}{dt} \tag{2-31}$$

We know that the instantaneous rate of change of volume, dV/dt, is given by the flow in (q_i), minus the flow out (q_o), or

$$\frac{dV}{dt} = q_i - q_o \tag{2-32}$$

Using Equations 2-31 and 2-32, we obtain

$$q_i - q_o = A\frac{dh}{dt} \tag{2-33}$$

If we assume turbulent flow from the tank, then $q_o = 2h/R$, or

$$q_i = A\frac{dh}{dt} + \frac{2h}{R} \tag{2-34}$$

Note that Equation 2-34 is a first-order linear differential equation that expresses liquid level as a function of time, with the fluid flow in (q_i) as the forcing function.

The example that follows shows how to determine the differential equation for a typical fluid flow system.

Thermal Systems

The final physical process we will examine to understand process time constants is a thermal system. The basic thermal processes encountered in the process industries are the mixing of hot and cold fluids, the exchange of heat through adjoining bodies, and the generation of heat by combustion or chemical reaction.

EXAMPLE 2-2

Problem: Determine the differential equation for flow (q_o) out of the process tank shown in Figure 2-11. Find the system time constant if the operating head is 5m, the steady state flow is 0.2m^3/s, and the surface area of the liquid is 10m^2.

Solution: Equation 2-33 states that

$$q_i - q_o = A\frac{dh}{dt}$$

To obtain the differential equation in terms of variable q_o, *dh*/*dt* must be expressed in terms of dq_o/dt. We know from the definition of hydraulic resistance (Equation 2-23) that

$$R_h = \frac{dh(t)}{dq(t)}$$

or $dh(t) = R_h\, dq(t)$. Taking the derivative of this equation with respect to time yields

$$\frac{dh}{dt} = R_h\frac{dq_o}{dt}$$

Therefore, the system equation becomes

$$AR_h\frac{dq_o}{dt} + q_o = q_i$$

To calculate the system time constant, remember that for our system, the liquid capacitance C_l is equal to the surface area of liquid in the tank; therefore, the system time constant τ is given by R_hC_l or R_hA. The hydraulic resistance is given by

$$R_h = \frac{2h}{q}$$

Since h = 5 m and q = 0.2 m^3/s in our example, then

$$R_h = \frac{2(5\text{m})}{0.2m^3/s} = 50\ s/m^2$$

The system time constant τ is given by

$$\tau = R_hA = (50\ s/m^2)(10\ m^2) = 500s.$$

Two laws of thermodynamics are used in the study of thermal systems. The first governs the way in which heat energy is produced and determines the amount generated. The second governs the flow of heat.

Temperature changes in an isolated body conform to the first law of thermodynamics. For a given body, heat input raises the internal energy, and the rate of change of body temperature will be proportional to the heat flow to the body. The constant that relates temperature change and heat flow is called the *thermal capacity* of the body.

$$C\frac{dT}{dt} = q \tag{2-35}$$

where

C = thermal capacitance (cal/°C)
dT/dt = the rate of change of temperature °C/s
q = heat flow (cal/s)

The thermal capacitance of a body is found by multiplying the specific heat of the material by the mass of the body.

$$C = MS \tag{2-36}$$

where

M = the mass of the body (g)
S = the specific heat of the material (cal/g)(°C)

Thermal capacitance is analogous to electric capacitance. For example, in Figure 2-12, heat flowing into a body with thermal capacitance C causes the temperature (T) to rise above the ambient value T_o.

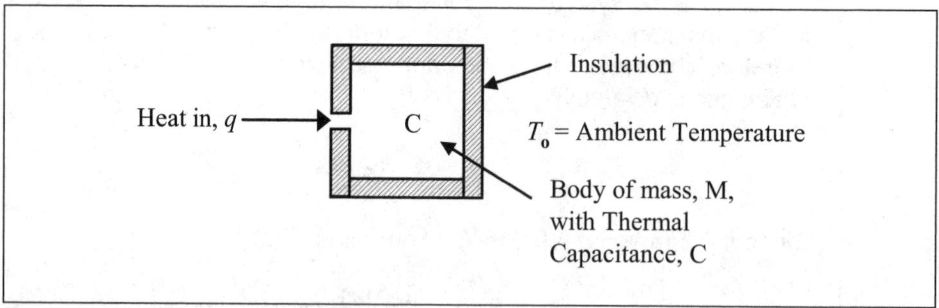

Figure 2-12. Thermal capacitance

Heat flow and charge flow as well as temperature and voltage are analogous quantities. Heat transmission takes place by *conduction, convection,* or *radiation*. Conduction involves transmission through adjoining bodies, convection involves transmission and mixing, and radiation uses electromagnetic waves to transfer heat.

The rate of heat flow through a body is determined by its *thermal resistance*. This is defined as the change in temperature that results from a unit change in heat flow rate. Thermal resistance is normally a linear function, in which case,

$$R_T = \frac{T_2 - T_1}{q} \tag{2-37}$$

where

R_T = thermal resistance (°C/cal/s)
$T_2 - T_1$ = temperature difference in (°C)
q = the heat flow (cal/s)

Thermal resistance is analogous to the resistance in an electrical circuit.

If the temperature of a body is considered to be uniform throughout, its thermal behavior can be described by a linear differential equation. This assumption is generally true for small bodies of gases or liquids where perfect mixing takes place. For such a system, thermal equilibrium requires that at any instant the heat added (q_i) to the system equals the heat stored (q_s) plus the heat removed (q_o). Thus,

$$q_i = q_s + q_o \tag{2-38}$$

Example 2-3 illustrates how to determine the differential equation for a typical thermal system.

Table 2-1 compares variables in the three different physical systems we have discussed. It is important to have a general idea of the physical meaning of process time constants so you can observe a process with some understanding of its capacity to store material or energy. In this way, you can gain some insight into the process dynamics of a system. Such an understanding is very important in process control design.

EXAMPLE 2-3

Problem: Derive the differential equation for the water temperature, T_W, for an insulated tank of water that is heated using an electric heater as shown in Figure 2-13. Assume that the rate of heat flow from the heating element is q_i and that the water is at a uniform temperature. Also assume that no heat is stored in the insulation.

Solution: Using Equation 2-35, the heat stored in the water is as follows:

$$q_w = C\frac{dT}{dt}$$

where C is the thermal capacitance of the water. From Equation 2-37, the heat loss through the insulation is the following:

$$q_o = \frac{T_W - T_o}{R}$$

where T_o is the temperature of the air outside the tank. Applying Equation 2-38 yields

$$q_i = q_s + q_o$$

or

$$q_i = C\frac{dT_W}{dt} + \frac{T_W - T_o}{R}$$

By multiplying this equation by R, we obtain

$$Rq_i = RC\frac{dT_W}{dt} + T_W - T_o$$

or

$$RC\frac{dT_W}{dt} + T_W = T_o + Rq_i$$

If we replace RC with the time constant symbol, τ, we obtain

$$\tau\frac{dT_W}{dt} + T_W = T_o + Rq_i$$

which is a first-order differential equation for water temperature T_W.

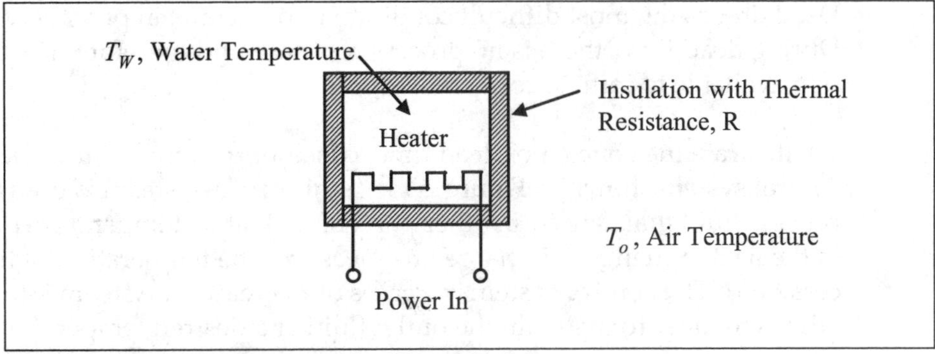

Figure 2-13. Thermal system for Example 2-3

Table 2-1. Comparison of Physical Units

Variable	System Type		
	Electrical	**Liquid**	**Thermal**
Quantity	Coulomb (c)	Cubic meter (m^3)	cal
Potential	Volt (v)	meter (m)	°C
Flow	Ampere (A)	m^3/s	cal/s
Resistance	Ohms (Ω)	m/m^3/s	°C/cal/s
Capacitance	Farad (F)	m/m^3	cal/°C
Time	Second (s)	Second (s)	Second (s)

Dead Time Lag

In general, not all processes can be neatly characterized by first-order lags. In some cases, a process will produce a response curve like that shown in Figure 2-14. In this process, the maximum rate of change for the output does not occur at time zero (t_o) but at some later time (t_1). This is called *dead time* in process control: the period of time (t_d) that elapses between the moment a change is introduced into a process and the moment the output of the process begins to change. Dead time is shown in Figure 2-14 as the time between t_1 and t_o, or $t_d = t_1 - t_o$.

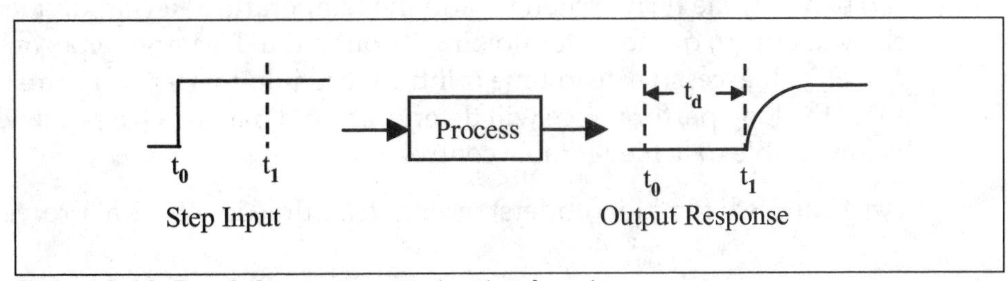

Figure 2-14. Dead time response to step input

Dead time is the most difficult condition to overcome in process control. During dead time, there is no process response and therefore no information available to initiate corrective action.

To illustrate the concept of dead time, consider the temperature feedback control system shown in Figure 2-15. In this process, steam is used to heat process fluid that is used in other parts of a plant. A temperature detector in the heat exchanger discharge line measures the temperature of the process fluid. The control system increases or decreases the steam into the heat exchanger to maintain the outlet fluid at a desired temperature.

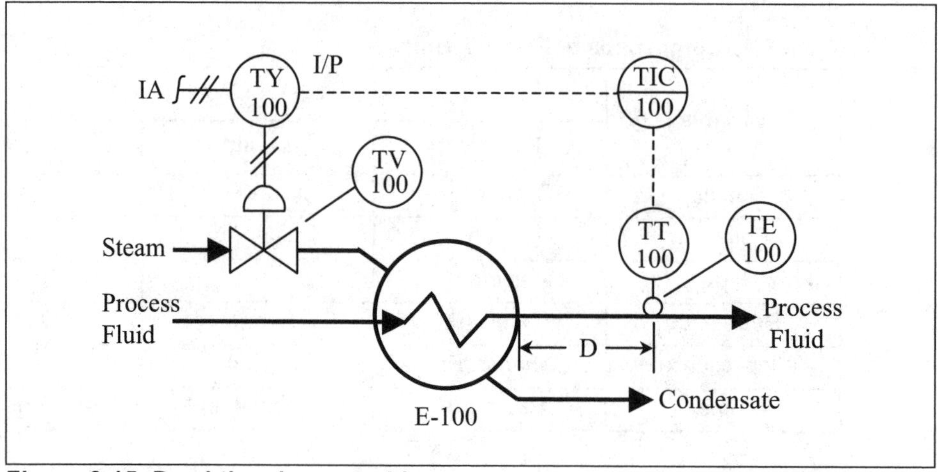

Figure 2-15. Dead time in control loop

In the design of this control loop, the location of the temperature detector is critical. It is tempting to say that the detector should be installed farther down the outlet pipe and closer to the point at which the process fluid is used. This seems correct because the temperature of the process fluid in the heat exchanger or at the discharge point of the exchanger is of little importance to the user of the heated process fluid. However, this line of reasoning can be disastrous because, as the detector is moved farther and farther down the line, a larger and larger dead time is introduced. In the example in Figure 2-15, the process dead time (t_d) is equal to the distance (D) between the heat exchanger and the temperature sensor, divided by the velocity (v) of the water flowing through the discharge pipe, or $t_d = D/v$. If excessive dead time is introduced into this temperature control loop, the loop performance will deteriorate and may reach a point where it is impossible to achieve stable control.

An example will aid in understanding dead time in a typical process.

EXAMPLE 2-4

Problem: Determine the dead time for the process shown in Figure 2-15 if the temperature detector is located 50 meters from the heat exchanger and the velocity of the process fluid in the discharge pipe is 10 m/s.

Solution: The dead time is given by $t_d = D/v$. Since D = 50m and v = 10m/s

$$t_d = \frac{D}{v} = \frac{50\text{m}}{10\text{m/s}} = 5\text{s}$$

You can detect process dead time from a process response curve by measuring the amount of time that elapses before any output response occurs after an input change is applied to a system. Process reaction curves are discussed in a later section of this chapter.

Advanced Control Loops

Up to this point, we have discussed only single-loop feedback control. This section will now expand the discussion to include *cascade, ratio,* and *feedforward* control loops.

Cascade Control Loops

The general concept of cascade control is to place one feedback loop inside another. In effect, one takes the process being controlled and finds some intermediate variable within the process to use as the set point for the main loop.

Cascade control exhibits its real value when a very slow process is being controlled. In such circumstances, errors can exist for a very long time, and when disturbances enter the process there may be a significant wait before any corrective action is initiated. Also, once corrective action is taken one may have to wait a long time for results. Cascade control allows the operator to find intermediate controlled variables and to take corrective action on disturbances more promptly. In general, cascade control offers significant advantages and is one of the most underutilized feedback control techniques.

An important question you may confront when implementing cascade control is how to select the most advantageous secondary controlled variable. Quite often, the designer has a large number of choices. The overall strategy or goal should be to get as much of the process lag into the outer loop as possible while, at the same time, getting as many of the disturbances as possible to enter the inner loop.

Figures 2-16 and 2-17 illustrate two different cascade control arrangements for a furnace that is used to increase the temperature of a fluid that is passing through it. In both cases, the primary controlled variable is the same, but in each case a different intermediate controlled variable has been selected. The question then is which type of cascade control is best.

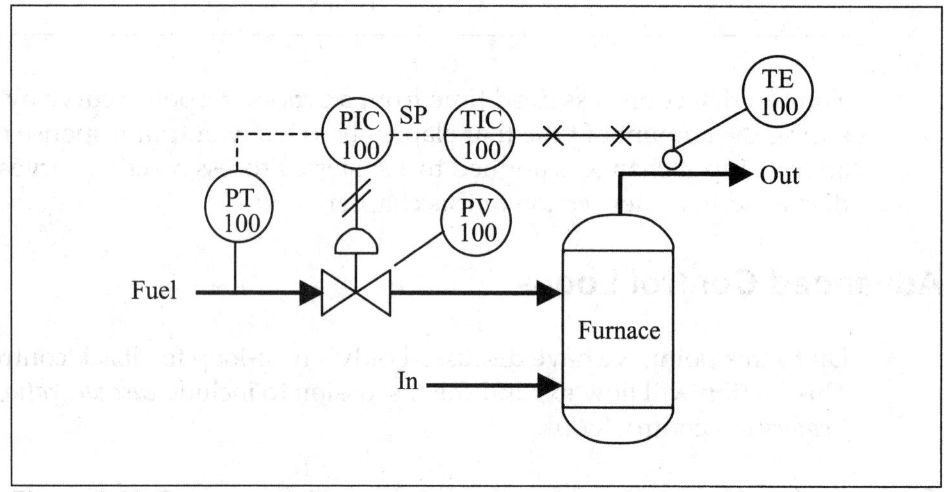

Figure 2-16. Pressure and temperature cascade control

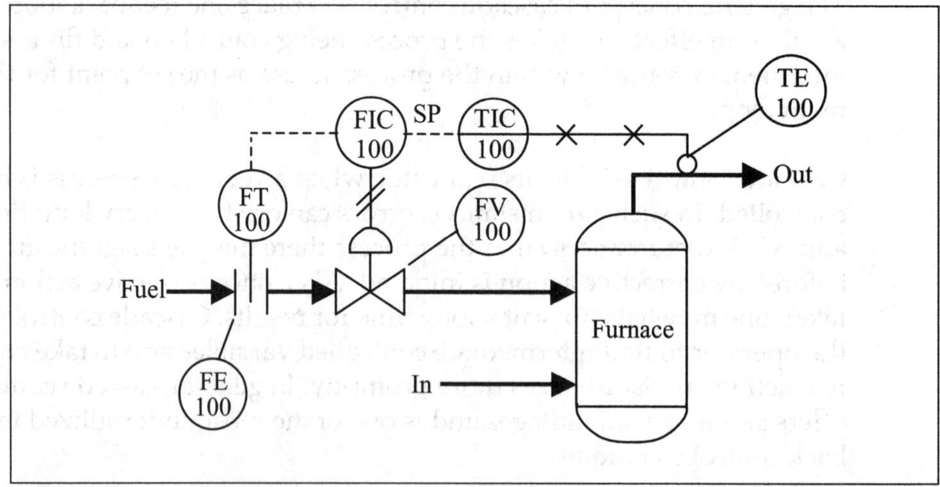

Figure 2-17. Flow and temperature cascade control

To determine the best cascade control arrangement, you must identify the most likely disturbances to the system. It is helpful to make a list of these in order of increasing importance. Once this has been done, the designer can review the various cascade control options available and determine which one best meets the overall strategy outlined earlier: to make the inner loop as fast as possible while at the same time receiving the bulk of the important disturbances.

If both controllers of a cascade control system are three-mode controllers, there are a total of six tuning adjustments. It is doubtful that such a system could ever be tuned effectively. Therefore, you should select with care the modes to be included in both the primary and secondary controllers of a cascade arrangement.

For the secondary (inner or slave) controller, it is standard practice to include the proportional mode. There is little need to include the reset mode to eliminate offset because the set point for the inner controller will be reset continuously by the outer or master controller. For the outer loop, the controller should contain the proportional mode. If the loop is sufficiently important to merit cascade control then you should probably include reset to eliminate offset in the outer loop. You should undertake rate or derivative control in either loop only if it has a very large amount of lag.

The tuning of cascade controllers is the same as the tuning of all feedback controllers, but the loop must be tuned from the inside out. The master controller should be put on manual (i.e., the loop broken), and then the inner loop can be tuned. Once the inner loop is properly tuned, the outer loop can be tuned. This allows the outer loop to "see" the tuned inner loop functioning as part of the total process or as the "all else" that is being controlled by the master controller. If you follow this general inside-first principle when tuning cascade controllers, you should encounter no special problems.

Ratio Control Loops

Another common type of feedback control system is *ratio control*. Based on hardware, ratio control is quite often confused with cascade control. The basic operation of ratio control, however, is quite different.

Ratio control is often associated with process operations in which two or more streams must be mixed together continuously to maintain a steady composition in the resulting mixture. A practical way to do this is to use a conventional flow controller on one stream and to control the other stream with a ratio controller that maintains flow in some preset ratio or fraction to the primary stream flow. A preset ratio regulates the flow of the controlled variable. For example, if the ratio is 10 to 1, then for every gallon per minute of the uncontrolled variable that is flowing, ten gallons per minute of the controlled variable are allowed to flow. A typical ratio control system is shown in Figure 2-18.

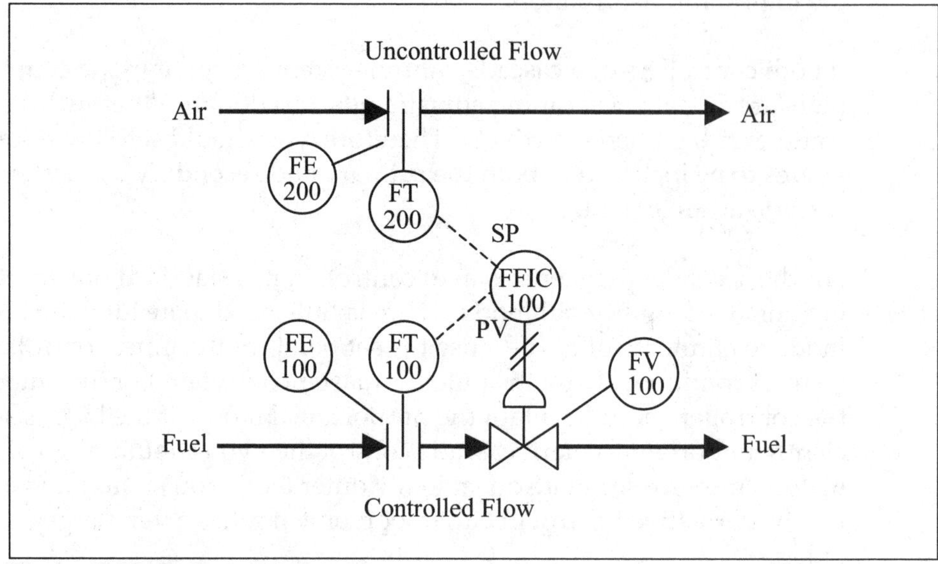

Figure 2-18. Ratio control

Use the signal from the uncontrolled flow transmitter, or *wild flow*, as the ratio input of the ratio controller. Multiply the value by an adjustable factor or ratio setting to determine the set point of the flow controller. The process variable to the controller is the flow of the controlled stream. The output from the ratio controller adjusts the control valve.

You can use a ratio control loop with any combination of suitably related process variables, and the control action selected is normally proportional plus integral. The response to process upsets of a control loop with ratio control is the same as the response found in a single-feedback loop.

Feedforward Control Loops

A feedback control loop is reactive in nature and represents a response to the effect of a load change or upset. A feedforward control loop, on the other hand, responds directly to load changes and thus provides improved control.

In *feedforward control*, a sensor is used to detect process load changes or disturbances as they enter the system. A block diagram of a typical feedforward control loop is shown in Figure 2-19. Sensors measure the values of the load variables, and a computer calculates the correct control signal for the existing load conditions and process set point.

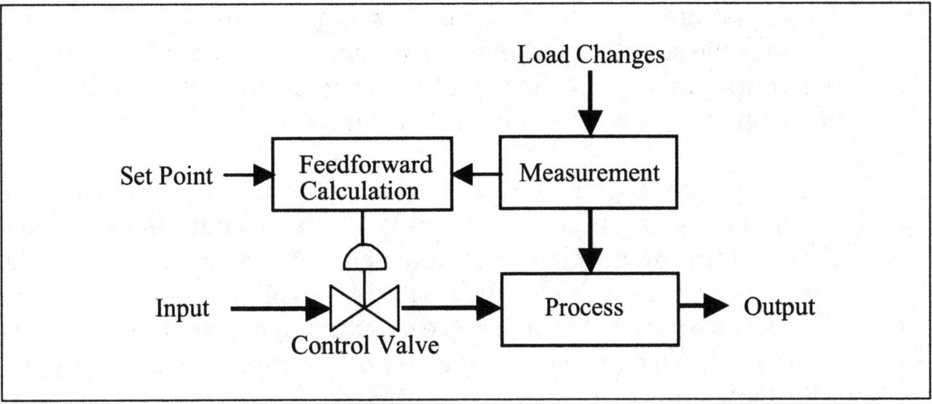

Figure 2-19. Feedforward control

Feedforward control poses some significant problems. Its configuration assumes that the disturbances are known in advance, that they will have sensors associated with them, and that no important undetected disturbances will occur.

Therefore, feedforward control is more complicated and more expensive, and it requires the operator to have a better understanding of the process than does a standard feedback control loop. So, feedforward control is generally reserved for well-understood and critical applications.

Tuning Control Loops

There are three important factors to consider when tuning a PID controller: the characteristics of the process, the selection of controller modes, and the performance criteria of the control loop.

Selecting Controller Modes

Which controller mode to select is an important consideration when tuning controllers. Four basic control combinations are used: proportional only, proportional plus integral (PI), proportional plus derivative (PD), and proportional plus integral plus derivative (PID).

The most basic continuous control mode is proportional control. Here, the controller output (m) is algebraically proportional to the error (e) input signal to a controller. The equation for proportional control is given by the following equation:

$$m = K_c e \tag{2-39}$$

This equation is also called the proportional control *algorithm*. This control action is the simplest and most commonly encountered of all the continuous control modes. In effect, there is a continuous linear relationship between the controller input and output.

The proportional gain of the controller is the term K_C, which is also referred to as the proportional sensitivity of the controller. K_C indicates the change in the manipulated variable per unit change in the error signal. In a true sense, the proportional sensitivity or gain is a multiplication term. It represents a parameter on a piece of actual hardware that must be adjusted by a technician or engineer; for example, the gain may be an adjustable knob on the process controller.

On many industrial controllers, this gain-adjusting mechanism is not expressed in terms of proportional sensitivity or gain but in terms of proportional band (*PB*). Proportional band is defined as the span of values of the input that corresponds to a full or complete change in the output. This is usually expressed as a percentage. It is related to proportional gain by the following equation:

$$\mathrm{PB} = \frac{100\%}{K_c} \tag{2-40}$$

Most controllers have a scale that indicates the value of the final controlled variable. Therefore, the proportional band can be conveniently expressed as the range of values of the controlled variable that corresponds to the full operating range of the final control valve. As a matter of practice, *wide bands* (high percentages of *PB*) correspond to less sensitive response, and *narrow bands* (low percentages) correspond to more sensitive response of the controller.

Proportional control is quite simple and is the easiest of the continuous controllers to tune since there is only one parameter to adjust. It also provides very rapid response and is relatively stable. Proportional control has one major disadvantage, however, at steady state, it exhibits offset. That is, there is a difference at steady state between the desired value or set point and the actual value of the controlled. To correct for this problem proportional control is combined with integral action.

Integral action or *reset action* is the integration of the input error signal *e* over a very small time period *dt*. In effect, this means that for integral action, the value of the manipulated variable *V* is changed at a rate that is proportional to the amount of error *e* that exists for a given duration. In other words, integral control responds to the duration of the error as well as to its magnitude and direction. When the controlled variable is at the set point, the final control element remains stationary. This indicates that, at steady state, when integral action is present there can be no offset; therefore, the steady-state error is zero.

This combination of proportional and integral action is called *PI control*. The control equation for PI control action is as follows:

$$V = K_c e + \frac{K_c}{t_i} \int_0^t e\,dt \tag{2-41}$$

where

K_c	=	the proportional gain
t_i	=	the *integral time*
e	=	the error signal
V	=	controller output

The advantage of including the integral mode with the proportional mode is that the integral action eliminates offset. Typically, there is some decreased stability due to the presence of the integral mode. In other words, the addition of the integral action makes the total loop slightly less stable. One significant exception to this is in liquid flow control. Liquid flow control loops are extremely fast and tend to be very noisy. As a result, plants often add integral control to the feedback controller in liquid flow control loops to provide a dampening or filtering action for the loop. Of course, the advantage of eliminating any offset is still present, but this is not the principal motivating factor in such cases.

Tuning a PI controller is more difficult than tuning a simpler proportional controller. Two separate tuning adjustments must be made, and each depends on the other. The difficulty of tuning a controller increases dramatically with the number of adjustments that must be made.

It is conceivable to have a control action that is based solely on the rate of change or derivative of the error signal *e*. Although this is possible, it is not practical because, while the error might be large, if it were unchanging the controller output would be zero. Thus, *rate control* or derivative control is usually found in combination with proportional control. The equation for a proportional-derivative (PD) controller is as follows:

$$V = K_c e + K_c t_d \frac{de}{dt} \tag{2-42}$$

where t_d is the derivative time.

By adding derivative action to the controller, lead time is added in the controller to compensate for lag around the loop. Almost any process has a time delay or lag around the loop; therefore, the theoretical advantages of lead in the controller are appealing. However, it is a difficult control action to implement and adjust, and its usage is limited to cases of extensive lag in the process. This often occurs with large temperature-control systems. Taking the derivative of the error signal has the side effect of producing upsets whenever the set point is changed, so most controllers take the derivative of the process signal.

Adding derivative control to the controller makes the loop more stable if it is correctly tuned. Since the loop is more stable, the proportional gain may be higher, and it can thus decrease offset better than proportional action alone. It does not, of course, eliminate offset.

The equation for a three-mode controller, or PID (proportional-integral-derivative) controller, is as follows:

$$V = K_c e + \frac{K_c}{t_i}\int_0^t e\,dt + K_c t_d \frac{de}{dt} \tag{2-43}$$

The three-mode control gives rapid response and exhibits no offset, but it is very difficult to tune because of the three terms that must be adjusted. As a result, it is used only in a very small number of applications, and operators often must adjust it extensively and continuously to keep it properly tuned. The PID mode, however, offers excellent control when proper tuning is used.

Five common types of control loops are encountered in process control: flow, level, temperature, analytical, and pressure. Table 2-2 provides basics guidelines for the controller modes that are generally used for each type of process control loop.

Table 2-2. Guidelines for Selecting Controller Modes

Control Loop	Controller Mode		
	Proportional	Integral	Derivative
Flow	Always	Usually	Never
Level	Always	Usually	Rarely
Temperature	Always	Usually	Usually
Analytical	Always	Usually	Sometimes
Pressure	Always	Usually	Sometimes

Performance Criteria

The final factor plants must consider before tuning a loop is the performance criteria that are to be used. There are four widely used criteria; each is based on a different method for minimizing the integral error of the deviation of measured control signal from the set point. The first three are the minimum integral of square error (ISE), the minimum integral of absolute error (IAE), and the minimum integral of absolute error multiplied by time (ITAE). In equation form, these three criteria are, respectively,

$$ISE = \int_0^\infty e^2 dt \tag{2-44}$$

$$IAE = \int_0^\infty |e| dt \tag{2-45}$$

$$ITAE = \int_0^\infty t|e| dt \tag{2-46}$$

These three integral criteria methods are best suited for computer-based control applications. They are recommended only for such applications, where they can give excellent tuning and control results.

The fourth and most commonly used performance criterion was presented by J. G. Ziegler and N. B. Nichols at the annual meeting of the American Society of Mechanical Engineers in New York City in December 1941. It is known as the "Ziegler and Nichols one-quarter wave decay" criterion and is the basis of their tuning methods. A decay ratio of one-quarter means that the ratio of the overshoot of the first peak in the process response curve to the overshoot of the second peak is 4:1 (this is illustrated in Figure 2-20). Using a decay ratio of one-quarter represents a compromise between a rapid initial response and a fast return to the set point.

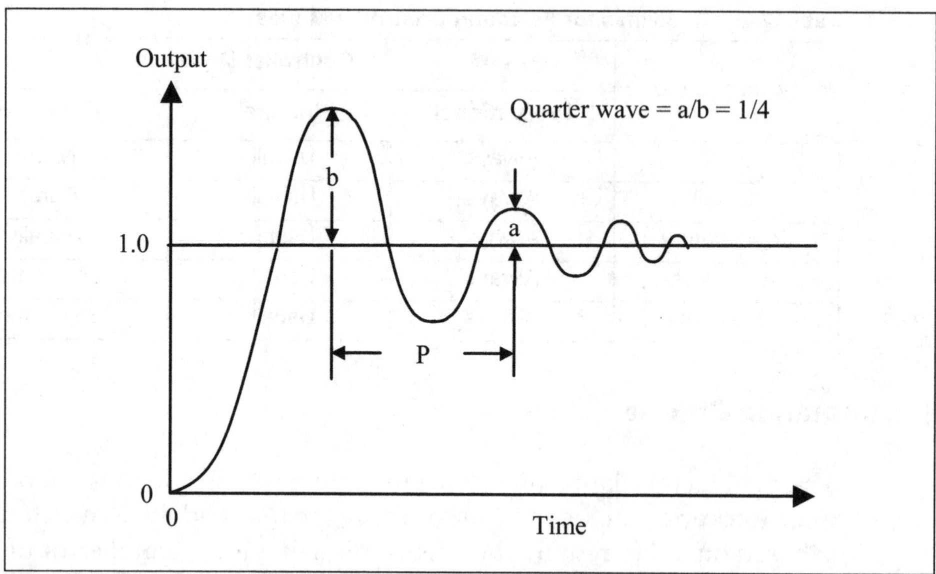

Figure 2-20. Process response curve for one-quarter decay ratio

Ziegler-Nichols Tuning Methods

Ziegler and Nichols called their tuning procedures the *ultimate* method because they required users to determine the ultimate gain and ultimate period for the closed control loop. The ultimate gain is the maximum allowable value of gain for a controller that has only a proportional mode in operation for which the closed-loop system shows a stable sine wave response to a disturbance.

Suppose we have a closed-loop feedback control system that has the controller in the automatic mode, as shown in Figure 2-21c. If we increase the proportional gain the loop will tend to oscillate. If the gain is increased more, continuous cycling or oscillation in the controller output variable occur, as shown in Figure 2-21b. This is the maximum gain at which the system can be operated before it becomes unstable. This gain is called the ultimate gain or sensitivity (S_u). The period of these sustained oscillations is called the ultimate period (P_u). If the gain is increased past this point, the system will become unstable, as shown in Figure 2-21a.

To determine the ultimate gain and the ultimate period, we perform the following steps:

1. Remove the reset and derivative action from the controller by setting the derivative time to zero, the reset time to infinity or the highest value possible, and the proportional gain to one.
2. Place the controller in automatic and make sure the loop is closed.

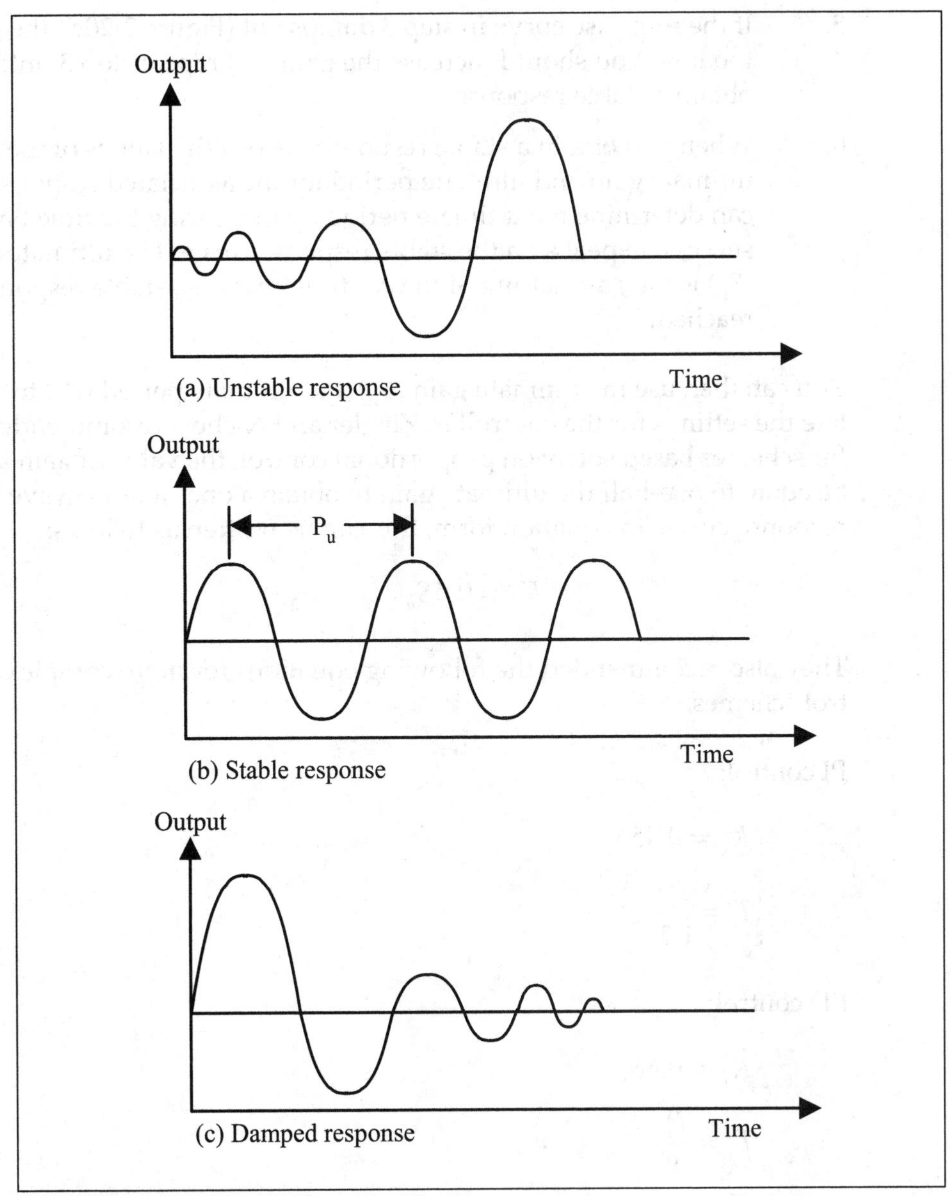

Figure 2-21. Typical process response curves

3. Impose an upset on the control loop and observe the response. The easiest way to impose an upset is to change the set point by a small amount.

4. If the response curve produced by step 3 does not damp out but is unstable (Figure 2-20a), the gain is too high. You should decrease gain and repeat step 3 until you obtain the stable response shown in Figure 2-20b.

5. If the response curve in step 3 damps out (Figure 2-20c), the gain is too low. You should increase the gain and repeat step 3 until you obtain a stable response.

6. When you obtain a stable response, record the values of the ultimate gain and ultimate period for the associated response. You can determine the ultimate period by measuring the time between successive peaks on the stable response curve. The ultimate gain (S_u) is the gain setting of the controller when a stable response is reached.

You can then use the ultimate gain (S_u) and ultimate period (P_u) to calculate the settings for the controller. Ziegler and Nichols recommended that for schemes based solely on proportional control, the value of gain should be equal to one-half the ultimate gain to obtain a one-quarter-wave response curve. In equation form, this can be written as follows:

$$K_c = 0.5S_u \tag{2-47}$$

They also recommended the following equations for more complex control schemes.

PI control:

$$K_c = 0.45S_u \tag{2-48}$$

$$T_i = \frac{P_u}{1.2} \tag{2-49}$$

PD control:

$$K_c = 0.6S_u \tag{2-50}$$

$$T_d = \frac{P_u}{8} \tag{2-51}$$

PID control:

$$K_c = 0.6S_u \tag{2-52}$$

$$T_i = 0.5P_u \tag{2-53}$$

$$T_d = \frac{P_u}{8} \tag{2-54}$$

These equations are empirical and are intended to achieve a decay ratio of one-quarter wave, which Ziegler and Nichols defined as good control. In many cases, this criterion is insufficient for specifying a unique combination of controller settings, each with a different period. (In two-mode or three-mode controllers an infinite number of settings will yield a decay ratio of one-quarter.) This illustrates the problem of defining what constitutes sufficient control.

In some cases, it is important that you tune the system so there is no overshoot. In other cases, a slow and smooth response is required. Some applications require a very rapid response in which high oscillations are not a problem. You will have to determine the proper control scheme for each specific loop.

The only parameter that needs to be adjusted in a single-mode controller is the proportional gain K_C. Two parameters must be adjusted in a two- mode *PI* controller: the proportional gain K_c and the integral time t_i. Three parameters must be adjusted in a PID controller: the controller gain K_c for the proportional mode, the integral time t_i for the integral mode, and the derivative time t_d for the derivative mode. When the controller is being adjusted, the gains around the loop will tend to dictate what the optimum gain in the controller should be. Similarly, the time constants and dead times that characterize the lag dynamics of the process will tend to dictate the optimum value of the reset time and the proper derivative time in the controller. In other words, before you can calculate or select the best values for the tuning parameters in the controller, you must obtain quantitative information about the overall gain and the process lags that are present in the balance of the feedback loop. This illustrates why controllers must be individually tuned at the process plant, rather than at the factory.

Example 2-5 illustrates the use of the Ziegler-Nichols method to determine controller settings.

Process Reaction Curve Method

Another method proposed by Ziegler and Nichols for tuning control loops is based on data from the process reaction curve for the system under control. The process reaction curve is simply the *reaction* of the process to a step change in its input signal. This process curve is the reaction of all components in the control system (excluding the controller) to a step change to the process. It is first-order process with a time delay, which is the most common process encountered in control applications.

EXAMPLE 2-5

Problem: The Ziegler-Nichols ultimate method was used to determine an ultimate sensitivity of 0.3 psi/ft and an ultimate period of 1 min for a level control loop. Determine the PID controller settings that are needed for good control.

Solution: Using the equations for PID control,

$$K_c = 0.6S_u = (0.6)(0.3 \text{ psi/ft}) = 0.18 \text{ psi/ft}$$

$$T_i = 0.5P_u = 0.5(1 \text{ min}) = 0.5 \text{ min}$$

$$T_d = \frac{P_u}{8} = \frac{1 \text{ min}}{8} = 0.125 \text{ min}$$

To obtain a graph of a process reaction curve, place a high-speed recorder in the control loop, as shown in Figure 2-22. To obtain a recording of the reaction curve, perform the following steps:

1. Allow the control system under study to come to a steady state.
2. Place the process controller for the system in the manual mode.
3. Manually adjust the controller output signal to the value at which it was operating automatically.
4. Wait until the control system comes to a steady state.
5. With the controller still in manual mode, introduce a step change in the controller output signal. Normally, this is done by making a small step change in the controller set point.
6. Record the response of the measured variable.
7. Then, return the process controller output signal to its previous value and place the controller in the automatic mode.

Once you have completed these steps, you can use the recorded process reaction curve to calculate the tuning parameters that are needed for the process controller. To use this process reaction curve method, you must determine only the parameters R_r and L_r. A sample determination of these parameters for a control loop is illustrated in Figure 2-23.

To obtain process information parameters in the process reaction curve method, draw a tangent to the curve at its point of maximum slope. This

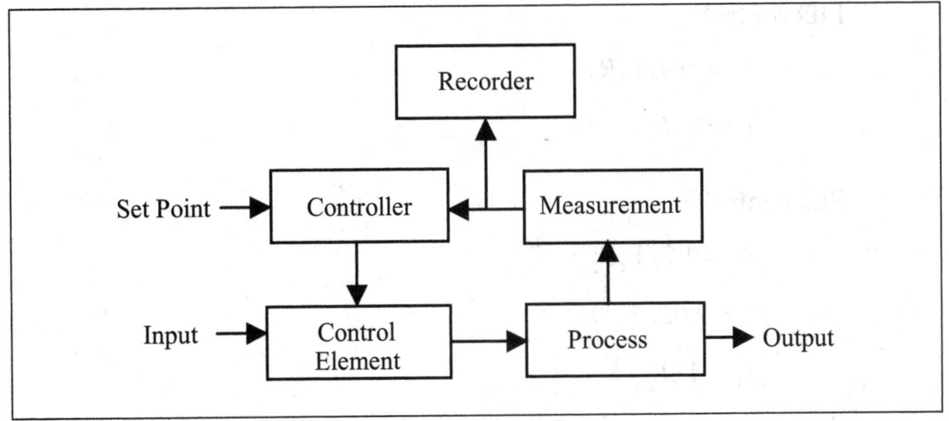

Figure 2-22. Determining process reaction curve

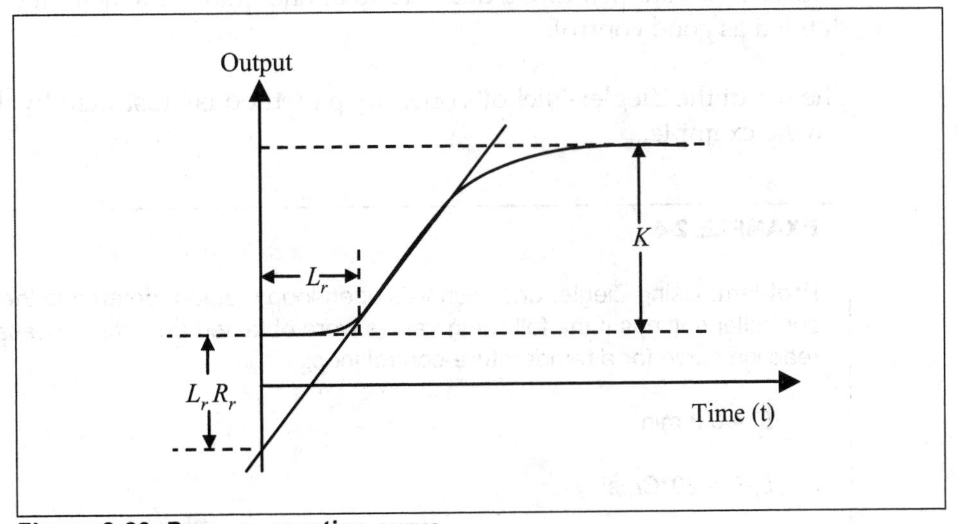

Figure 2-23. Process reaction curve

slope, R_r, is the process reaction rate. The intersection of this tangent line with the original baseline gives an indication of L_r, the process lag. L_r is really a measure of overall dead time for the control valve, the measurement transducer, and the process. If you extrapolate this tangent drawn at the point of maximum slope to a vertical axis that is drawn when the step is imposed, then the amount by which this is below the horizontal baseline will represent the product L_rR_r. Using these parameters, Ziegler and Nichols recommend that the following equations be used to calculate appropriate controller settings for optimal control.

Proportional-only control:

$$K_c = 1/L_rR_r \tag{2-55}$$

PID control:

$$K_c = 0.9/L_r R_r \tag{2-56}$$

$$t_i = 3.3L_r \tag{2-57}$$

PID control:

$$K_c = 1.2/\mathrm{L_r R_r} \tag{2-58}$$

$$t_i = 2.0L_r \tag{2-59}$$

$$t_d = 0.5L_r \tag{2-60}$$

Ziegler and Nichols indicated that using these equations to tune a PID controller should produce a decay ratio of one-quarter, which they defined as good control.

The use of the Ziegler-Nichol's open-loop method is illustrated by the following example.

EXAMPLE 2-6

Problem: Using Ziegler and Nichols's open-loop method, determine the PID controller settings if the following values were obtained from the process reaction curve for a temperature-control loop:

L_r = 0.8 min

$L_r R_r$ = 20°C/psi

Solution: Using the equations for PID control,

$K_c = 1.2/L_r R_r = 1.2/(20°C/psi) = 0.06$ psi/°C

$t_i = 2.0L_r = 2.0(0.8 \text{ min}) = 1.6$ min

$t_d = 0.5L_r = 0.5(0.8 \text{ min}) = 0.4$ min

We have presented several practical and theoretical approaches for tuning controllers in this chapter. You can use many other mathematical equations to predict the ideal proportional, integral, and derivative settings for a given process. Some process personnel tune controllers by relying on their experience or trial-and-error methods. However, it is generally best to use one of the methods presented here until you have gained some experience in tuning control loops.

EXERCISES

2.1 Explain the operation of the single-loop feedback control loop shown in Figure 2-1.

2.2 Define the term *lag* in the context of process control.

2.3 A sensor measures temperature linearly with a transfer function of 22 mv/°C and a 0.5 s time constant. Find the sensor output 1 s after the input changes rapidly from 20°C to 22°C.

2.4 Show that the time constant (τ) in an electrical resistance and capacitance circuit system has the units of time.

2.5 Show that the time constant (τ) in a thermal system has the units of time.

2.6 The tank shown in Figure 2-11 has an operating head (h) of 5 ft and a normal output flow of 5 ft^3/s. The cross-sectional area of the tank is 16 ft^2. Find the system time constant (τ).

2.7 Explain the concept of *dead time* in a process.

2.8 Determine the dead time for the process shown in Figure 2-15 if the temperature detector is located 5 m from the water tank and the velocity of the water in the discharge pipe is 5 m/s.

2.9 Describe the type of process that typically uses Ratio Control.

2.10 Discuss the typical method used to tune a Cascade Control Loop.

2.11 Explain why proportional plus integral (PI) action is effective in most flow control applications.

2.12 The Ziegler-Nichols ultimate method was used to determine an ultimate sensitivity of 0.6 psi/ft and an ultimate period of 2 min for a level control loop. Determine the PID controller settings needed for a one-quarter damping response.

2.13 Using the Ziegler-Nichols open-loop method, determine the PID controller settings if the following values are obtained from the reaction curve for a temperature control loop: $L_r = 1.0$ min, $L_r R_r = 15°C/psi$, and $K_c = 26°C/psi$.

3

Electrical and Electronic Fundamentals

Introduction

The design of programmable control systems requires a thorough knowledge of the basic principles of electricity and electronics. This chapter will discuss the fundamentals of electricity and magnetism and then examine the electrical and electronic circuits that are commonly encountered in process control. We will also discuss the operation and purpose of electro-magnetic and electronic devices, such as transformers, power supplies, electromagnetic relays, solenoid valves, contactors, control switches, and indicating lights.

Fundamentals of Electricity

Electricity is a fundamental force in nature that can produce heat, motion, light, and many other physical effects. The force is an attraction or repulsion between electric charges called *electrons* and *protons*. An electron is a small atomic particle having a negative electric charge. A proton is a basic atomic particle with a positive charge. It is the arrangement of electrons and protons that determines the electrical characteristics of all substances. As an example, this page of paper contains electrons and protons, but you cannot detect any evidence of electricity because the number of electrons equals the number of protons. In this case, the opposite electrical charges cancel, making the paper electrically neutral.

If we want to use the electrical forces that are associated with positive and negative charges, work must be performed that separates the electrons and protons. For example, an electric battery can do such work because its chemical energy separates electrical charges to produce an excess of electrons at its negative terminal and an excess of protons at its positive terminal.

The basic terms encountered in electricity are electric *charge* (Q), *current* (I), *voltage* (V), and *resistance* (R). For most common applications of electricity a charge of a very large number of electrons and protons is required. It is therefore convenient to define a practical unit called the *coulomb* (C) that is equal to the charge of 6.25×10^{18} electrons or protons. The symbol for electric charge is Q, which stands for *quantity of charge*, and a charge of 6.25×10^{18} electrons or protons is stated as $Q = 1$ C. Voltage is the potential difference or force produced by the differences in opposite electrical charges. Fundamentally, the volt is a measure of the work required to move an electric charge. When 1 *joule* of work is required to move 1 C between two points, the potential difference is 1 V.

When the potential difference between two different charges forces a third charge to move, the charge in motion is called *electric current*. If the charges move past a given point at the rate of 1 C per second, the amount of current is defined as 1 *ampere* (*A*). In equation form, $I = dQ/dt$, where I is the instantaneous current in amperes and dQ is the differential amount charge in coulombs passing a given point during the time period (dt) in seconds. If the current flow is constant, it is simply given by the following equation:

$$I = \frac{Q}{t} \tag{3-1}$$

where Q is the amount of current in coulombs flowing past a given point in t seconds.

Example 3-1 shows how to calculate current flow.

EXAMPLE 3-1

Problem: A steady flow of 12 C of charge passes a given point in an electrical conductor every 3 seconds. What is the current flow?

Solution: Since constant current flow is defined as the amount of charge (Q) that passes a given point per period of time (t) in seconds, we have

$$I = \frac{Q}{t} = \frac{12 \text{ coulombs}}{3 \text{ seconds}} = 4 \text{ amps}$$

The fact that a conductor carrying electric current can become hot is evidence that the work done by the applied voltage in producing current must be meeting some form of opposition. This opposition, which limits current flow, is called *resistance.*

Conductivity, Resistivity, and Ohm's Law

An important physical property of some material is *conductivity*, that is, the ability to pass electric current. Suppose we have an electric wire (conductor) of length L and cross-sectional area A, and we apply a voltage V between the ends of the wire. If V is in volts and L is in meters, we can define the voltage gradient (E), as follows:

$$E = \frac{V}{L} \tag{3-2}$$

Now, if a current I in amperes flows through a wire of area A in meters squared (m^2), we can define the current density J as follows:

$$J = \frac{I}{A} \tag{3-3}$$

The conductivity C is defined as the current density per unit voltage gradient E or, in equation form, we have

$$C = \frac{J}{E} \tag{3-4}$$

Using the definitions for current density J and voltage gradient E, we obtain conductivity in a different form

$$C = \frac{I/A}{V/L} \tag{3-5}$$

Resistivity (r) is defined as the inverse of conductivity, or

$$r = \frac{1}{C} \tag{3-6}$$

The fact that resistivity is a natural property of certain materials leads to the basic principle of electricity called *Ohm's law.*

Consider a wire of length L and cross-sectional area A. If the wire has resistivity, r, then its resistance, R, is

$$R = r\frac{L}{A} \tag{3-7}$$

The unit of resistance is the ohm, which is denoted by the Greek letter omega, Ω. Since resistivity, r, is the reciprocal of conductivity, we obtain the following:

$$r = \frac{V/L}{I/A} \tag{3-8}$$

When Equation 3-8 is substituted into Equation 3-7, we obtain the following:

$$R = \frac{V}{I} \tag{3-9}$$

This relationship in its three forms—$R = V/I$ or I = V/R or $V = IR$—is called Ohm's law. It assumes that the resistance of the material that is used to carry the current flow will have a linear relationship between the voltage applied and the current flow. That is, if the voltage across the resistance is doubled, the current through it also doubles. The resistance of materials such as carbon, aluminum, copper, silver, gold, and iron is linear and follows Ohm's law. Carbon is the material most commonly used to manufacture a device that has fixed resistance. This device is called a *resistor*.

Wire Resistance

To make it possible to compare the resistance and size of one conductor with another, the U.S. established a standard or unit size for conductors. The standard unit of measurement for the diameter of a circular wire or conductor is the *mil* (0.001 inch), and the standard unit of wire length is the *foot*. The standard unit of wire size in the U.S. is the *mil-foot*. That is, a wire is said to have a unit size if it has a diameter of 1 mil and a length of 1 foot.

The *circular mil* is the standard unit of cross-sectional area for wire used in U.S. wire-sizing tables. Because the diameter of circular wire is normally only a small fraction of an inch, it is convenient to express these diameters in mils, to avoid the use of decimals. For example, the diameter of a 0.010-inch conductor is expressed as 10 mils instead of 0.010 inch. The circular mil area is defined as the square of the mil diameter of a conductor, or area (cmil) = d^2 (mil^2). Example 3-2 illustrates how to calculate conductor area in circular mils.

EXAMPLE 3-2

Problem: Calculate the area in circular mils of a conductor with a diameter of 0.002 inch.

Solution: First, we must convert the diameter in inches to a diameter in mils. Since we defined 0.001 in. = 1 mil, this means that 0.002 in. = 2 mils, so the circular mil area is (2 mil)^2 or 4 cmil.

A *circular-mil per foot* is a unit conductor 1 foot in length that has a cross-sectional area of 1 circular mil. The unit cmil per foot is useful for comparing the resistivity of various conductors. Table 3-1 gives the specific resistance (r) in cmil-ohms per foot of some common solid metals at 20°C.

Conductors that are used to carry electric current are normally manufactured out of copper because of its low resistance and relatively low cost. We can use the specific resistance (r) given in Table 3-1 to find the resistance of a conductor of length (L) in feet and cross-sectional area (A) in cmil by using Equation 3-6, $R=rL/A$.

Material	*r*, resistivity, cmil-Ω/ft
Silver	9.8
Copper (drawn)	10.37
Gold	14.70
Aluminum	17.02
Tungsten	33.20
Brass	42.10
Steel	95.80

Table 3-1. Specific Resistivity

Example 3-3 illustrates how to find the resistance of copper wire with a fixed conductor area in circular mils.

The equation for the resistance of a conductor, $R = rL/A$, can be used in many applications. For example, if you know R, r, and A, you can determine the length by a simple mathematical transformation of the equation for the resistance of a conductor. A typical application is locating a problem ground point in a telephone line. To find ground faults, the

EXAMPLE 3-3

Problem: Find the resistance of 1,000 feet of copper (drawn) wire that has a cross-sectional area of 10,370 cmil and a wire temperature of 20°C.

Solution: From Table 3-1, the specific resistance of copper wire is 10.37 cmil-ohms/ft. By inserting the known values into Equation 3-6, the resistance is determined as follows:

$$R = r\frac{L}{A}$$

$$R = (10.37\ \text{cmil} - \Omega/\text{ft})\left(\frac{1000\ \text{ft}}{10,400\ \text{cmil}}\right)$$

$$R = 1\Omega$$

telephone company uses specially designed test equipment. This equipment operates on the principle that the resistance of a conductor varies in direct relation to distance, so the distance between a test point and a fault can be measured accurately with properly designed equipment.

Example 3-4 illustrates how to find the distance to a short to ground in a telephone line.

EXAMPLE 3-4

Problem: The resistance to ground on a faulty underground telephone line is 5Ω. Calculate the distance to the point where the wire is shorted to ground if the line is a copper conductor that has a cross-sectional area of 1,020 cmil and the ambient temperature of the conductor is 20°C.

Solution: To calculate the distance to the point where the wire is shorted to ground, we transform Equation 3-6 as follows:

$$L = \frac{RA}{r}$$

Since R = 5Ω, A = 1,020 cmils, and r = 10.37 cmil-Ω/ft, we have

$$L = \frac{(5\Omega)(1020 cmil)}{10.37 cmil - \Omega / ft} = 492 ft$$

Wire Gauge Sizes

Electrical conductors are manufactured in sizes that are numbered according to a system known as the American Wire Gauge (AWG). As Table 3-2 shows, the wire diameters become smaller as the gauge numbers increase, and the resistance per one thousand feet increases as the wire diameter decreases. The largest wire size listed in the table is "08," and the smallest is "24." This is the normal range of wire sizes typically encountered in process control applications. The complete AWG table goes from 0000 to 40; the larger and smaller sizes not listed in Table 3-2 are manufactured but are not commonly encountered in process control.

AWG Number	Diameter, mil	Area, cmil	Ω/1000 ft at 25°C	Ω/1000 ft at 65°C
08	128.0	16500	0.641	0.739
10	102.0	10400	1.020	1.180
12	81.0	6530	1.620	1.870
14	64.0	4110	2.58	2.97
16	51.0	2580	4.09	4.73
18	40.0	1620	6.51	7.51
20	32.0	1020	10.4	11.9
22	25.3	642	16.5	19.0
24	20.1	404	26.2	30.2

Table 3-2. American Wire Gauge for Copper Wire

Example 3-5 shows how to calculate the resistance of a typical conductor.

EXAMPLE 3-5

Problem: Determine the resistance of 2,500 feet of 14 AWG copper wire. Assume the wire temperature is 25°C.

Solution: Using Table 3-2, we see that 14 AWG wire has a resistance of 2.58Ω per 1,000 ft at 25°C. So the resistance of 2,500 ft is calculated as follows:

$$R = (2.58\Omega/1000 \text{ ft})(2500 \text{ ft}) = 6.5\Omega$$

Direct and Alternating Current

Basically, two types of voltage signals are encountered in process control and measurement: direct current (DC) and alternating current (AC). In direct current, the flow of charges is in just one direction. A battery is one example of a DC power source. Figure 3-1a shows a graph of a DC voltage over time, and Figure 3-1b shows the symbol for a battery.

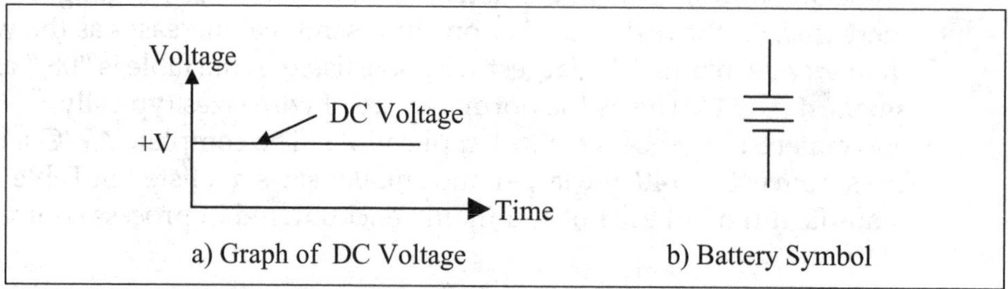

Figure 3-1. DC voltage graph and symbol

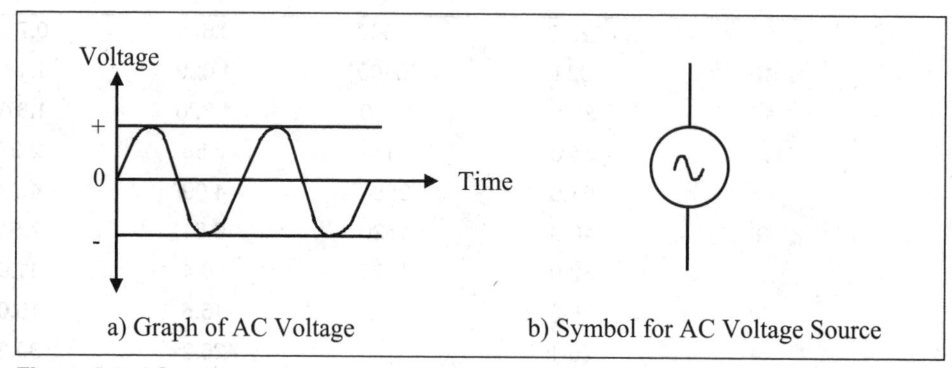

Figure 3-2. AC voltage graph and symbol

An alternating voltage source periodically reverses its polarity. Therefore, the resulting current flow in a closed circuit will reverse direction periodically. Figure 3-2a shows a sine-wave example of an AC voltage signal over time. Figure 3-2b shows the schematic symbol for an AC power source.

The 60-cycle AC power that is used in homes and industry in the U.S. is a common example of AC power. The term *60 cycle* means that the voltage polarity and current direction go through 60 reversals or changes per second. The signal is said to have a frequency of 60 cycles per second. The unit for 1 cycle per second (cps) is called 1 hertz (Hz). Therefore, a 60-cycle-per-second signal has a frequency of 60 Hz.

Series Resistance Circuits

The components in a circuit form a series circuit when they are connected in successive order with the end of a component that is joined to the end of the next element. Figure 3-3 shows an example of a series circuit.

In this circuit, the current (I) flows from the negative terminal of the battery through the two resistors (R_1 and R_2) and back to the positive terminal. According to Ohm's law, the amount of current (I) flowing between two points in a circuit equals the potential difference (V) divided by the resistance (R) between these points. If V_1 is the voltage drop across R_1, V_2 is defined as the voltage drop across R_2, and the current (I) flows through both R_1 and R_2, then from Ohm's law we obtain the following:

$$V_1 = IR_1 \text{ and } V_2 = IR_2$$

so that

$$V_t = IR_1 + IR_2$$

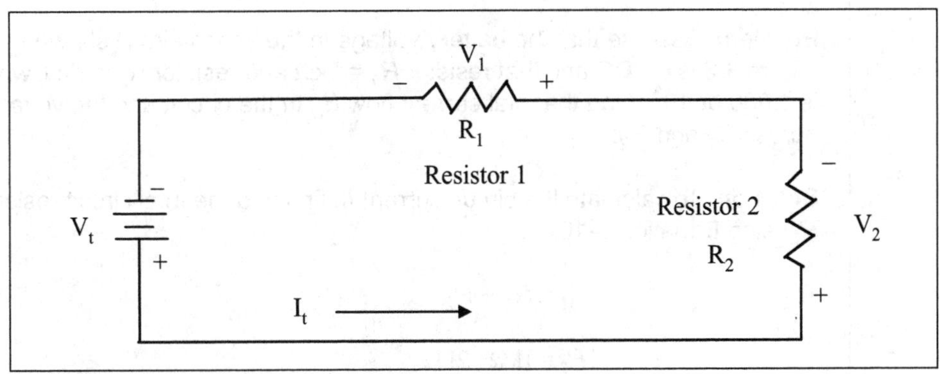

Figure 3-3. Series resistance circuit

If we divided both sides of this equation by the current, I, in the series circuit, we obtain the following:

$$V_t/I = R_1 + R_2$$

Ohm's law defines the total resistance of the series circuit as the total voltage applied divided by the current in the circuit or $R_t = V_t/I$. As a result, the total resistance of the circuit is given by the following equation:

$$R_t = R_1 + R_2$$

We can derive a more general equation for any number of resistors in series by using the classical law of conservation of energy. According to this law, the energy or power supplied to a series circuit must equal the power dissipated in the resistors in the circuit. Thus,

$$P_t = P_1 + P_2 + P_3 \ldots\ldots P_n$$

since power in a resistive circuit is given by $P = I^2R$, we obtain the following:

$$I^2R_t = I^2R_1 + I^2R_2 + I^2R_3 + \ldots.. + I^2R_n$$

If we divide this equation by I^2, we obtain the series resistance formula for a circuit with n resistors:

$$R_t = R_1 + R_2 + R_3 + \ldots.. + R_n \tag{3-10}$$

Example 3-6 illustrates calculations for a typical series DC circuit.

EXAMPLE 3-6

Problem: Assume that the battery voltage in the series circuit shown in Figure 3-3 is 6 VDC and that resistor $R_1 = 1\text{k}\Omega$ and resistor $R_2 = 2\text{k}\Omega$, where k is 1,000 or 10^3. Find the total current flow (I_t) in the circuit and the voltage across R_1 and R_2.

Solution: To calculate the circuit current I_t, first find the total circuit resistance R_t using Equation 3-10:

$$R_t = R_1 + R_2$$

$$R_t = 1\text{k}\Omega + 2\text{k}\Omega$$

$$R_t = 3\text{k}\Omega$$

Now, according to Ohm's law, we have:

$$I_t = V_t / R_t$$

$$I_t = 6V/3k\Omega = 2 \text{ mA}$$

Parallel Resistance Circuits

When two or more components are connected across a power source, they form a parallel circuit. Each parallel path is called a *branch,* and each branch has its own current. In other words, parallel circuits have one common voltage across all the branches but individual currents in each branch. These characteristics contrast with series circuits, which have one common current but individual voltage drops.

Figure 3-4 shows a parallel resistance circuit with two resistors across a battery. You can find the total resistance (R_t) across the power supply by

EXAMPLE 3-6 continued

The voltage across R_1 (i.e., V_1) is given by

$$V_1 = R_1 I_t$$

$$V_1 = (1k\Omega)(2\ mA) = 2\ V$$

Thus, the voltage across R_2 (i.e., V_2) is obtained as follows:

$$V_2 = R_2 I_t$$

$$V_2 = (2k\Omega)\ (2\ mA) = 4\ V$$

dividing the voltage across the parallel resistance by the total current into the two branches. In the circuit of Figure 3-4, total current is given by $I_t = I_1 + I_2$, where the current in branch 1 is given by $I_1 = V_t/R_1$ and the current in branch 2 is given by $I_2 = V_t/R_2$. Thus, $I_t = V_t/R_1 + V_t/R_2$. Since according to Ohm's law total current is given by $I_t = V_t/R_t$, we obtain Equation 3-11 for total resistance, R_t, of two resistors in parallel:

$$\frac{1}{R_t} = \frac{1}{R_1} + \frac{1}{R_2} \tag{3-11}$$

Transforming this equation into a more useful form, we obtain the following:

$$R_t = \frac{R_1\ x\ R_2}{R_1 + R_2} \tag{3-12}$$

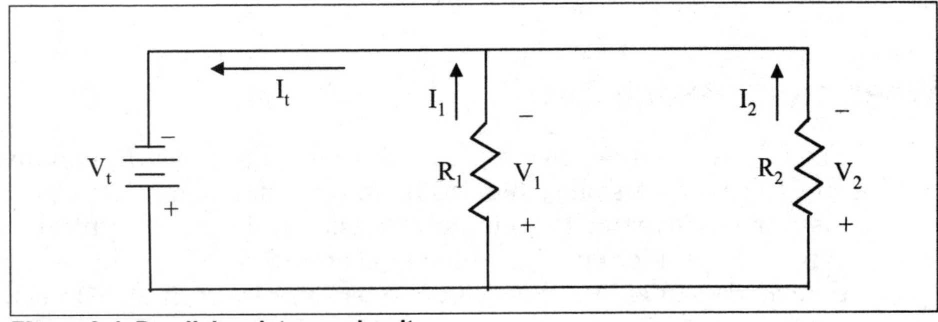

Figure 3-4. Parallel resistance circuit

We can derive the general reciprocal resistance formula for any number of resistors in parallel from the fact that the total current I_t is the sum of all the branch currents, or

$$I_t = I_1 + I_2 + I_3 + \ldots + I_n$$

Since current flow is defined as $I = dQ/dt$, this is simply the law of electrical charge conservation:

$$\frac{dQ_t}{dt} = \frac{dQ_1}{dt} + \frac{dQ_2}{dt} + \frac{dQ_3}{dt} + \ldots\ldots + \frac{dQ_n}{dt}$$

This law implies that, since no charge accumulates at any point in the circuit, the differential charge, dQ, from the power source in the differential time period, dt, must appear as charges $dQ_1, dQ_2, dQ_3 \ldots dQ_n$ through the resistors $R_1, R_2, R_3 \ldots R_n$ at the same time. Since the voltage across each branch is the applied voltage V_t and $I = V/R$, we obtain the following:

$$\frac{V_t}{R_t} = \frac{V_t}{R_1} + \frac{V_t}{R_2} + \frac{V_t}{R_3} + \ldots\ldots + \frac{V_t}{R_n}$$

If we divide both sides of this equation by V_t, we obtain the total resistance of n resistors in parallel:

$$\frac{1}{R_t} = \frac{1}{R_1} + \frac{1}{R_2} + \frac{1}{R_3} + \ldots\ldots + \frac{1}{R_n} \qquad (3\text{-}13)$$

To illustrate the basic concepts of parallel resistive circuits, let's look at Example 3-7.

Wheatstone Bridge Circuit

The Wheatstone bridge circuit, named for English physicist and inventor Sir Charles Wheatstone (1802-1875), was one of the first electrical instruments invented to accurately measure resistance. A typical Wheatstone bridge circuit is shown in Figure 3-5. The circuit has two parallel resistance branches with two series resistors in each branch and a galvanometer (G) connected across the branches. The galvanometer is a very sensitive electric-current-measuring instrument that is connected across the bridge between points "a" and "b" when switch 2 is closed. If these points are at the same potential, the meter will not deflect. The purpose of the circuit is to balance the voltage drops across the two

EXAMPLE 3-7

Problem: Assume that the voltage for the parallel circuit shown in Figure 3-4 is 12 V and we need to find the currents I_t, I_1, and I_2 and the parallel resistance R_t, given that R_1 = 30 kΩ and R_2 = 30 kΩ.

Solution: We can find the parallel circuit resistance R_t by using Equation 3-12:

$$R_t = \frac{R_1 \, x \, R_2}{R_1 + R_2}$$

$$R_t = \frac{(30\text{k})(30\text{k})}{30\text{k} + 30\text{k}} \Omega = 15\text{k}\Omega$$

The total current flow is given by the following equation:

$$I_t = \frac{V_t}{R_t} = \frac{12\text{V}}{15\text{k}\Omega} = 0.8 \text{ mA}$$

The current branch 1 (I_1) is obtained as follows:

$$I_1 = \frac{V_t}{R_1} = \frac{12\text{V}}{30\text{k}\Omega} = 0.4 \text{ mA}$$

Since $I_t = I_1 + I_2$, the current flow in branch 2 is given by

$$I_2 = I_t - I_1 = 0.8 \text{ mA} - 0.4 \text{ mA} = 0.4 \text{ mA}$$

parallel branches to obtain 0 V across the meter and zero current through the meter. In the Wheatstone bridge, the unknown resistance R_x is balanced against a standard accurate resistor R_s to make possible precise measurement of resistance.

In the circuit shown in Figure 3-5, the switch S_1 is closed to apply the voltage V_t to the four resistors in the bridge, and switch S_2 is closed to connect the galvanometer (G) across the bridge. To balance the circuit, the value of R_x is varied until a zero reading is obtained on the meter with switch S_2 closed.

A typical application of the Wheatstone bridge circuit is to replace the unknown resistor R_x with a resistance temperature detector (RTD). An RTD is a device that varies its resistance in response to a change in temperature so the balancing resistor dial can be calibrated to read out a process temperature. In a typical application, an electronic sensor that is

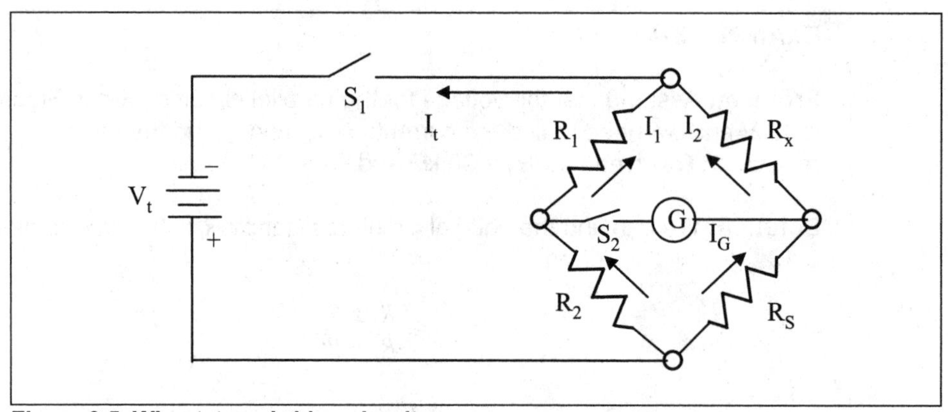

Figure 3-5. Wheatstone bridge circuit

designed to convert the ambient temperature into resistance is connected to a Wheatstone bridge circuit to display the temperature.

When the bridge circuit is balanced (i.e., there is no current through the galvanometer G), the bridge circuit can be analyzed as two series resistance strings in parallel. The equal voltage ratios in the two branches of the Wheatstone bridge can be stated as follows:

$$\frac{I_2 R_x}{I_2 R_s} = \frac{I_1 R_1}{I_1 R_2} \tag{3-14}$$

Note that I_1 and I_2 can be canceled out in the previous equation, so we can invert R_s to the right side of the equation to find R_x as follows:

$$R_x = R_s \frac{R_1}{R_2} \tag{3-15}$$

To illustrate the use of a Wheatstone bridge, let's consider Example 3-8.

Instrumentation Current Loop

The brief discussion on resistive circuits earlier in this chapter served as a starting-off point for this section on instrumentation current loops. The DC current loop shown in Figure 3-6 is used extensively in the instrumentation field to transmit process variables to indicators and controllers. It is also used to send control signals to field devices to manipulate process variables such as temperature, level, and flow. The standard current range used in these loops is 4 to 20 mA, and this value is normally converted to 1 to 5 VDC by a 250 Ω resistor at the input to controllers and indicators. These instruments are normally high-input-

EXAMPLE 3-8

Problem: Assume a Wheatstone bridge with R_1 = 1 kΩ and R_2 = 10kΩ. If R_s is adjusted to read 50Ω when the bridge is balanced (i.e., the galvanometer reading is at zero), calculate the value of R_x.

Solution: We can use Equation 3-15 to determine R_x:

$$R_x = R_s \frac{R_1}{R_2}$$

$$R_x = 50\Omega \frac{1\text{K}\Omega}{10\text{K}\Omega} = 5\Omega$$

impedance (Zin > 10 MΩ) electronic devices that draw virtually no current from the instrument loop.

There are two main advantages to using the 4-to-20 mA current loop. First, only two wires are required for each remotely mounted field transmitter, so a cost savings is realized on both labor and wire when installing field devices. The second advantage is that the current loop is not affected by electrical noise or changes in lead wire resistance caused by temperature changes.

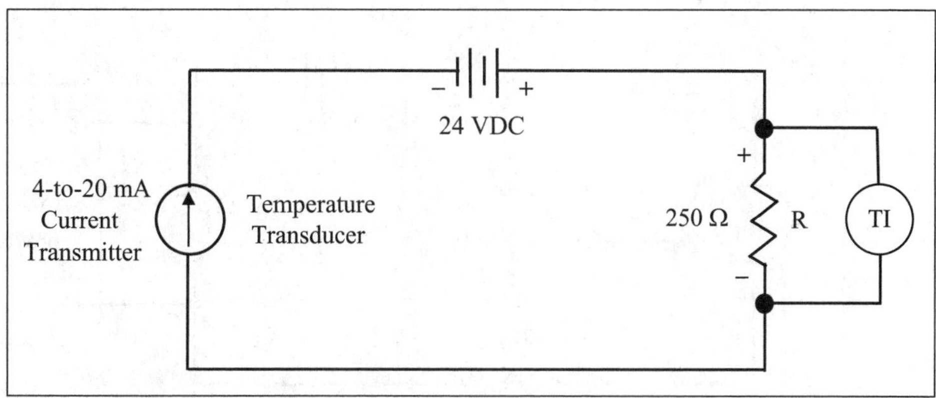

Figure 3-6. Typical 4-to-20mA current loop

Selecting Wire Size

You must consider several factors when selecting the wire size for a programmable controller application. One factor is the permissible power loss ($P = I^2R$) in the electrical line. This power loss is electrical energy being converted into heat. If the heat produced is excessive, the conductors or system components may be damaged. Using large-diameter (low-wire-gauge) conductors will reduce the circuit resistance and, therefore, the power loss. However, larger-diameter conductors are more expensive than smaller ones and they are more costly to install, so you will need to make design calculations to select the proper wire size.

A second factor when selecting wire is the resistance of the wires in the circuit. For example, let's assume that we are sending a full-range 20 mA DC signal from a field instrument to a programmable controller input module that is located 2,000 feet away, as shown in Figure 3-7.

Assuming that we are using 20 AWG wire, we can easily calculate the wire resistance (R_w). Using data from Table 3-2, we see that 20 AWG wire has a resistance of 10.4Ω per 1,000 ft at 25°C. So the resistance for the 4,000 feet of wire is calculated as follows:

$$R_W = (10.4\Omega/1000\ \text{ft})(4000\ \text{ft}) = 41.6\Omega$$

This resistance increases the load on the 24 VDC power supply that is used to drive the current loop. You can calculate the voltage drop V_w in the wire when the field instrument is sending 20 mA of current to the programmable controller analog input module by using Ohm's law:

$$V_W = IR_W = (20\ \text{mA})(41.6\Omega) = 0.832\ \text{V}$$

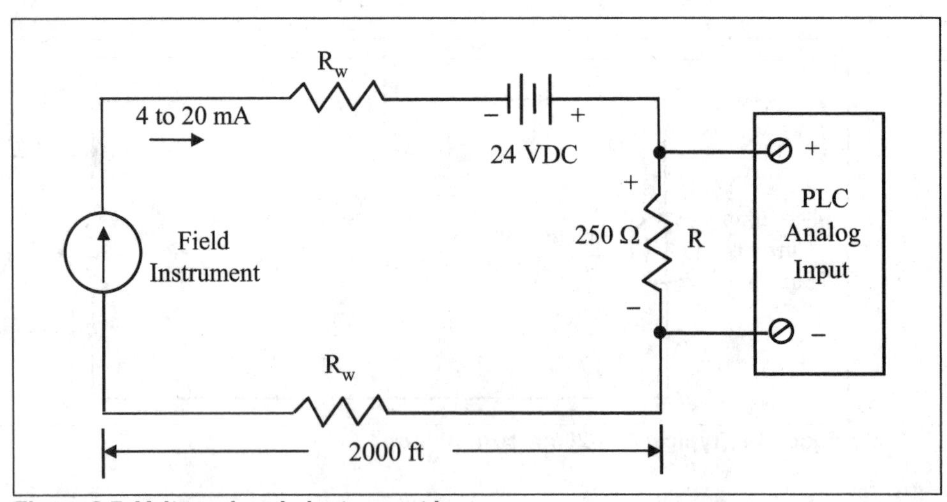

Figure 3-7. Voltage drop in instrument loop

A third factor when selecting wire is the current-carrying capacity of the conductor. When current flows in a wire, heat is generated. The temperature of the wire will rise until the heat that is radiated away, or otherwise dissipated, is equal to the heat generated in the conductor. If the conductor is insulated, the heat produced is not dissipated so quickly as it would if the conductor were not insulated. So, to protect the insulation from excessive heat, you must keep the current flowing in the wire below a certain value. Rubber insulation will start to deteriorate at relatively low temperatures. Teflon™ and certain plastic insulations retain their insulating properties at higher temperatures, and insulations such as asbestos are effective at still higher temperatures.

You could install electrical cables in locations where the ambient temperature is relatively high. In these cases, the heat produced by external sources adds to the total heating of the electrical conductor. When designing a programmable controller system, you must therefore make allowances for the ambient heat sources encountered in industrial environments. The maximum allowable operating temperature for insulated wires and cables is specified in manufacturers' electrical design tables.

Table 3-3 gives an example of the maximum allowable current-carrying capacities of copper conductor with three different types of insulation. The current ratings listed in the table are those permitted by the National Electrical Code. This table can be used to determine the safe and proper wire size for any electrical wiring application. Example 3-9 will help to illustrate a typical wire-sizing calculation.

EXAMPLE 3-9

Problem: Calculate the proper wire size for the 120-VAC feed to the programmable controller system shown in Figure 3-8.

Solution: The total AC feed current is the sum of all the branch currents, or

$$I_t = I_1 + I_2 + I_3 + I_4$$

$$I_t = 2\text{ A} + 8\text{ A} + 5\text{ A} + 5\text{ A}$$

$$I_t = 20\text{ A}$$

Using Table 3-3, we can see that the main power conductors (hot, neutral, and ground) must have a minimum size of 14 AWG.

Fundamentals of Magnetism

Magnetism, like electricity, is a fundamental force in nature. As we discussed earlier, electrical effects exist in two forms, voltage and current.

Conductor Size (AWG)	60°C (140°F) Types: TW, UF	75°C (167°F) Types: FEPW, RH, RHW, THHW, THW, THWN, XHHW, USE, ZW	85°C (185°F) Type: V18
14	20	20	25
12	25	25	30
10	30	35	40
8	40	50	55
6	55	65	70

Table 3-3. Ampacities of Insulated Copper Conductor (National Electrical Code®)

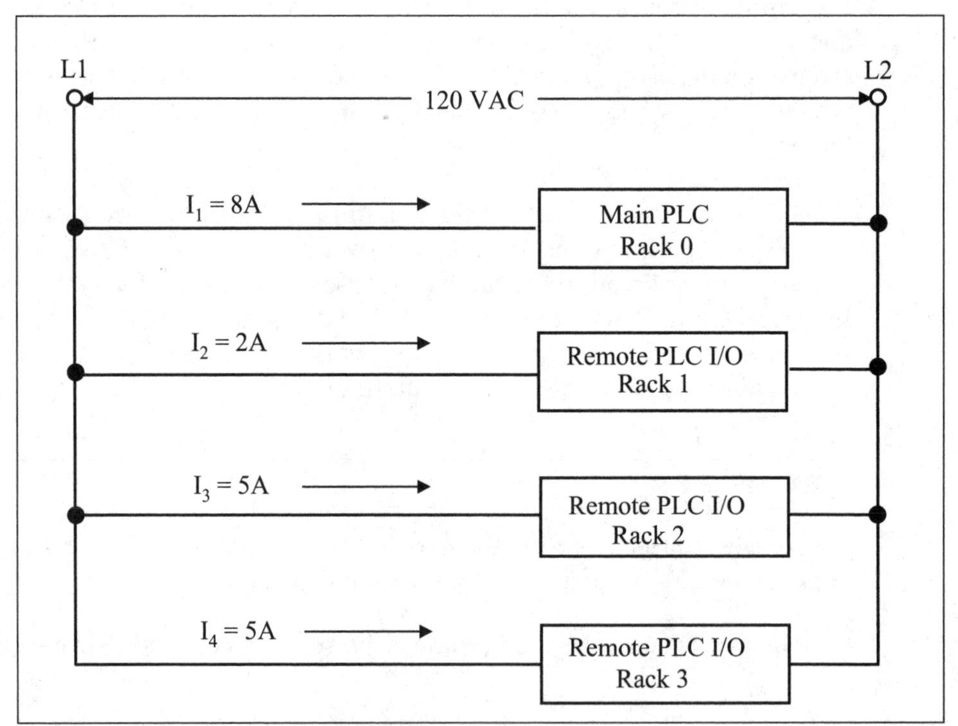

Figure 3-8. PLC wire-sizing application

Separated electric charges or voltage have the potential to do mechanical work in attracting or repelling charges. Similarly, a changing electric current has an associated magnetic field that can do work of attraction or repulsion. Some materials made of iron, nickel, and cobalt can concentrate their magnetic effects on opposite ends, where the magnetic material meets a nonmagnetic medium such as air.

The term magnetism is derived from the iron oxide mineral, *magnetite*, a naturally occurring magnet first used in navigational compasses. Ferromagnetism refers specifically to the magnetic properties of iron.

There are two types of magnets: permanent and temporary. The natural magnet, *magnetite*, and the man-made magnets, such as horseshoe, compass, and bar magnets, are examples of permanent magnets. An example of a man-made temporary magnet is shown in Figure 3-9. It consists of a soft iron core wound with an electrical wire that is connected to a DC power supply through an on/off switch.

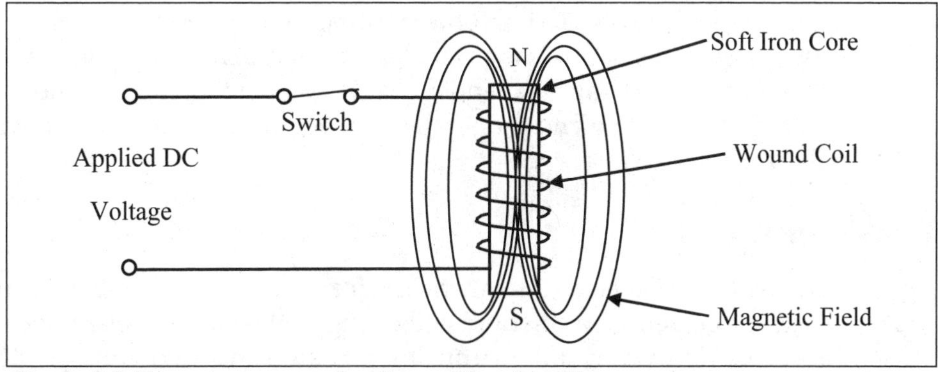

Figure 3-9. Temporary magnet using a solenoid coil

If the switch is closed, current will flow through the coil and a magnetic field is produced as shown by the magnetic lines of force in Figure 3-9.

Danish physicist Hans C. Oersted discovered this relationship between electricity and magnetism in 1819. He found that when current flows through an electrical conductor, a magnetic field is created around that conductor. At about the same time another physicist, A. M. Ampere, found that when the conductor is tightly wound in a cylindrical coil shape as shown in Figure 3-8, the magnetic field is intensified. This was the first time that magnetism other than that produced by the rock *magnetite* was observed. A. M. Ampere found that there are three ways to increase the magnetic effect of a coil of wire. First, you can increase the amount of electrical current by increasing the applied voltage. Second, you can increase the number of turns of wire in the coil. Third, you can insert an iron core through the center of the coil and it will concentrate the magnetic lines of force. The softer the iron core the more intense the magnetic field.

The result of the discoveries and studies of early scientists like Oersted and Ampere was the development of electromagnetic industries that rely on magnetic coils in many devices such as motors, electrical generators, transformers, electromagnetic relays, and solenoids.

Solenoids

A solenoid is simply an electromagnet consisting of a coil of wire and an applied voltage as shown in Figure 3-9. However, in practical applications solenoids are combined with a moving armature that transmits the force created by the solenoid into a useful function. Since, the solenoid is used in a variety of electromagnetic devices and control applications, the term solenoid can be confusing unless the full device name or application for the solenoid is also stated. In other words, a complete or proper description of a solenoid might include terms such as solenoid valve, solenoid operated valve, solenoid clutch, solenoid switch, solenoid operated relay, or solenoid operated contactors. Most of theses devices will be discussed later.

Transformers

Another common electromagnetic device encountered is the transformer. When the current in a solenoid coil changes, the varying magnetic flux cuts across any other coil nearby and produces induced voltage in the other coil. In Figure 3-10, the primary coil L_P is connected to an AC power source that produces varying current in the turns of the primary coil. The secondary coil L_S is not directly connected to L_P, but the turns are linked by the magnetic field produced by the varying AC current in the primary. This arrangement is called a *transformer*. A soft iron core is generally used to concentrate the magnetic field to improve the efficiency of the transformer. One purpose of the transformer is to transfer power from the primary, where the AC power source is connected, to the secondary, where the induced secondary voltage can produce current in the load resistance that is connected across the secondary coil.

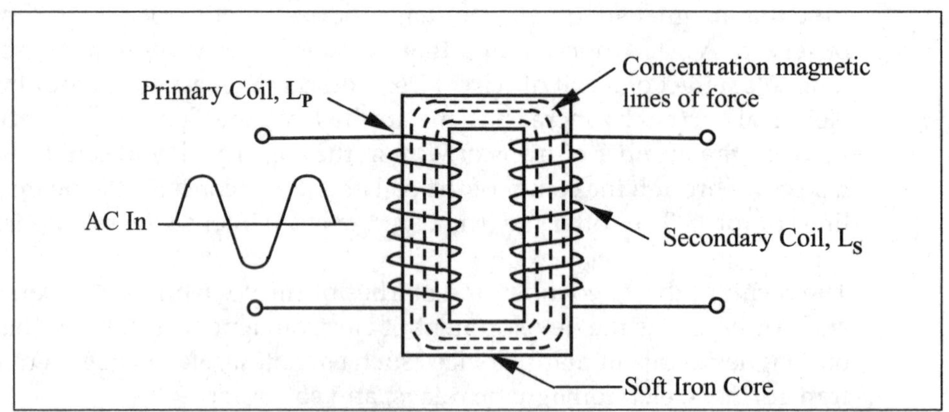

Figure 3-10. Soft iron core transformer

Another purpose of a transformer is to isolate the primary circuit from the secondary load, since the primary and secondary are not electrically connected to each other but are coupled to each other by a magnetic field. A final function of a transformer is to step up or step down the input AC voltage to provide a different value of AC voltage to the output load.

A common application for transformers is their use in power supplies used by the control industry.

Power Supplies

It is important for process control users to have a basic understanding of the design and operation of DC power supplies because of their wide use in control systems applications. DC power supplies use either half-wave or full-wave rectification, depending on the power supply application in question. Figure 3-11 shows the output waveforms produced by half-wave and full-wave rectification. A schematic diagram of a half-wave rectifier is shown in Figure 3-12. In the circuit in Figure 3-12, the positive and negative cycles of the AC voltage across the secondary winding are in phase with the signal at the primary of the transformer. This is indicated by the dots on each side of the transformer symbol.

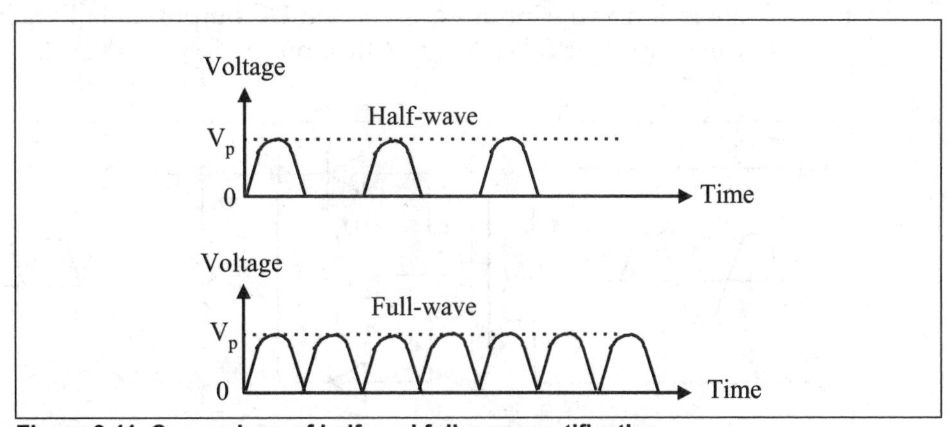

Figure 3-11. Comparison of half- and full-wave rectification

If we assume that the top of the transformer secondary is positive and the bottom is negative during a positive cycle of the input, then the diode is forward biased and current flow is permitted through the load resistor, R_L, as indicated.

During the negative input cycle, the top of the transformer will be negative and the bottom will be positive. This voltage polarity reverse biases the diode. A reverse-biased diode represents an extremely high

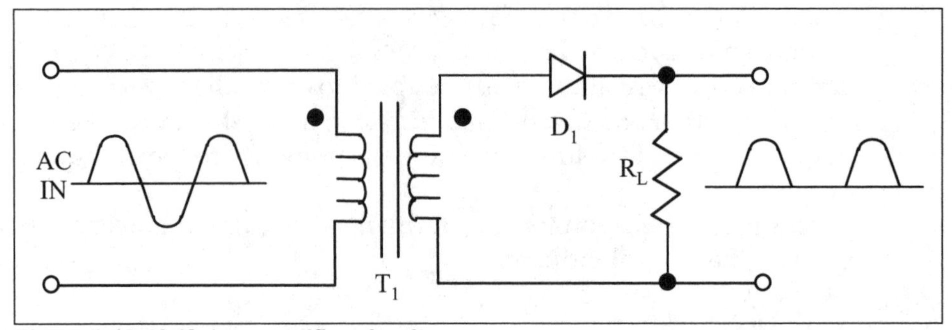

Figure 3-12. Half-wave rectifier circuit

resistance and serves as an open circuit. Therefore, with no current flowing through R_L, the output voltage will be zero for this cycle.

The basic full-wave rectifier using two diodes is shown in Figure 3-13. The cathodes of both diodes are connected to obtain a positive output. The anodes of each diode are connected to opposite ends of the transformer secondary winding. The load resistor (R_L) and the transformer center are connected to ground to complete the circuit.

In a full-wave rectifier, the current through R_L goes in the same direction for each alteration of the input, so we obtain DC output for both halves of the sine-wave input, or full-wave rectification.

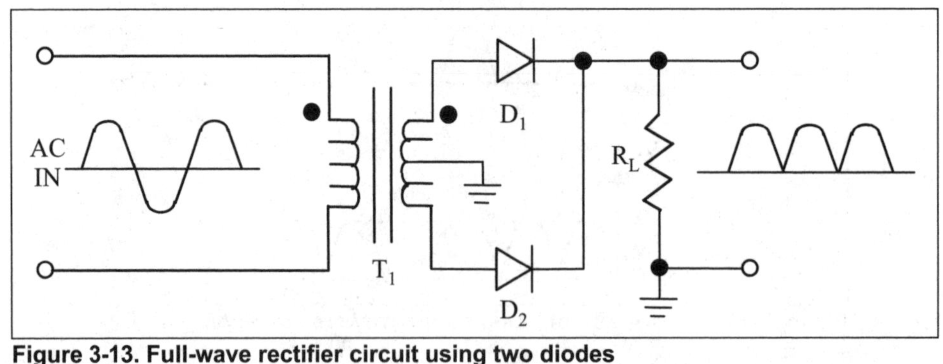

Figure 3-13. Full-wave rectifier circuit using two diodes

The ripple frequency of the full-wave rectifier is also different from that of a half-wave circuit. Since each cycle of the input produces an output across R_L, the ripple frequency will be twice the input frequency. The higher ripple frequency and characteristic output of the full-wave rectifier is easier to filter than is the corresponding output of a half-wave rectifier. To reduce the amount of ripple, a filter capacitor is placed across the output. A capacitor consists of two metal plates separated by a dielectric material.

A bridge structure of four diodes is used in most power supplies to achieve full-wave rectification. In the bridge rectifier shown in Figure 3-14, two diodes will conduct during the positive alteration and two will conduct during the negative alteration. A bridge rectifier does not require a center-tapped transformer as is used in a two-diode, full-wave rectifier.

The DC output that appears across R_L of the bridge circuit has less ripple amplitude because a filter capacitor has been placed across the output. A capacitor opposes any changes in the voltage signal. If the filter capacitor is properly sized and placed across the output, it will prevent the output signal from returning to zero at the end of each half cycle, as shown in Figure 3-14.

Bridge rectifiers are commonly used in the electronic power supplies of instruments because of their simple operation and desirable output. The diode bridge is generally housed in a single enclosure that has two input and two output connections.

The DC output voltage signal from a bridge rectifier is filtered to produce the smooth DC signal that most instrumentation circuits and devices need.

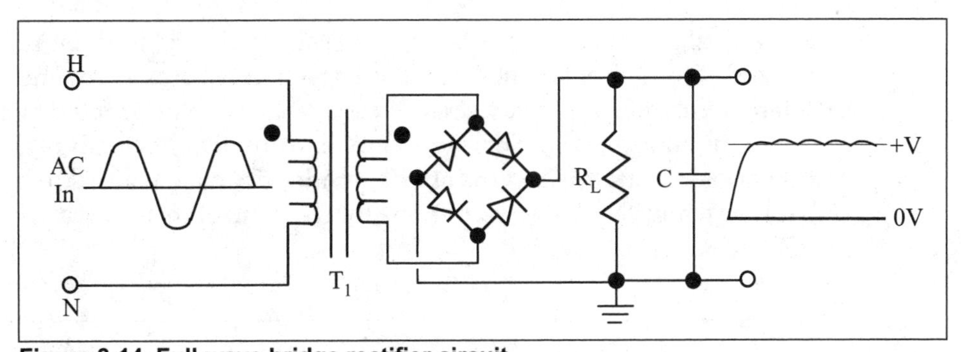

Figure 3-14. Full-wave bridge rectifier circuit

We have now presented the most common types of electronic devices and circuits encountered in programmable controller applications. We now need to discuss some of the common electrical control devices.

Electrical Control Devices

A wide variety of electrically operated devices can be found in control applications. In this section, we will discuss the most common devices (as well as the electrical symbol used in drawings to represent each device).

Electrical Relays

The most common control device encountered in control applications is the *electromechanical relay*. It is called electromechanical because it consists both of electrically operated parts and mechanical components. One electrical part is a long thin wire that is wound into a close-packed helix around an iron core. When a changing electric current is applied, a strong magnetic field is produced. The resulting magnetic force moves the iron core, which is connected to a set of contacts. This transfers the common contact from the normally closed contact to the normally open contact.

A simplified representative diagram of an electrical relay is shown in Figure 3-15, with two sets of relay contacts. Each set consists of a common (COM) contact, a normally open (NO) contact, and a normally closed (NC) contact. *Normally open* and *normally closed* refer to the status of contacts when no electrical power is applied to the relay or when the relay is in the shelf position. For example, if there is no electrical continuity between a common contact and another contact in the set when the relay is on a storage shelf, then this set of contacts is termed the "normally open" set of contacts. These contacts in the relay are used to make or break electrical connections in control circuits.

The electrical schematic symbol for a NO set of contacts is given in Figure 3-16a, and the symbol for the NC set is shown in Figure 3-16b. The standard symbol for the relay solenoid is a circle with two connection dots as shown in Figure 3-16c. This symbol will normally also include a combination of letter(s) and number(s) to identify each control relay on a control drawing. Typical designations for the control relays on a schematic are CR1, CR2, . . . CRn.

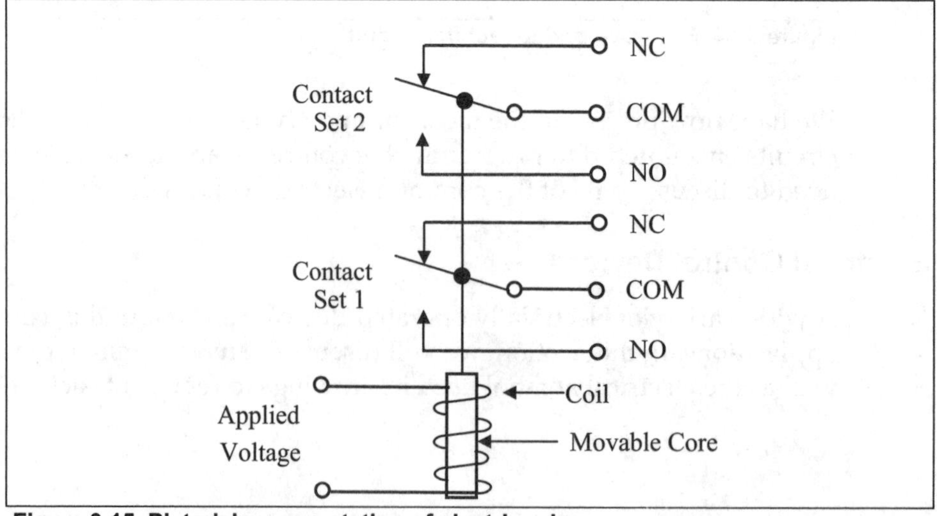

Figure 3-15. Pictorial representation of electric relay

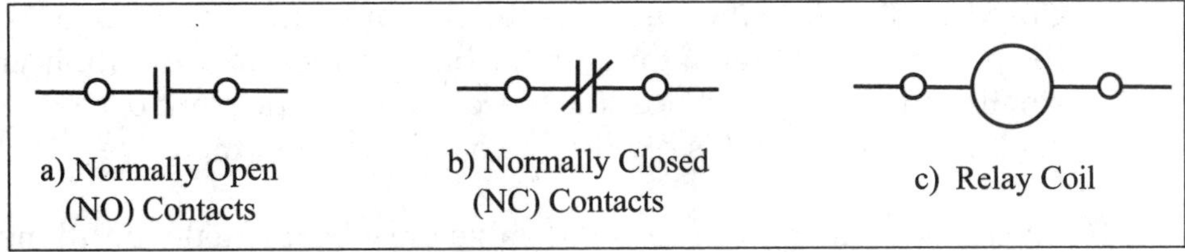

Figure 3-16. Electrical relay symbols

Electric Solenoid Valves

Another electrically operated device that is commonly encountered in process control is the *solenoid valve*. The solenoid valve is a combination of two basic functional units: an electromagnetic solenoid with its core and a valve body that contains one or more orifices. The flow through an orifice can be allowed or prevented by the action of the core when the solenoid is energized or deenergized, respectively.

Solenoid valves normally have a solenoid coil that is mounted directly on the valve body. The coil is enclosed and free to move in a sealed tube called a *core tube*, which thus gives it a compact assembly. The two different electrical schematic symbols that are used to represent solenoid valve coils on electrical and control drawings are given in Figure 3-17.

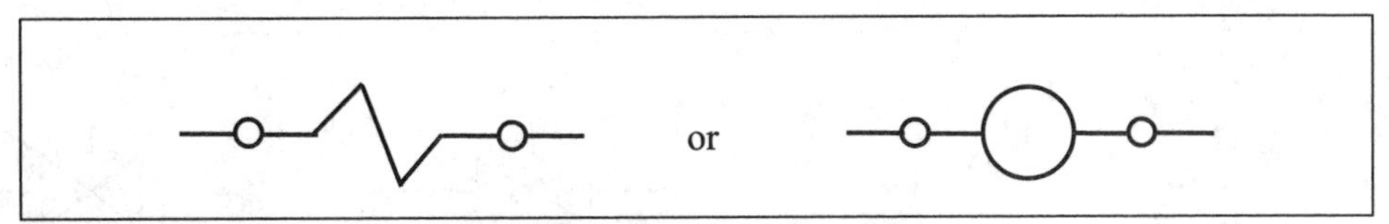

Figure 3-17. Schematic symbols for solenoid valve coils

The three most common types of solenoid valves are direct acting, internal pilot operated, and manual reset. In *direct-acting valves*, the solenoid core directly opens or closes the orifice, depending on whether the solenoid is energized or deenergized, respectively. The valve will operate from 0 psi inlet pressure to its maximum rated inlet pressure. The force that is required to open a solenoid valve is proportional to the orifice size and the pressure drop across the valve. As the orifice size increases, the force needed to operate the valve increases. To keep the solenoid coil small and open large orifices at the same time, some plants use internal pilot-operated solenoid valves.

Internal pilot-operated solenoid valves have a pilot and bleed orifice and use the line pressure to operate. When the solenoid is energized, the core opens the outlet side of the valve. The imbalance of pressure causes the line pressure to lift the piston or diaphragm off the main valve orifice, thus opening the valve. When the solenoid is deenergized, the pilot orifice is closed, and to close the valve, the full line pressure is applied to the top of the piston or diaphragm through the bleed orifice. In some applications, a small manual valve replaces the bleed orifice so the plant personnel can control the speed that is required to open and close the valve.

The *manual reset* type of solenoid valve must be manually positioned (latched). It will return to its original position when the solenoid is energized or deenergized, depending on the valve design.

Three configurations of solenoid valves are normally available for process control applications: two-, three-, and four-way. Figure 3-18 shows the process diagram symbols that are used to represent two-, three-, and four-way solenoid valves. Two-way valves have one inlet and one outlet process connection. They are available in either NO or NC configurations. Three-way solenoid valves have three process piping connections and two orifices (one orifice is always open and the other is always closed). These valves are normally used in process control to alternately apply air pressure to and exhaust air from diaphragm-operated control valves or single-acting cylinders. Four-way solenoid valves have four pipe connections: one for supplying air pressure, two for process control, and one for air exhaust. These valves are normally used to control double-acting actuators on control valves.

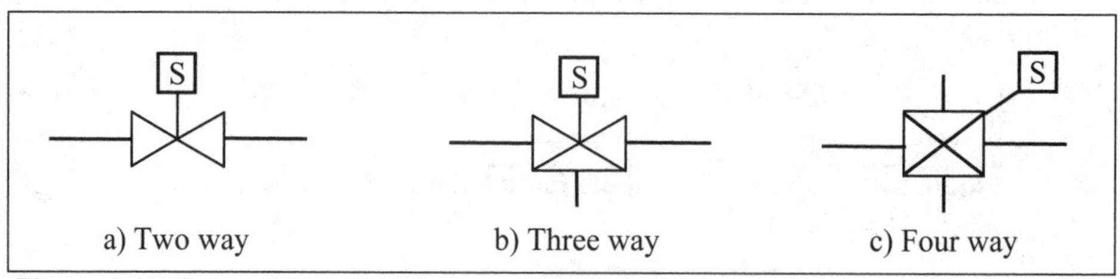

Figure 3-18. Process diagram symbols for solenoid valves

Electrical Contactors

Electrical contactors are relays that can switch high-current loads that are normally greater than 10 amps. They are used to repeatedly make and break electrical power circuits. An electrically operated solenoid is the most common operating mechanism for contactors. The solenoid is powered by low-ampere voltage signals from control circuits. Instead of opening or closing a valve, the linear action of the solenoid coil is used to open or close sets of contacts with high-power ratings. The higher-rated sets of electrical contacts are used in turn to control loads, such as motors, pumps, and heaters. The schematic symbol for a typical electrical contactor is shown in Figure 3-19.

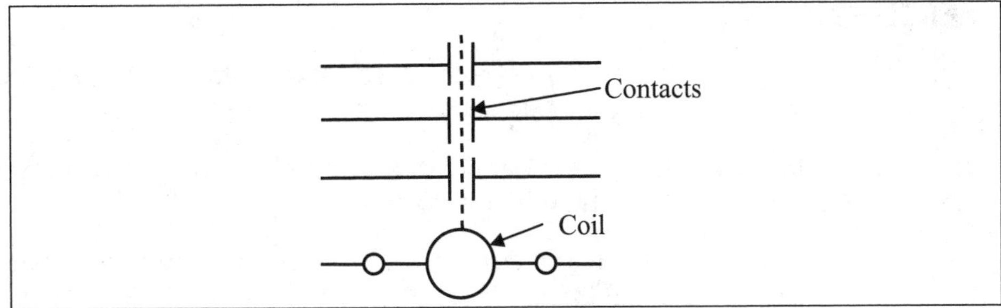

Figure 3-19. Schematic symbol for an electrical contactor

Other Control Devices

Hand switches, instrument switches, control switches, pushbuttons, and indicating lights are some of the other devices commonly encountered in process control. Switches and pushbuttons are used to control other devices, such as solenoid valves, electric motors, motor starters, heaters, and other process equipment.

Hand switches come in a variety of designs and styles. They can be mounted on control panels or in the field near the process equipment. Instrumentation switches are generally field-mounted and are used to sense pressure, temperature, flow, liquid level, and position.

To make it easier to read control drawings, a set of standardized symbols is defined by IEEE 315-1975, Graphic Symbols for Electrical and Electronics Diagrams (Including Reference Designation Letters). The most common symbols used in control diagrams are shown in Figure 3-20.

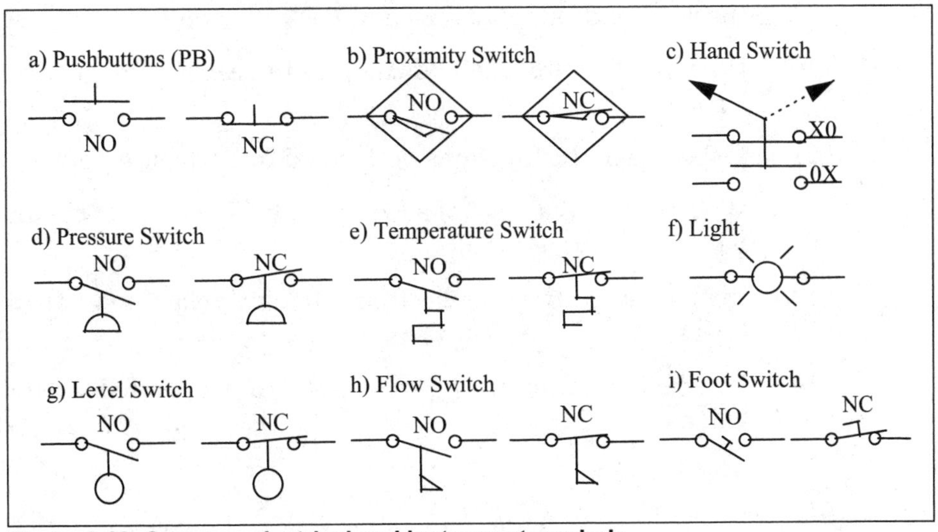

Figure 3-20. Common electrical and instrument symbols

EXERCISES

3.1 A charge of 15 coulombs moves past a given point every second. What is the current flow?

3.2 Calculate the current in a conductor, if 30×10^{18} electrons pass a given point in the wire every second.

3.3 Calculate the area in circular mils (cmil) of a conductor that has a diameter of 0.003 inch.

3.4 Calculate the resistance of 500 feet of copper wire that has a cross-sectional area of 4110 cmil, if the ambient temperature is 20°C.

3.5 Calculate the resistance of 500 feet of silver wire with a diameter of 0.001 inch, if the ambient temperature is 20°C.

3.6 The resistance to ground on a faulty underground telephone line is 20 Ω. Calculate the distance to the point where the wire is shorted to ground, if the line is a 22 AWG copper conductor and the ambient temperature is 25°C.

3.7 Find the total current (I_t) in a circuit that has two 250 Ω resistors in series with a voltage source of 24 VDC. Also find the voltage drop across each resistor, where V_1 is the voltage across the first resistor and V_2 is voltage across the other resistor.

3.8 In the parallel resistance circuit shown in Figure 3-4, assume that $R_1 = 100\ \Omega$, $R_2 = 200\Omega$, and $V_t = 100$ VDC. Find I_1, I_2, and I_t.

3.9 Assume that the Wheatstone bridge circuit shown in Figure 3-5 is balanced (i.e., $I_G = 0$) and that $R_1 = 2\text{k}\ \Omega$, $R_2 = 10\text{k}\ \Omega$, $R_s = 2\text{k}\ \Omega$, and $V_t = 12$ VDC. Calculate the value of I_1, I_2, and R_x. Also, what are the voltage drops across R_1 and R_2?

3.10 What are the three methods used to increase the magnetic effect of a coil of wire?

3.11 List some of the functions performed by a transformer.

3.12 What is the advantage of a full-wave bridge rectifier circuit over a half-wave bridge circuit?

3.13 Explain the purpose of electromechanical relays and how they operate.

3.14 List the three common types of electrically operated solenoid valves.

4

Digital System Fundamentals

Introduction

Plant personnel who are applying digital techniques to process control applications must understand the fundamentals of digital systems. The design and implementation of computer-based control systems also requires a working knowledge of digital principles. In this chapter, we will discuss the basic concepts of digital signals, numbering systems, binary logic functions, and relay ladder logic design.

Since the data in process control computers is in binary form (zero or one), we will turn our attention to binary signals and codes first.

Binary Signals and Codes

If digital techniques are to be used in process control, the measurement and control signals must be encoded into binary form. Binary signals are simply two-state signals (on/off, start/stop, high voltage/low voltage, etc.). The simplest approach to encoding analog data as binary-based digital words is provided by the American Standard Code for Information Interchange (ASCII), which uses a pattern of seven bits (ones and zeros) to represent letters and numbers. Sometimes an extra bit (parity bit) is used to check that the correct pattern has been transmitted. For example, the number one is represented by 011 0001, and the number two is represented by 011 0010. On the other hand, the capital letter *A* is represented by 100 0001 and the capital letter B by 100 0010.

Many other methods are used to encode digital numbers in programmable controllers and digital computers. The most common method is the simple binary code that we will discuss later in this chapter. First, however, we will look briefly at numbering systems.

Numbering Systems

The most commonly used numbering system in computers is the binary system, but the octal and hexadecimal numbering systems can also be encountered in these applications. Let's first briefly review the traditional decimal numbering system before discussing the three other systems.

Decimal Numbering System

The decimal numbering system is based on the number ten. It is in common use, probably because man first used his fingers to count. However, the decimal system is not easy to implement electronically. A ten-state electronic device would be quite costly and complex. It is much easier and more efficient to use the binary (two-state) numbering system when manipulating numbers through the logic circuits found in computers.

A decimal number, N_{10}, can be written mathematically as follows:

$$N_{10} = d_n R^n + \ldots\ldots + d_2 R^2 + d_1 R^1 + d_0 R^0 \tag{4-1}$$

where R is equal to the number of digital symbols used in the system. R is called the *radix* and is equal to ten in the decimal system. The subscript *10* on the number N in Equation 4-1 indicates that N is a decimal number. However, the common practice is to omit this subscript when writing out decimal numbers. The decimal digits, d_n ..... d_2, d_1, d_0, can assume the values of 0, 1, 2, 3, 4, 5, 6, 7, 8, or 9 in the decimal numbering system.

For example, the decimal number 1,735 can be written as follows:

$$1735 = 1\text{x}10^3 + 7\text{x}10^2 + 3\text{x}10^1 + 5\text{x}10^0$$

$$1735 = 1000 + 700 + 30 + 5$$

When written as "1735," the powers of ten are implied by positional notation. The value of the decimal number is computed by multiplying each digit by the weight of its position and summing the result. As we will see, this is true for all numbering systems. The decimal equivalent of any number can be calculated by multiplying the digit times its base raised to the power of the digit's position. The general equation for numbering systems is the following:

$$N_b = Z_n R^n + \ldots\ldots + Z_2 R^2 + Z_1 R^1 + Z_0 R^0 \tag{4-2}$$

Where Z is the value of the digit, and R is the radix or base of the numbering system.

Binary Numbering System

The binary numbering system has a base of two, and the only allowable digits are zero (0) or one (1). This is the basic numbering system for computers and programmable controllers, which are basically electronic devices that manipulate zeros and ones to perform math and control functions.

It was easier and more convenient to design digital computers that operated on two entities or numbers rather than on the ten numbers used in the decimal world. Furthermore, most physical elements in the process environment have only two states, such as a pump on or off, a valve open or closed, a switch on or off, and so on.

A binary number follows the same format as a decimal one, that is, the value of a digit is determined by its position in relation to the other digits in a number. In the decimal system, a one (1) by itself is worth one; placing it to the left of a zero makes the one worth ten (10), and putting it to the left of two zeros makes it worth one hundred (100). This simple rule is the foundation of the numbering systems. For example, when numbers are to be added or subtracted by hand they are first arranged so their place columns line up.

In the decimal system, each position to the left of the decimal point indicates an increasing power of ten. In the binary system, each place to the left signifies an increased power of two, that is, 2^0 is one, 2^1 is two, 2^2 is four, 2^3 is eight, and so on. So, finding the decimal equivalent of a binary number is simply a matter of noting which place columns the binary 1s occupy and adding up their values. A binary number also uses standard positional notation. The decimal equivalent of a binary number can be found using the following equation:

$$N_2 = Z_n 2^n + ... + Z_2 2^2 + Z_1 2^1 + Z_0 2^0 \qquad (4\text{-}3)$$

where the radix or base equals 2 in the binary system, and each binary digit (bit) can have the value of 0 or 1 only.

Thus, the decimal equivalent of the binary number 10101_2 can be found as follows:

$$10101_2 = 1 \times 2^4 + 0 \times 2^3 + 1 \times 2^2 + 0 \times 2^1 + 1 \times 2^0$$

or

$$(1x16)+(0x8)+(1x4)+(0x2)+(1x1)= 21 \text{ (decimal)}$$

Example 4-1 shows how to convert a binary number into its decimal equivalent.

EXAMPLE 4-1

Problem: Convert the binary number 101110_2 into its decimal equivalent.

Solution: Using Equation 4-3,

$$N_2 = Z_n 2^n + ... + Z_2 2^2 + Z_1 2^1 + Z_0 2^0$$

we have

$$N_2 = 1 \times 2^5 + 0 \times 2^4 + 1 \times 2^3 + 1 \times 2^2 + 1 \times 2^1 + 0 \times 2^0$$

$$= (1 \times 32) + (0 \times 16) + (1 \times 8) + (1 \times 4) + (1 \times 2) + (0 \times 1)$$

$$= 32 + 0 + 8 + 4 + 2 + 0$$

$$= 46$$

To this point, we have discussed only positive binary numbers. Several common methods are used to represent negative binary numbers in computer systems. The first is *signed-magnitude binary*. This method places an extra bit (sign bit) in the left-most position and lets this bit determine whether the number is positive or negative. The number is positive if the sign bit is 0 and negative if the sign bit is 1. For example, suppose that in a 16-bit machine we have a 12-bit binary number, $000000010101_2 = 21_{10}$. To express the positive and negative values, we would manipulate the left-most or most significant bit (MSB). So, using the signed magnitude method, 0000000000010101 = + 21 and 1000000000010101 = –21.

Another common method used to express negative binary numbers is called *two's complement binary*. To complement a number means to change it to a negative number. For example, the binary number 10101 is equal to decimal 21. To get the negative using the two's complement method, you complement each bit and then add 1 to the least significant bit. In the case of the binary number 010101 = 21, its two's complement would be 101011 = – 21.

Octal Numbering System

The binary numbering system requires substantially more digits to express a number than does the decimal system. For example, $130_{10} = 10000010_2$, so it takes 8 or more binary digits to express a decimal number larger than 127. It is difficult for people to read and manipulate large numbers without making errors. To reduce such errors when manipulating binary numbers, some computer manufacturers began using the octal numbering system. This system uses the number eight (8) as a base or radix with the eight digits 0, 1, 2, 3, 4, 5, 6, and 7.

Like all other number systems, each digit in an octal number has a weighted value according to its position. For example,

$$1301_8 = 1 \times 8^3 + 3 \times 8^2 + 0 \times 8^1 + 1 \times 8^0$$

$$= 1 \times 512 + 3 \times 64 + 0 \times 8 + 1 \times 1$$

$$= 512 + 192 + 0 + 1$$

$$= 705_{10}$$

The octal system is used as a convenient way of writing or manipulating binary numbers in computer systems. A binary number that has a large number of ones and zeros can be represented with fewer digits by using its equivalent octal number. As Table 4-1 shows, one octal digit can be used to express three binary digits, so the number is reduced by a factor of three.

Table 4-1. Binary and Octal Equivalent Numbers

Binary	Octal
000	0
001	1
010	2
011	3
100	4
101	5
110	6
111	7

For example, the binary number 11010101010_2 can be converted into an octal number by grouping binary bits in sets of three starting with the least significant bit (LSB), as follows:

$$11\ 010\ 101\ 010 = 3252_8$$

Example 4-2 illustrates how to represent a binary number in octal.

EXAMPLE 4-2

Problem: Represent the binary number 101011001101111_2 in octal.

Solution: To convert a number from binary to octal, we simply divide the binary number into groups of three bits, starting with the least significant bit (LSB). Then we use Table 4-1 to convert the three-bit groups into their octal equivalent.

To solve, we place binary numbers in groups of three: $101\ 011\ 001\ 101\ 111_2$. Since $101_2 = 5_8$, $011_2 = 3_8$, $001_2 = 1_8$, $101_2 = 5_8$, and $111_2 = 7_8$,

we obtain $101011001101111_2 = 53157_8$.

We can convert a decimal number into an octal number by successively dividing the decimal number by the octal base number 8. This is best illustrated in Example 4-3.

EXAMPLE 4-3

Problem: Convert the decimal number 370_{10} into an octal number.

Solution: Decimal-to-octal conversion is achieved by successively dividing by the octal base number 8, as follows:

Division	Quotient	Remainder
370/8	46	2 (LSD)
46/8	5	6
5/8	0	5 (MSD)

Thus, $370_{10} = 562_8$.

Hexadecimal Numbering System

The hexadecimal numbering system provides an even shorter notation than the octal system and is commonly used in computer applications. The hexadecimal system has a base of sixteen (16), and four binary bits are used to represent a single symbol. The sixteen symbols are 0, 1, 2, 3, 4, 5, 6,

7, 8, 9, A, B, C, D, E, and F. The letters A through F are used to represent the binary strings 1010, 1011, 1100, 1101, 1110, and 1111, which correspond to the decimal numbers 10 through 15. The hexadecimal digits and their binary equivalents are given in Table 4-2.

Table 4-2. Hexadecimal and Binary Equivalent Numbers

Hexadecimal	Binary	Hexadecimal	Binary
0	0000	9	1001
1	0001	A	1010
2	0010	B	1011
3	0011	C	1100
4	0100	D	1101
5	0101	E	1110
6	0110	F	1111
7	0111	10	10000
8	1000	11	10001

To convert a binary number into a hexadecimal number, we use Table 4-2. For example, the binary number 0110 1111 1000 is 6F8. Again, the hexadecimal numbers follow the standard positional convention H_n ..... H_2, H_1, H_0, where the positional weights for hexadecimal numbers are powers of sixteen, with 1, 16, 256, and 4096 being the first four decimal values.

To convert from hexadecimal into decimal numbers, we use the following equation:

$$N_{16} = H_n 16^n + ... + H_2 16^2 + H_1 16^1 + H_0 16^0 \tag{4-4}$$

where the radix equals 16 in the hexadecimal system, and each digit can take on the value of zero through nine and the letters *A*, *B*, *C*, *D*, *E*, and *F*.

Example 4-4 illustrates how to convert a hexadecimal number into its decimal equivalent.

To convert a decimal number into a hexadecimal number, we use the following procedure:

1. Divide the decimal number by 16 and record the quotient and the remainder.
2. Divide the quotient from the division in step 1 by 16 and record the quotient and the remainder.
3. Repeat step 2 until the quotient is zero.

EXAMPLE 4-4

Problem: Convert the hexadecimal number 1FA into its decimal equivalent.

Solution: We use Equation 4-4:

$$N_{16} = H_n 16^n + ... + H_2 16^2 + H_1 16^1 + H_0 16^0$$

Since $H_2 = 1 = 1_{10}$, $H_l = F = 15_{10}$, and $H_0 = A = 10_{10}$, we obtain the following:

$$N_{16} = 1 \times 16^2 + 15 \times 16^1 + 10 \times 16^0$$

$$= \quad 256 + 240 + 10 = 506_{10}$$

4. The hexadecimal equivalents of the remainders generated by the previous divisions are the digits of the hexadecimal number. The first remainder is the least significant digit (LSD), and the last remainder is the most significant digit (MSD).

Example 4-5 will illustrate the process of converting from a decimal into hexadecimal number.

EXAMPLE 4-5

Problem: Convert the decimal number 610_{10} into a hexadecimal number.

Solution:

Division	Quotient	Remainder
610/16	38	2 (LSD)
38/16	2	6
2/16	0	2 (MSD)

Therefore, $610_{10} = 262_{16}$.

It is important to note that the octal and hexadecimal systems are used for human convenience only. The computer system actually converts the octal and hexadecimal numbers into binary strings and then operates on binary digits.

Data Codes

Data codes translate information (alpha, numeric, or control characters) into a form that can be transferred electronically and then converted back into its original form. A code's efficiency is a measure of its ability to utilize the maximum capacity of the bits and to recover from error. As the various codes have evolved, their efficiency at transferring data has steadily increased. In this section, we discuss the most commonly used codes.

Binary Code

It is possible to represent 2^n different symbols in a purely binary code of *n* bits. The binary code is a direct conversion of the decimal number into the binary. This is illustrated in Table 4-3.

Table 4-3. Binary to Decimal Code

Decimal	Binary	Decimal	Binary
0	00000	11	01011
1	00001	12	01100
2	00010	13	01101
3	00011	14	01110
4	00100	15	01111
5	00101	16	10000
6	00110	17	10001
7	00111	18	10010
8	01000	19	10011
9	01001	20	10100
10	01010	21	10101

Binary code is the most commonly used code in computers because it is a systematic arrangement of the digits. It is also a weighted code in that each column has a magnitude of 2^n associated with it, and it is easy to translate. In Table 4-3, note that the least significant bit (LSB) alternates every time, whereas the second LSB repeats every two times, the third LSB bit repeats every four times, and so on.

Baudot Code

The Baudot code was the first successful data communications code. It is also known as the International Telegraphic Alphabet #2 (ITA#2). The code was intended primarily for text transmission. It has only uppercase letters and is used with punched-paper tape units on teletypewriters. The Baudot code uses five consecutive bits and an additional start/stop bit to represent a data character, as shown in Figure 4-1. Because it relies on the teletypewriter, it transfers asynchronous data at a very slow rate (ten characters per second).

Most early teletypewriters used basically the same circuit as the telegraph but combined with the mechanics of a typewriter. As with the telegraph, the teletypewriter required a technology that would enable the receiving end to know when the other end wanted to transmit. When there was no transmission, a mark signal (current) would be sent as a "line idle" signal. Since a mark is the idle condition, the first element or bit of any code would have to be a space (no current). This bit, shown in Figure 4-1, is known as the "start space." After the character code pulses had been sent, there also had to be a "current on" (mark) condition, so the receiving device would know when the character was complete and thus would separate this transmission character from the next character to be transmitted. This period of current is known as the "stop mark" and is either 1, 1.42, or 2 elements in duration. The bit time or duration is determined by the teletype's motor speed.

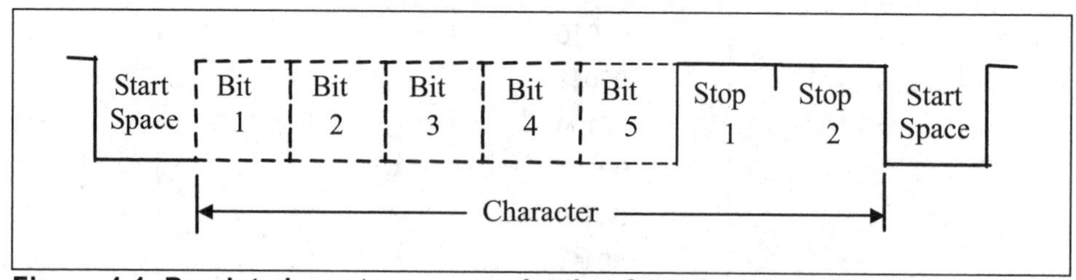

Figure 4-1. Baudot character communication format

The Baudot or teletypewriter code is given in Table 4-4. There are twenty-six uppercase letters, ten numerals, and various elements of punctuation and teletype control. This code uses five bits (two to a fifth power) or thirty-two patterns. However, the code developers used the mechanical shift of the teletypewriter to produce twenty-six patterns out of a possible thirty-two for letters and twenty-six patterns shifted for numbers and punctuation. Only twenty-six were available in either shift because, as Table 4-4 shows, six patterns were the same for both. These six common patterns are carriage return, line feed, shift up (figures), shift down (letters), space, and blank (no current).

Table 4-4. Baudot Code

Character Case		Bit Pattern	Character Case		Bit Pattern
Lower	**Upper**	**54321**	**Lower**	**Upper**	**54321**
A		00011	Q	1	10111
B	?	11001	R	4	01010
C	:	01110	S	'	00101
D	$	01001	T	5	10000
E	3	00001	U	7	00111
F	!	01101	V	;	11110
G	&	11010	W	2	10011
H	#	10100	X	/	11101
I	8	00110	Y	6	10101
J	Bell	01011	Z	"	10001
K	(	01111	Letters Shift Down		11111
L	)	10010	Figures Shift Up		11011
M	.	11100	Space		00100
N	,	01100	Carriage Return		01000
O	9	11000	Line Feed		00010
P	0	01101	Blank or null		00000

In terms of transmission overhead, the Baudot code is still the most efficient code for narrative text because it requires very little machine operation or error detection. While the code is no longer widely used, at one time it was the most common binary transmission code.

One disadvantage of the Baudot code is that it can represent only the fifty-eight characters shown in Table 4-4. Other limitations are the sequential nature of the code, the high overhead, and the lack of error detection.

BCD Code

As computer and data communications technology improved, more efficient codes were developed. The BCD, or binary coded decimal, code was first used to perform internal numeric calculations within data processing devices. The BCD code is commonly used in programmable controllers to code data to numeric light-emitting diode (LED) displays and from panel-mounted digital thumb wheel units. Its main disadvantages are that it has no alpha characters and no error-checking capability. Table 4-5 lists the BCD code for decimal numbers from 0 to 19.

Note that four-bit groups are used to represent the decimal numbers zero through nine. To represent higher numbers, such as ten through nineteen, another four-bit group is used. It is placed to the left of the first four-bit group.

Table 4-5. BCD Code

Decimal	BCD Code	Decimal	BCD Code
0	0000	10	0001 0000
1	0001	11	0001 0001
2	0010	12	0001 0010
3	0011	13	0001 0011
4	0100	14	0001 0100
5	0101	15	0001 0101
6	0110	16	0001 0110
7	0111	17	0001 0111
8	1000	18	0001 1000
9	1001	19	0001 1001

Example 4-6 shows how to convert a decimal number into a BCD code.

EXAMPLE 4-6

Problem: Convert the following decimal numbers into BCD code:

(a) 276,

(b) 567,

(c) 719, and

(d) 4,500.

Solution: Based on Table 4-5, the decimal numbers can be expressed in BCD code as follows:

(a) 276 = 0010 0111 0110,

(b) 567 = 0101 0110 0111,

(c) 719 = 0111 0001 1001, and

(d) 4500 = 0100 0101 0000 0000

ASCII Code

The most widely used code is ASCII, the American Standard Code for Information Interchange. Developed in 1963, this code has seven bits for data (allowing 128 characters), as Table 4-6 shows.

The ASCII code can operate synchronously or asynchronously with one or two stop bits. ASCII format has thirty-two control characters, the legend for which is given in Table 4-7. These control codes are used to indicate, modify, or stop a control function in the transmitter or receiver. Seven of the ASCII control codes are called *format effectors* because they pertain to the control of a printing device. Format effectors increase code efficiency and speed because they replace frequently used character combinations with a single code. The format effectors used in the ASCII code are BS (backspace), HT (horizontal tab), LF (line feed), VT (vertical tab), FF (form feed), CR (carriage return), and SP (space).

Table 4-6. ASCII Code

Bits	7 6 5	0 0 0	0 0 1	0 1 0	0 1 1	1 0 0	1 0 1	1 1 0	1 1 1
4321	HEX	0	1	2	3	4	5	6	7
0000	0	NUL	DLE	SP	0	@	P	‘	p
0001	1	SOH	DC1	!	1	A	Q	a	q
0010	2	STX	DC2	“	2	B	R	b	r
0011	3	ETX	DC3	#	3	C	S	c	s
0100	4	EOT	DC4	$	4	D	T	d	t
0101	5	ENQ	NAK	%	5	E	U	e	u
0110	6	ACK	SYN	&	6	F	V	f	v
0111	7	BEL	ETB	’	7	G	W	g	w
1000	8	BS	CAN	(	8	H	X	h	x
1001	9	HT	EM	)	9	I	Y	i	y
1010	A	LF	SUB	*	:	J	Z	j	z
1011	B	VT	ESC	+	;	K	[	k	{
1100	C	FF	FS	,	<	L	\	l	|
1101	D	CR	GS	-	=	M	]	m	}
1110	E	SO	RS	.	>	N	^	n	~
1111	F	SI	US	/	?	O	-	o	DEL

Table 4-7. Legend for ASCII Control Characters

Mnemonic	Meaning	Mnemonic	Meaning
NUL	Null	DLE	Data Link Escape
SOH	Start of heading	DC1	Device Control 1
STX	State of Text	DC2	Device Control 2
ETX	End of Text	DC3	Device Control 3
EOT	End of Transmission	DC4	Device Control 4
ENQ	Enquiry	NAK	Negative Acknowledge
ACK	Acknowledge	SYN	Synchronous Idle
BEL	Bell	ETB	End of Transmission Block
BS	Backspace	CAN	Cancel
HT	Horizontal Tabulation	EM	End of Medium
LF	Line Feed	SUB	Substitute
VT	Vertical Tabulation	ESC	Escape
FF	Form Feed	FS	File Separator
CR	Carriage Return	GS	Group Separator
SO	Shift Out	RS	Record Separator
SI	Shift In	US	Unit Separator
		DEL	Delete

Example 4-7 shows how to express a simple message in ASCII code.

EXAMPLE 4-7

Problem: Express the words *PUMP 100 ON* using ASCII code. Use hexadecimal notation for brevity.

Solution: Using Table 4-6, we obtain the following:

MESSAGE:	PUMP 100 0N
ASCII (Hex) String:	50 55 4D 50 20 31 30 30 20 4F 4E

Binary Logic Functions

In control system applications, the binary numbers one (1) and zero (0) are represented by voltage levels, relay contact status, switch position, and so on. For example, in transistor-transistor logic (TTL) gates, a binary one is represented by a voltage signal in the range of 2.4 to 5.0 volts, and a binary zero is represented by a voltage level between 0 and 0.8 volt. Solid-state electronic circuits are available that manipulate digital signals so they perform a variety of logical functions, such as NOT, AND, OR, NAND, and NOR. In hardwired electrical logic systems, electrical relays are used to implement the logic functions.

NOT Function

The most basic binary logic function is the NOT or inversion function. The NOT, or logic inverter as it is also known, produces an output that is opposite to that of the input. An inversion bar is drawn over a logic variable to indicate the NOT function. For example, if a NOT operation is performed on a logic variable *A*, it is designated by $Z = \bar{A}$.

Table 4-8 shows a binary logic truth table for the NOT function which lists the results of the NOT function on the Input A. In relay-based logic circuits, a normally closed (NC) set of contracts is used to perform the NOT function, as shown in Figure 4-2. If the electric relay A is not energized, there is electrical current flow or logic continuity through the normally closed contacts, so output relay coil is energized or On. In other words, if input A is logic 0 or not On then the output Z is logic 1 or On. If input A is logic 1 or On, the normally closed contacts are opened, there is no current flow or logic flow in the circuit, relay Z is off, and output Z is 0.

Table 4-8. NOT function Binary Logic Truth Table

Input	Output
A	Z
0	1
1	0

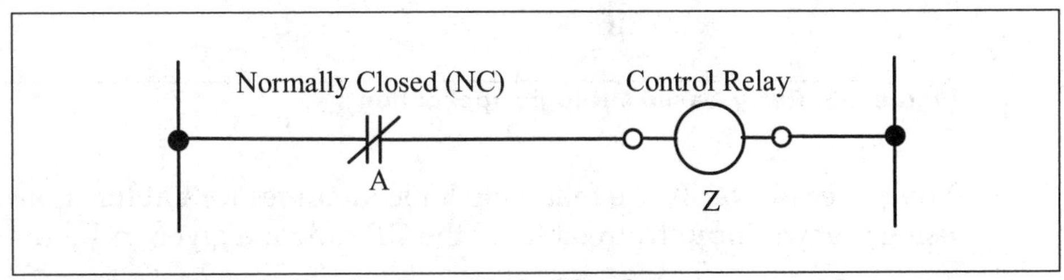

Figure 4-2. NOT function implemented with relay

OR Function

A logical OR function that has two or more inputs and a single output operates in accordance with the following definition: *The output of an OR function assumes the 1 state if one or more inputs assume the 1 state.*

The inputs to a logic function OR gate can be designated by $A, B, \ldots, N$ and the output by Z. We make the assumption that the inputs and outputs can take only one of two possible values, either 0 or 1. The logic expression for this function is $Z = A + B + \ldots + N$. Table 4-9 gives a two-input truth table for a two-input OR function.

Table 4-9. Two-input OR Function Truth Table

Input	Output
A B	Z
0 0	0
0 1	1
1 0	1
1 1	1

Here is an example of OR logic in process control: If the water level in a hot water heater is low, or the temperature in the tank is too high, a logic system can be designed that will turn off the heater in the system using logic circuits or relays. Figure 4-3 shows this application, which uses relays to perform the logic function.

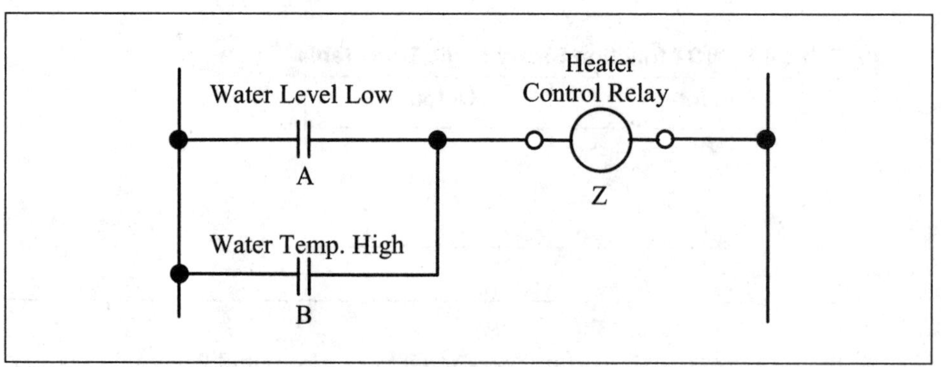

Figure 4-3. Relay-based OR logic application

You can easily verify the following logic identities for OR functions by using the two-input truth table for the OR function given in Table 4-9:

$$A + B + C = (A + B) + C = A + (B + C) \tag{4-5}$$

$$A + B = B + A \tag{4-6}$$

$$A + A = A \tag{4-7}$$

$$A + 1 = 1 \tag{4-8}$$

$$A + 0 = A \tag{4-9}$$

Remember that *A*, *B*, and *C* can only assume the value of 0 or 1.

AND Function

An AND function has two or more inputs and a single output, and it operates according to the following rule: *The output of an AND gate assumes the 1 state if and only if all the inputs assume the 1 state.* The general equation for the AND function is given by *ABC ..... N = Z*. Table 4-10 is a two-input AND function truth table.

Table 4-10. Two-input AND Truth Table

Inputs		Output
A	B	Z
0	0	0
0	1	0
1	0	0
1	1	1

Consider the following example of AND logic being used in process control. Suppose the liquid level in a process tank is high and the inlet feed pump to the tank is running. Design a logic circuit for opening the tank outlet valve using electric relays. Figure 4-4 shows the relay ladder logic diagram that will perform the required AND function for this example.

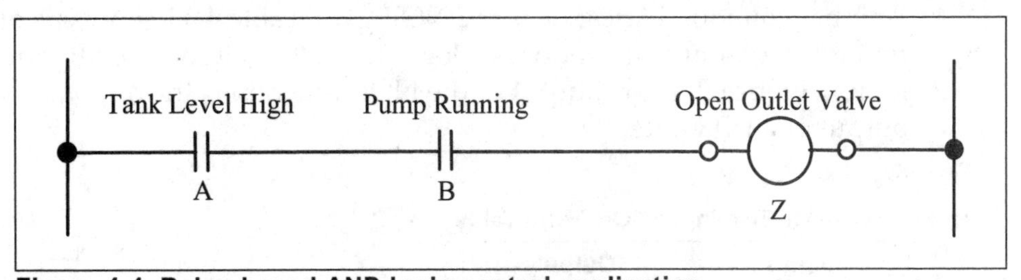

Figure 4-4. Relay-based AND logic control application

The logic expressions for the AND function are as follows:

$$ABC = (AB)C = A(BC) \tag{4-10}$$

$$AB = BA \tag{4-11}$$

$$AA = A \tag{4-12}$$

$$A1 = A \tag{4-13}$$

$$A0 = 0 \tag{4-14}$$

You can verify these identities by referring to the definitions of the AND gate and by using a truth table for the AND gate. For example, you can verify Equation 4-13 (A1 = A) as follows: let B = 1 and tabulate AB = Z or A1 = Z in a truth table such as the following:

Inputs	Output
A B	Z
0 1	0
1 1	1

Note that A = Z in this truth table, so A1 = A for both values (0 or 1) of *A*.

Some important auxiliary identities used in logic design are as follows:

$$A + AB = A \tag{4-15}$$

$$A + \bar{A}B = A + B \tag{4-16}$$

$$(A + B)(A + C) = A + BC \tag{4-17}$$

These identities are important because they help reduce the number of logic elements or gates that are required to implement a logic function.

NOR Function

Another common logic gate is the NOR gate. Table 4-11 shows its operation for two inputs. It produces a logic 1 result if and only if all inputs are logic 0. Notice that the output of the NOR function is the opposite of the output of an OR gate.

Table 4-11. Two-input NOR Truth Table

Inputs	Output
A B	Z
0 0	1
0 1	0
1 0	0
1 1	0

NAND Function

Another logic operation that is relevant to process control is the NAND gate. Table 4-12 summarizes its operation for a two-input NAND function.

Note that the output of the NAND function is the exact opposite of the AND function's output. When both inputs to the NAND are 1, the output is 0. In all other configurations, the NAND function output is 1.

Table 4-12. Two-input NAND Function Truth Table

Inputs	Output
A B	Z
0 0	1
0 1	1
1 0	1
1 1	0

Logic Function Symbols

Two common sets of symbols are used in process control applications to represent logic function: graphic and ladder. The graphic symbols are generally used on engineering drawings to convey the overall logic plan for a discrete or batch control system. They are based on the standard ISA-5.2-1976 (R1992) - Binary Logic Diagrams for Process Operations. The ladder logic symbols are used to describe a logic control plan if relays are being used to implement the control system or if programmable controller ladder logic is used to implement the control plan. Figure 4-5 compares these two sets of symbols for the most common logic functions encountered in logic control design.

Ladder Diagrams

Ladder diagrams are a traditional method for describing electrical logic controls. These circuits are called ladder diagrams because they look like ladders with rungs. Each rung of a ladder diagram is numbered so we can easily cross-reference between the sections on the drawing that describe the control.

The utility of ladder diagrams can best be appreciated by investigating a simple control example. Figure 4-6 shows a typical process-level control application. In this application, let us assume that the flow into the tank is random. Let us also assume that we need to control the level in the tank by opening or closing the outlet valve (LV-1) based on the tank's level as it is sensed by a level switch (LSH-1). We will also provide the operator with a three-position switch so he or she can maintain the proper level in the tank by manually turning the valve on or off or by selecting automatic control position. The hand switch (HS-1) is designated "HOA," where "H" repre-

Logic Function	Logic Symbol	Ladder Symbol
NOT, $Z = \overline{A}$	A —○— $Z = \overline{A}$	A Z
OR, Z =A + B	A, B — OR — Z	A B Z
AND, Z = A• B	A, B — A — Z	A B Z

Figure 4-5. Comparison of logic symbols

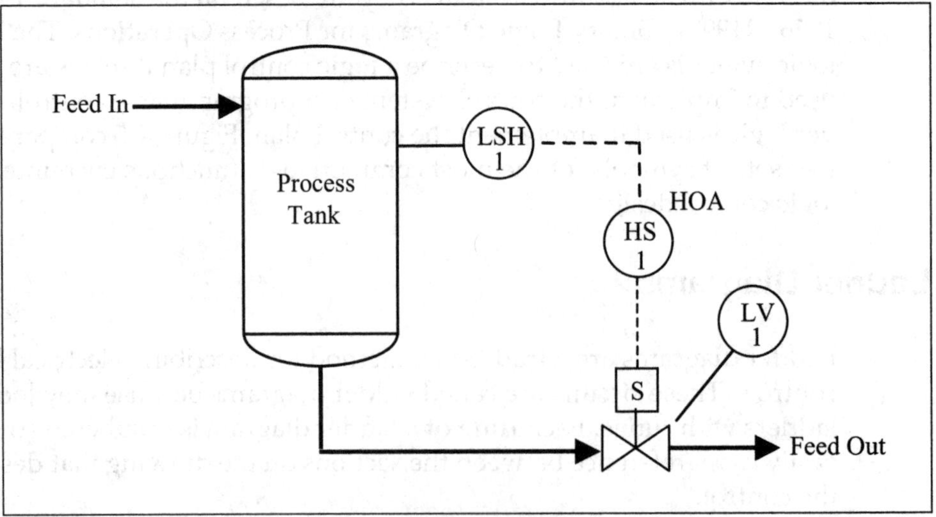

Figure 4-6. On/off control of process tank level

sents hand or manual position, the letter "O" is for the off position, and the "A" represents the automatic operation of the solenoid valve based on some control logic.

The ladder logic design for this application is shown in Figure 4-7. If the panel-mounted HOA switch is in the automatic position and the level switch is closed, the solenoid will be energized. This is a simple example of a logical AND function in process control. If the HOA switch is in the hand or manual position, the valve will also be turned on.

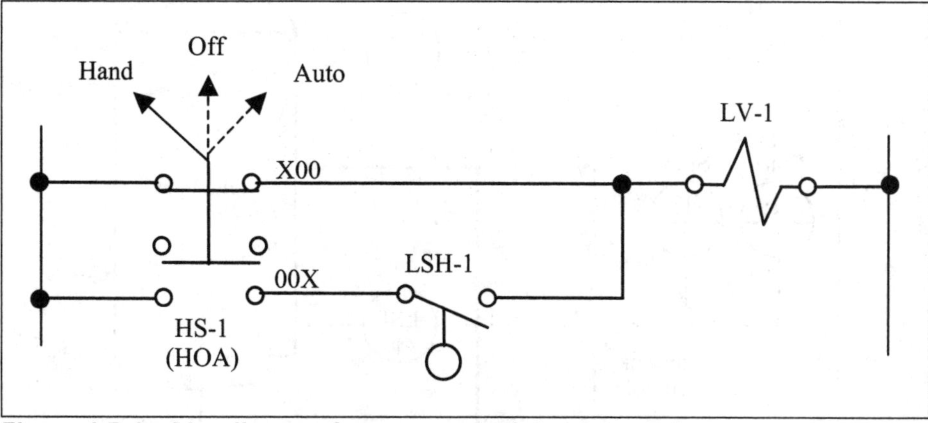

Figure 4-7. Ladder diagram for process tank level control

Suppose that we designate the logic variables for the ladder diagram shown in Figure 4-7 as follows: A for hand position of HOA switch, B for automatic position of the HOA switch, C for the status of level switch LSH-1, and Z for the output to turn on solenoid valve LV-1. In this case, the logic equation for the control circuit is Z = A + BC. We can implement this control circuit with logic gates, as shown in Figure 4-8.

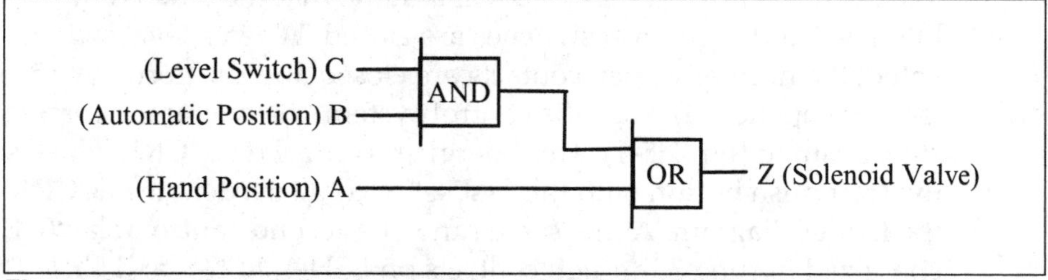

Figure 4-8. Logic gate implementation of tank level control

A more complex application might be to control the liquid level of the tank between two level switches: a level switch high (LSH) and a level switch low (LSL). In this application, a pump supplying liquid to a tank is turned on and off to maintain the liquid level in the tank between the two level switches. The process, shown in Figure 4-9, uses a regulated flow of steam to boil down a liquid so as to produce a more concentrated solution. This solution is drained off periodically by opening a manual valve on the bottom of the tank. Note that a small diamond symbol is used to indicate that the pump is interlocked with the level switches on the tank.

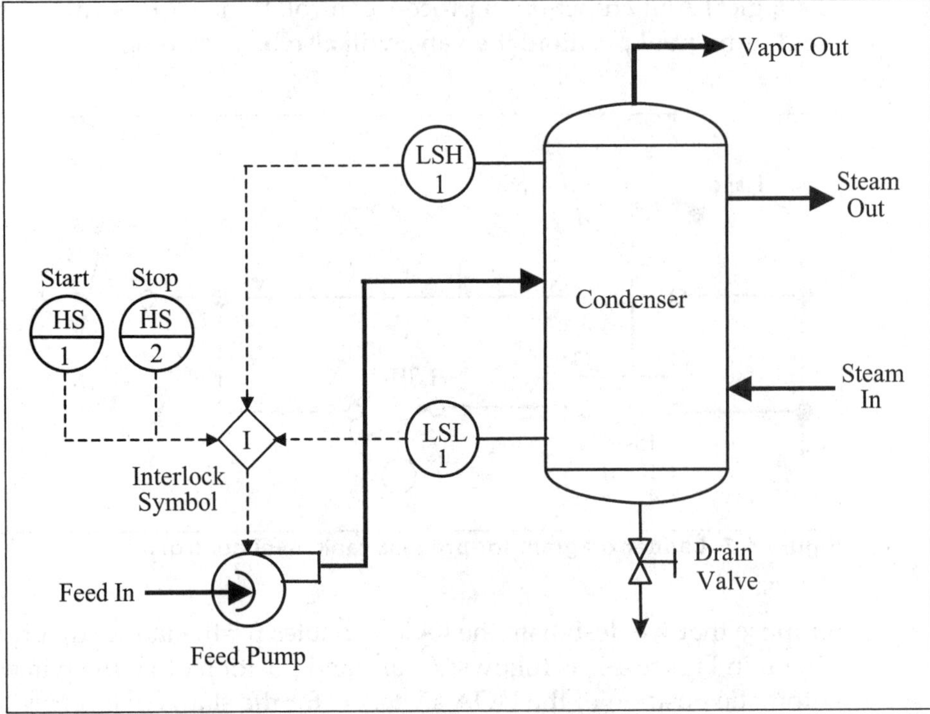

Figure 4-9. Pump control of condenser liquid level

Figure 4-10 shows the electrical ladder diagram that is used to control the feed pump and hence the liquid level in the process tank. To explain the logic of the control system, we will assume that the tank is empty and the low-level and high-level switches are closed. When a level switch is activated, the normally open contacts are closed and the normally closed contacts are opened. To start the control system, the operator depresses the start push button (HS-1). This energizes control relay CR1, which seals in the start push button with the first set of contacts, denoted as CR1(1) on the ladder diagram. At the same time, the second control relay (CR2) is energized in rung 3 through contacts on LSH-1, LSL-1, and CR1. This turns on the pump motor starter MS-1. The first set of contacts on CR2(1) is used to seal in the low-level switch contacts. As a result, when the level in the tank rises above the position of the low-level switch on the tank, the pump will stay on until the liquid level reaches the high-level switch. After the high-level switch is activated, the pump will be turned off. The system will now cycle on and off between the high and low levels until the operator depresses the stop push button (HS-2).

This pump control application is a typical example of logic control in the process industries. Plants can implement the logic control system by using a hardwired relay-based logic system, a programmable logic controller, or a distributed control system.

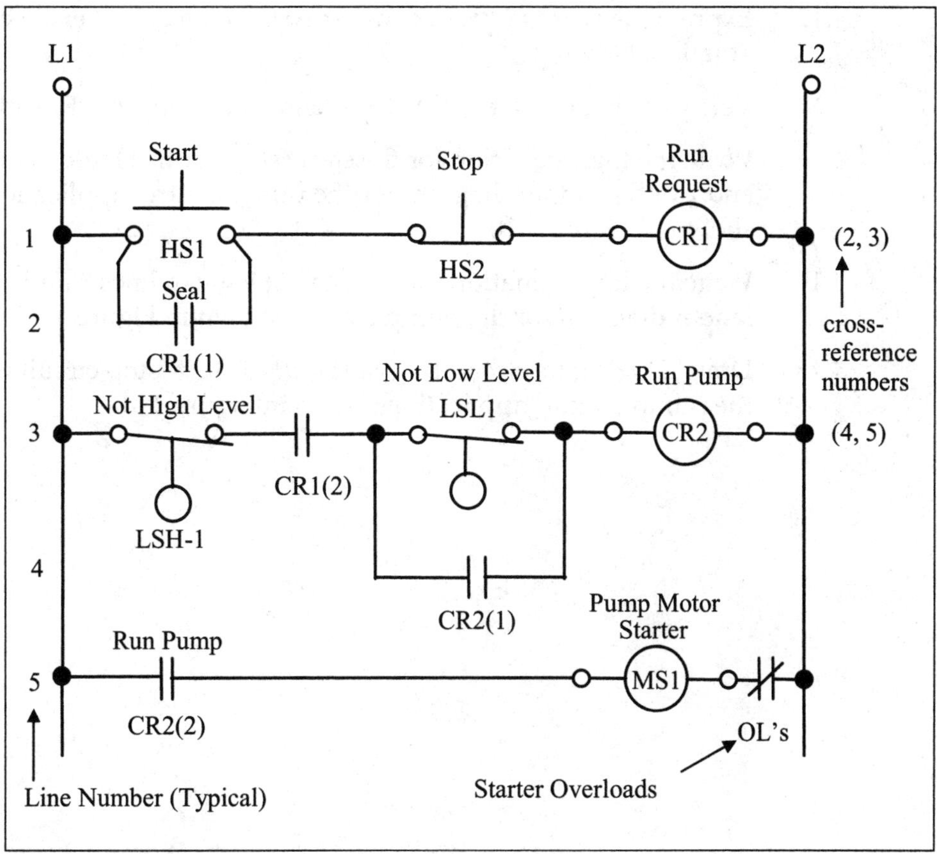

Figure 4-10. Ladder diagram for pump control

EXERCISES

4.1 Convert the binary number 1010 into a decimal number.

4.2 Convert the binary number 10011 into a decimal number.

4.3 Find the octal equivalent of the binary number 10101011.

4.4 Find the octal equivalent of the binary number 101111101000.

4.5 Convert the decimal number 33_{10} into its octal equivalent.

4.6 Convert the decimal number 451_{10} into its octal equivalent.

4.7 Convert the hexadecimal number 47_{16} into its decimal equivalents.

4.8 Convert the hexadecimal number 157_{16} into its decimal equivalents.

4.9 Convert the decimal number 56_{10} into its hexadecimal equivalent.

4.10 Convert the decimal number 37_{10} into its equivalent BCD code.

4.11 Convert the decimal number 270_{10} into its equivalent BCD code.

4.12 Express the words *Level Low* using ASCII code. Use hexadecimal notation.

4.13 Verify the logic identity $A + 1 = 1$ using a two-input OR truth table.

4.14 Write the logic equation for the start/stop control logic in lines 1 and 2 of the ladder diagram for the pump control application shown in Figure 4-10.

4.15 Write the logic equation for the control logic in lines 3 and 4 of the ladder diagram for the pump control shown in Figure 4-10.

4.16 Draw a logic gate circuit for the standard start/stop circuit used in the pump control application shown in Figure 4-10.

5

Pressure Measurement

Introduction

This chapter discusses the principles of pressure measurement and the pressure sensors commonly found in industrial control applications. Many methods are used to measure pressure. Which method your plant chooses depends on several factors, including the specific requirements of the measuring system, the accuracy required, and the static, dynamic, and physical properties of the process being measured or controlled.

A pressure sensor is always used to convert pressure into a mechanical or electrical signal. This signal is then used for process indication or to generate a scaled transmitted signal that a controller can use to control a process. Before discussing pressure elements or sensor types, we need to define pressure and other terms used in pressure measurement.

Definition of Pressure

A major concern in process control applications is the measurement of fluid pressure. The term *fluid* means a substance that can flow; hence, the term applies to both liquids and gases. Both will occupy the container in which they are placed. However, if a liquid does not completely fill the container it will have a free liquid surface, whereas a gas will always fill the volume of a container. If a gas is confined in a container, its molecules strike the container walls. The collision of molecules against the walls results in a force against the surface area of the container. The pressure is equal to the force applied to the walls divided by the area that is perpendicular to the force. For a liquid at rest, the pressure exerted by the fluid at any point is perpendicular to the boundary of the liquid. Pressure is defined as a force applied to, or distributed over, a surface area, or

$$P = \frac{F}{A} \tag{5-1}$$

where P is pressure, F is force, and A is area. The standard unit of pressure in the English or foot-pound-second (fps) system is pounds per square inch (psi), where the force is expressed in pounds and the area in square inches. This is the most common pressure unit encountered in process control applications and will be used extensively throughout this book. In the International System of Units (SI), the pressure units are Pascal (Pa) or Newton per square meter (N/m^2). Table 5-1 lists several pressure-unit conversion factors.

Table 5-1. Pressure-unit Conversion Factors

1 inches water (inH_2O) = 0.0360 psi = 0.0737 inches mercury (inHg)
1 foot water = 0.4320 psi = 0.8844 inHg
1 psi = 27.7417 inH_2O = 2.0441 inHg
1 kg/cm^2 = 14.22 psi = 98.067 kilopascals (kPa)
Note: All fluids at a temperature of 22°C.

Pressure in a Fluid

In our basic definition of pressure, we neglected the weight and density of the fluid and assumed that the pressure was the same at all points. It is a familiar fact, however, that atmospheric pressure is greater at sea level than at mountain altitudes and that the pressure in a lake or in the ocean increases the greater the depth below the surface. Therefore, we must more clearly define pressure at *any point* as the ratio of the normal force ΔF exerted on a small area ΔA (including the point) as follows:

$$P = \frac{\Delta F}{\Delta A} \tag{5-2}$$

If the pressure is the same at all points of a finite plane surface that has area A, this equation reduces to Equation 5-1.

Let's find the general relation between the pressure at any point in a fluid in a gravitational field and the elevation of the point y. If the fluid is in equilibrium, every volume element is in equilibrium. Consider a section of a homogeneous fluid with density ρ (rho) in the form of a thin slab, shown in Figure 5-1.

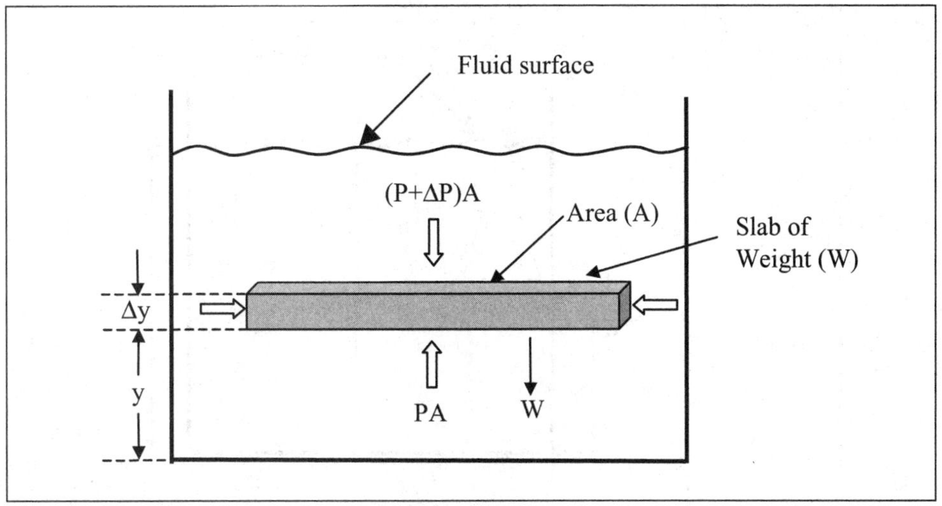

Figure 5-1. Forces on slab of fluid at equilibrium

Since density is defined as mass (m) per unit volume (V) or $\rho = m/V$, the mass of the thin slab is $\rho A\Delta y$ and its weight (W) is $\rho gA\Delta y$. The force exerted on the element by the surrounding fluid is everywhere perpendicular to the surface. By symmetry, the sum of the horizontal forces on the slab's vertical sides is equal to zero. The upward force on its lower face is PA, the downward force on its upper face is $(P + \Delta P)A$, and the weight of the slab is $\rho gA\Delta y$. Since the slab is in equilibrium, all three forces must equal zero, or

$$PA-(P+\Delta P)A-\rho gA\Delta y=0$$

Solving this equation for the change in pressure, ΔP, gives us

$$\Delta P = -\rho g\Delta y \tag{5-3}$$

Since density ρ and gravity g are both positive numbers, it follows that a positive Δy (increase of elevation) is accompanied by a negative ΔP (decrease of pressure). If P_1 and P_2 are the pressures at elevations y_1 and y_2 above some reference level (datum), then

$$P_2 - P_1 = -\rho g(y_2 - y_1) \tag{5-4}$$

Let's apply this equation to a given point in the liquid of the open vessel shown in Figure 5-2.

Take point 1 at a given level, and let P represent the pressure at that point. Select point 2 at the top of the liquid in the tank, where the pressure is atmospheric pressure, represented by P_a. Then Equation 5-4 becomes the following:

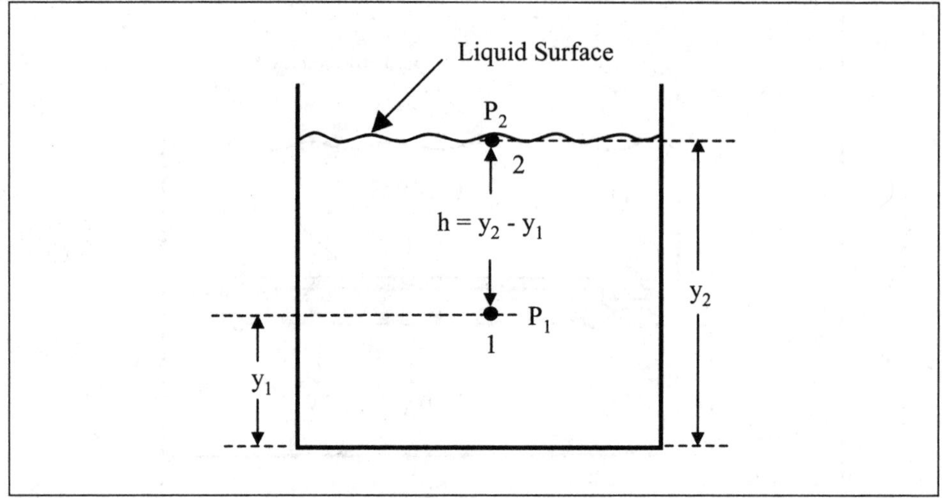

Figure 5-2. Pressure in a vessel open to atmosphere

$$P_a - P = -\rho g(y_2 - y_1) \tag{5-5}$$

And since $h = y_2 - y_1$, we obtain,

$$P = P_a + \rho gh \tag{5-6}$$

Equation 5-6 indicates that the shape of the open tank does not affect the pressure and that the pressure is influenced only by the depth (h) and the density (ρ). This fact was recognized in 1653 by the French scientist Blaise Pascal (1623-1662) and is called Pascal's law. This law states that whenever an external pressure is applied to any confined fluid at rest, the pressure is increased at every point in the fluid by the amount of the external pressure.

Since water is the most common fluid encountered in process control, the pressure properties of water are important. Figure 5-3 shows water in a cubic foot container. The weight of this cubic foot (ft^3) of water is 62.3 lbs at 20°C, and this weight is exerted over the surface area of 1 ft^2, or 144 in^2. Since pressure is defined as $P = F/A$ (Equation 5-1), the total pressure on the surface of 1 ft^2 is given by the following:

$$P = \frac{F}{A} = \frac{62.3 \text{ lbs}}{144 \text{ in}^2} = 0.433 psi$$

This means a 1-ft column of water exerts a pressure of 0.433 lb/in^2. Conversely, a pressure of 0.433 psi will cause a column of water to be raised 1 ft.

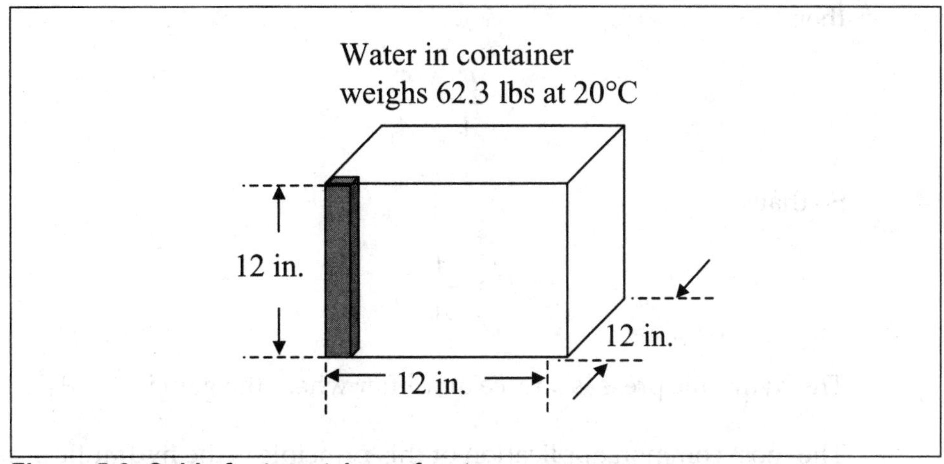

Figure 5-3. Cubic foot container of water

A practical application of Pascal's law is the hydraulic press shown in Figure 5-4. A small force (F_1) is applied to the small position (area A_1). The following relationship exists in the hydraulic press because, according to Pascal's law, the pressure at every point is equal:

$$P_1 = P_2$$

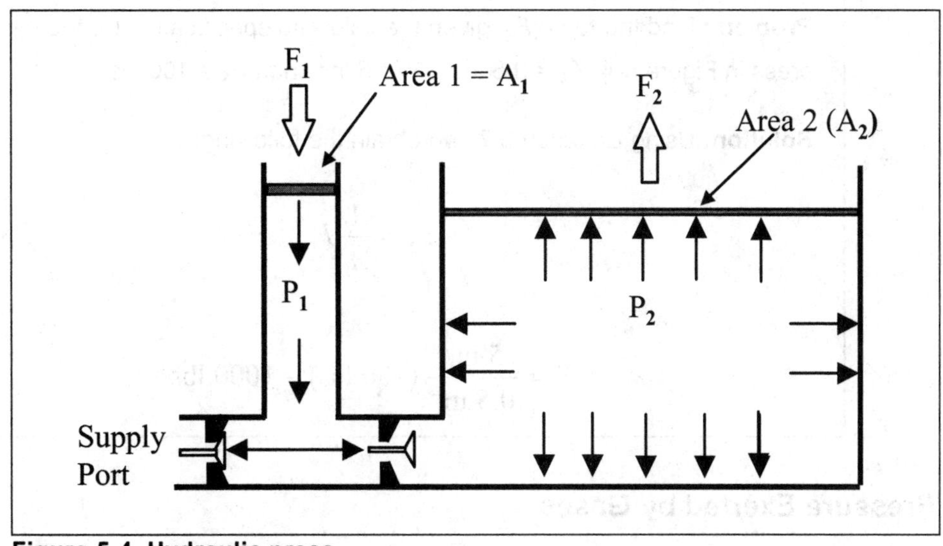

Figure 5-4. Hydraulic press

Since

$$P_1 = \frac{F_1}{A_1} \text{ and } P_2 = \frac{F_2}{A_2}$$

then,

$$\frac{F_1}{A_1} = \frac{F_2}{A_2}$$

So that

$$F_2 = \frac{A_2}{A_1} F_1 \tag{5-7}$$

The hydraulic press is a force amplifier where the gain is A_2/A_1.

The most common application of this principle is the hydraulic brake system in automobiles. It is also used very extensively in pneumatic and hydraulic instruments as well as in equipment used to amplify process signals.

Example 5-1 illustrates how the concept expressed by Pascal's law can be utilized in practical applications.

EXAMPLE 5-1

Problem: Find the force F_2, given the following specification for the hydraulic press in Figure 5-4: A_1 = 0.5 in.^{2}, A_2 = 5 in.^{2}, and F_1 = 100 lbs.

Solution: Using Equation 5-7, we obtain the following:

$$F_2 = \frac{A_2}{A_1} F_1$$

$$F_2 = \frac{5 \text{ in.}^2}{0.5 \text{ in.}^2}(100 \text{ lbs}) = 1000 \text{ lbs}$$

Pressure Exerted by Gases

Liquids conform to the shape of the vessel they are contained in and form a surface layer if the liquid does not fill the vessel. Gases, on the other hand, assume no definite shape. They will expand to fill the entire vessel that contains them. Therefore, a gas will exert an equal amount of pressure on all surfaces of the container. Two factors that affect the pressure that a gas exerts on the surface of the container are the volume of the vessel and the temperature of the gas. The relationship between the pressure (P) exerted by the gas and the volume (V) of the vessel is expressed by Boyle's law:

$$V = \frac{K}{P} \tag{5-8}$$

where K is a constant, and the gas in the closed system is held at a constant temperature.

Boyle's law states that if temperature is held constant, the pressure exerted by the gas on the walls of a container varies inversely with the volume of the vessel. This means that if the volume of the container is decreased, the pressure exerted by the gas increases. Conversely, if the volume of the vessel is increased, the pressure exerted by the gas decreases. For example, suppose that the pressure exerted by a gas in a 2 ft^3 container is 10 psi. If the container size is increased to 4 ft^3, the pressure will decrease to 5 psi if the temperature of the gas is held constant. This relationship can be expressed mathematically as $P_1 \times V_1 = P_2 \times V_2$, where P_1 and V_1 are the initial pressure and volume of the gas, respectively, and P_2 and V_2 are the final pressure and volume. To find the final pressure (P_2) if the volume changes and the temperature remains constant, we use the following equation:

$$V_2 = \frac{P_1}{P_2} V_1 \tag{5-9}$$

Example 5-2 illustrates the use of Boyle's Law.

EXAMPLE 5-2

Problem: A gas sample occupies a volume of 3 ft^3 at a pressure of 4 psi and a temperature of 60°F. What will be its volume if the pressure increases to 8 psi and the temperature remains at 60°F?

Solution: Using Equation 5-9, we obtain the following:

$$V_2 = \frac{P_1}{P_2} V_1$$

$$V_2 = \frac{4 \text{ psi}}{8 \text{ psi}} \times 3 \text{ ft}^3 = 1.5 \text{ ft}^3$$

The relationship between temperature and pressure for a gas at constant volume is expressed by Charles' law. This law states that if the volume of the vessel holding a gas is constant, then the pressure exerted by the gas on the walls of the vessel will vary directly with the temperature in Kelvin (K) of the gas.

Charles' law can be expressed as follows:

$$P = KT \tag{5-10}$$

where K is a constant and the volume of the gas in the closed system is constant. This relationship means that if the temperature in K is doubled, the pressure exerted by the gas on the walls of the container will double. This concept can be expressed mathematically as follows:

$$\frac{P_1}{T_1} = \frac{P_2}{T_2} \tag{5-11}$$

where P_1 and T_1 are the initial pressure and temperature of the gas, respectively, and P_2 and T_2 are the final pressure and temperature in a fixed volume.

Example 5-3 illustrates a typical application of Charles' law.

EXAMPLE 5-3

Problem: A sample of gas occupies a fixed volume of 10 ft^3 at a pressure of 4 psi and a temperature of 25°C. What will its pressure be if the temperature of the gas is increased to 50°C?

Solution: From Charles' law we know that $P_1/T_1 = P_2/T_2$ if temperature is in Kelvin. Thus, we must first convert the Celsius temperatures into Kelvin: T_1 (K) = 25° + 273.14° = 298.14° and T_2 (K) = 50° + 273.14° = 323.14°. Then, using Equation 5-11 and solving for P_2, we obtain

$$P_2 = P_1 \frac{T_2}{T_1}$$

$$P_2 = (4 \text{ psi}) \frac{323.14}{298.14} = 4.33 \text{ psi}$$

Gauge and Absolute Pressure

Absolute pressure is the pressure measured above total vacuum or zero absolute, where zero absolute represents a total lack of pressure. Gauge pressure is the pressure measured above atmospheric or barometric pressure. It represents the positive difference between measured pressure and existing atmospheric pressure.

Most pressure gauges and other pressure-measuring devices indicate a zero reading when the measuring point is exposed to the atmosphere. This point is called *zero psi*. In fact, most pressure instruments actually measure a difference in pressure. However, some instruments are designed to produce a reading that is referenced to absolute zero and to indicate a reading near 14.7 psi at sea level when the pressure point is exposed to atmospheric pressure. This reading is generally termed *psi*. Figure 5-5 illustrates the relationship between absolute and gauge pressure.

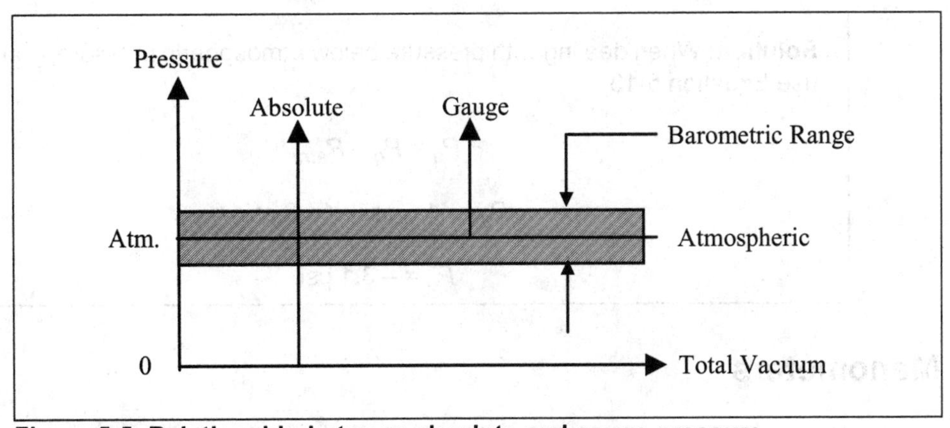

Figure 5-5. Relationship between absolute and gauge pressure

The equation for converting from gauge pressure (P_g) in psi to absolute pressure (P_a) in psi is given by the following:

$$P_a = P_g + P_{atm} \text{ (when } P_g > P_{atm}) \tag{5-12}$$

$$P_a = P_g - P_{atm} \text{ (when } P_g < P_{atm}) \tag{5-13}$$

where P_{atm} is atmospheric pressure.

It should be noted that a change in atmospheric pressure will cause a change in gauge pressure. Therefore, a change in barometric pressure will cause a change in the reading of a gauge-pressure-measuring instrument.

This principle can be best illustrated by Examples 5-4 and 5-5.

EXAMPLE 5-4

Problem: If a pressure instrument has a reading of 30 psi, find the absolute pressure if the local barometric reading is 14.6 psi.

Solution: Since $P_g > P_{atm}$, use Equation 5-12 to find the absolute pressure:

$$P_a = P_g + P_{atm}$$

$$P_a = 30 \text{ psi} + 14.6 \text{ psi}$$

$$P_a = 44.6 \text{ psi}$$

EXAMPLE 5-5

Problem: Find the absolute pressure if a vacuum gauge reads 11.5 psi and the atmospheric pressure is 14.6 psi.

Solution: When dealing with pressure below atmospheric pressure, you must use Equation 5-13:

$$P_a = P_g - P_{atm}$$

$$P_a = 11.5 \text{ psi} - 14.6 \text{ psi}$$

$$P_a = -3.1 \text{ psi}$$

Manometers

The manometer is one of the first pressure-measuring instruments ever designed. Today, it is mainly used in the laboratory or to calibrate pressure instruments in the process industries. A manometer consists of one or two transparent tubes and two liquid surfaces. Pressure applied to the surface of one tube causes an elevation of the liquid surface in the other tube. The amount of elevation is read from a scale that is usually calibrated to read directly in pressure units.

In general, the manometer works as follows: An unknown pressure is applied to one tube, and a reference (known) pressure is applied to the other tube. The difference between the known pressure and the unknown pressure is balanced by the weight per unit area of the displaced manometer liquid. Mercury and water are two commonly used liquids in manometers; however, any fluid can be used. In fact, some manometers now use a fill liquid that has a specific gravity of 2.95 to avoid the environmental problems associated with mercury. The formula for the pressure reading of a manometer is given by the following:

$$P = K_m (SG)h \tag{5-14}$$

where

P	=	pressure in psi
h	=	the inches of displaced liquid
SG	=	the specific gravity of the manometer liquid
K_m	=	0.03606 psi/in

The *specific gravity* of a material is the ratio of its density to that of water. It is therefore a pure number with no dimensions. *Specific gravity* is a very poor term since it has nothing to do with gravity. Relative density would describe the concept more precisely. But, the term *specific gravity* (SG) is widely used in engineering to reduce the number of dimensions in equations.

Example 5-6 shows how to calculate the amount of displacement of liquid in a manometer that is connected to a process tank.

EXAMPLE 5-6

Problem: Find the displacement in inches of a liquid that has a specific gravity of 2.95 in the manometer shown in Figure 5-6, if the pressure of the gas in the process tank is 3 psi.

Solution: Using Equation 5-14, we obtain

$$P = (0.03606)(SG)h$$

Therefore,

$$h = P/((0.03606)(SG))$$

$$h = (3\ \text{psi})/(0.03606)(2.95) = 28.2\ \text{inches}$$

Manometers can provide a very accurate measurement of pressure and are often used as calibration standards for other pressure-measurement devices. The pressure-measurement range of most manometers is usually from a few inches to about 30 inches. The range depends on the physical length and arrangement of the tubes and the specific gravity of the fill fluid.

EXAMPLE 5-7

Problem: Calculate the pressure detected by a mercury manometer if the mercury is displaced 10 inches and the specific gravity of the mercury in the manometer is 13.54.

Solution: Using Equation 5-14, we obtain

$$P = (0.03606)(SG)h$$

$$P = (0.03606)(13.54)(10 \text{ inches})$$

$$P = 4.9 \text{ psi}$$

The most common type of manometer is the *U-tube*. The open tube manometer shown in Figure 5-6, for example, is used to measure the pressure in a process tank. It consists of a U-shaped tube that contains a liquid. One end of the tube is open to the atmosphere, and the other end is connected to the system whose pressure we want to measure.

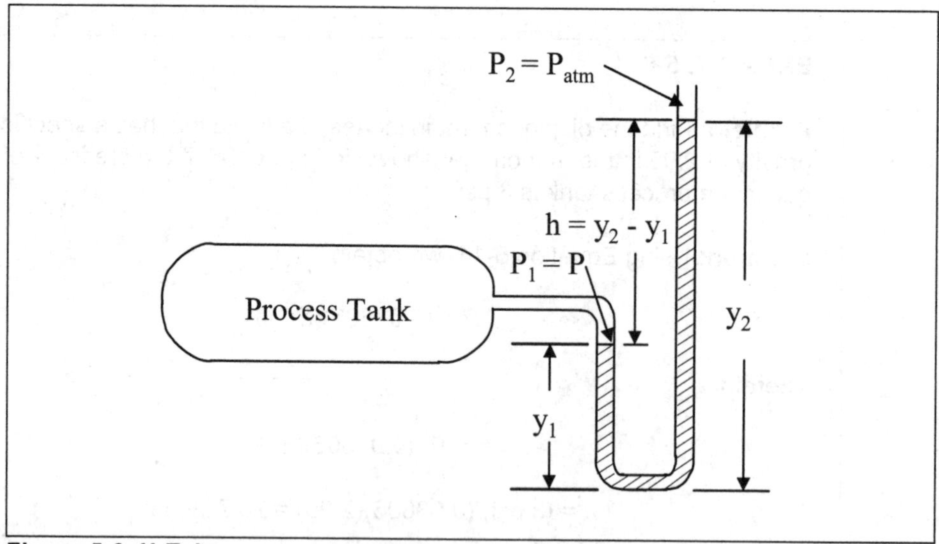

Figure 5-6. U-Tube manometer used on a process tank

Pressure Gauges

Pressure gauges are used for local indication and are the most common type of pressure-measurement instrument used in process industries. Pressure gauges consist of a dial or indicator and a pressure element. A pressure element converts pressure into a mechanical motion.

Most mechanical pressure elements rely on the pressure that acts on a surface area inside the element to produce a force that causes a mechanical deflection. The common elements used are *Bourdon tubes diaphragms* and *bellows elements.*

Figure 5-7 shows one of the most common and least expensive pressure gauges used in the process industries. This pressure gauge uses a "C" type Bourdon tube. In this device, a section of tubing that is closed at one end is partially flattened and coiled, as shown in Figure 5-7. When a pressure is applied to the open end, the tube uncoils. This movement provides a displacement that is proportional to the applied pressure. The tube is mechanically linked to a pointer on a pressure dial to give a calibrated reading.

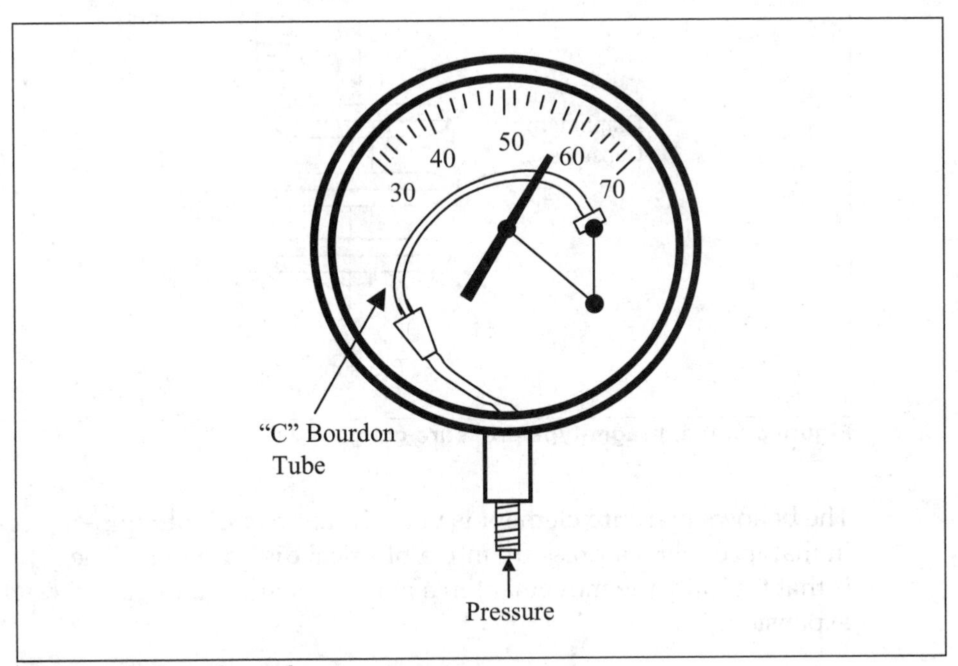

Figure 5-7. Bourdon tube pressure gauge

A diaphragm is another device that is commonly used to convert pressure into a physical movement. A diaphragm is a flexible membrane. When two are fastened together they form a container called a *capsule*. In pressure-measuring instruments, the diaphragms are normally metallic. Pressure applied inside the diaphragm capsule causes it to expand and produce motion along its axis. A diaphragm acts like a spring and will extend or contract until a force is developed that balances the pressure difference force. The amount of movement depends on how much spring there is in the type of metal used. A wide variety of materials are used, including brass, phosphor bronze, beryllium-copper, stainless steel, Ni-Span-C, Monel, Hastelloy, titanium, and tantalum.

To amplify the motion that a diaphragm capsule produces, several capsules are connected end-to-end, as shown on Figure 5-8. You can use diaphragm-type pressure gauges to measure gauge, absolute, or differential pressure. They are normally used to measure low pressures of 1 inch of Hg, but they can also be manufactured to measure higher pressures in the range of 0 to 330 psi. They can also be built for use in vacuum service.

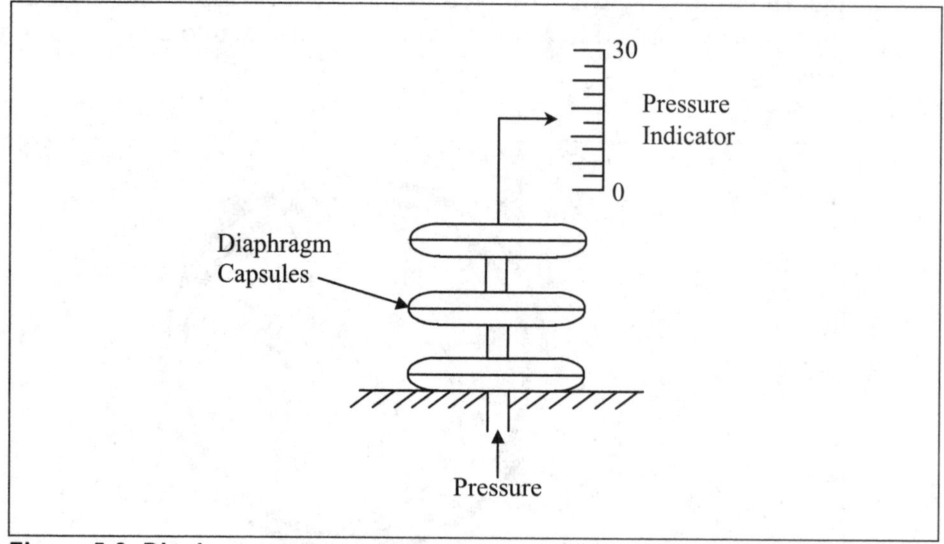

Figure 5-8. Diaphragm-type pressure gauge

The bellows pressure element is very similar to a diaphragm-type gauge in that it converts a pressure into a physical displacement. The difference is that typically the movement in a bellows is much more of a straight-line expansion.

A typical bellows-type pressure gauge is manufactured by forming many accordion-like pleats into a cylindrical tube, as shown in Figure 5-9. Bellows pressure cells are low-pressure cells of as little as 0 to 1 inch of Hg and as much as 0.5 to 30 psi. In vacuum service, they can be used to 30 inches of Hg. You can use the motion that the pressure input signal produces to position a pointer, recorder pen, or the wiper of a potentiometer.

Pressure Transmitters

In some control applications, the pressure value must be transmitted some distance, such as to a central control room, where it is then converted into

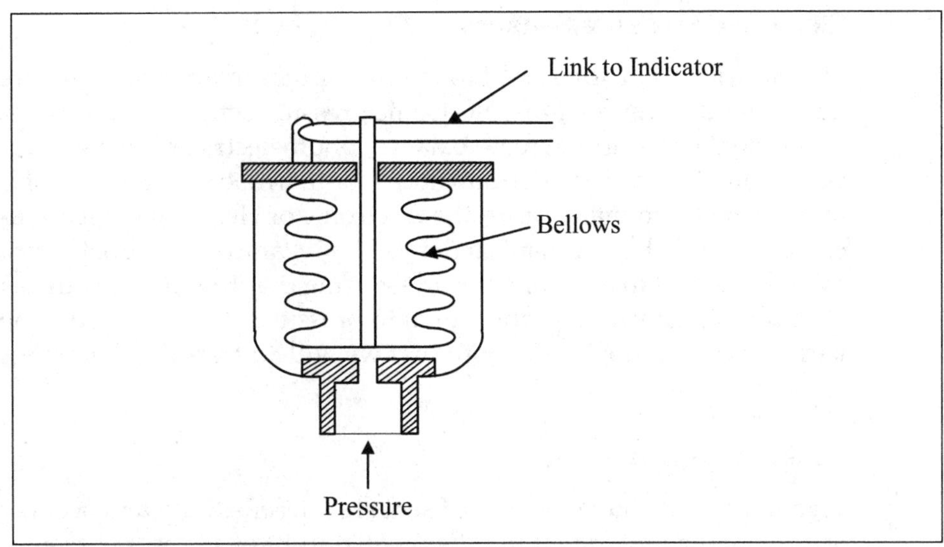

Figure 5-9. Pressure bellows

a usable pressure reading. Pressure transmitters are designed to convert a pressure value into a scaled signal, electric, pneumatic, or mechanical.

Pneumatic Pressure Transmitter

A typical pneumatic pressure transmitter is called a *force-balance pneumatic transmitter*. In this type of instrument, the pressure that is to be measured is applied to a metal diaphragm that is welded to the sides of a chamber. The force developed on the diaphragm is brought out of the chamber by a rigid rod called a *force bar*, which is attached to the diaphragm. A balancing force developed by a pneumatic feedback bellows opposes this force. Imbalance between the capsule force and the feedback bellows force is sensed by a pneumatic nozzle-baffle. This simple pneumatic servomechanism is responsive to nozzle pressure and reestablishes the balance. As a result, pneumatic pressure is maintained so it is exactly proportional to applied pressure and is used as an output signal (usually 3 to 15 psi).

It should be noted that most pneumatic pressure transmitters actually measure a differential pressure (ΔP) that is applied to the high and low inputs of the transmitter. Pressure measurements are always made with respect to a reference point. Gauge pressure, for example, is referenced to atmospheric pressure. Absolute pressure measurement represents a pressure level above a complete vacuum, which is the absence of pressure or 0 psi. In either case, a measurement represents the difference in pressure between a value and the reference level. In a strict sense, all pressure measurements are differential pressure measurements.

Electronic Pressure Sensors

The electrical principles used to measure pressure displacement are numerous and varied. Most electronic pressure sensors employ capacitive, differential transformer, force balance, photoelectric, piezoelectric, potentiometric, resistive, strain gauge, or thermoelectric means of measurement. In many cases, the electronic or electrical device is used in conjunction with a mechanical device. For example, a piezoelectric crystal can be attached to a metal pressure-sensing diaphragm to produce an electrical signal that is proportional to pressure. We will discuss the various types in this section, but first we will address the characteristics of sensors.

Sensor Characteristics

Two important characteristics of sensors are sensitivity and accuracy. *Sensitivity* is a measure of the change in output of a sensor in response to a change in input. In general, high sensitivity is desirable in a sensor because a large change in output for a small change in input makes it easier to take a measurement. The value of sensitivity is generally indicated by the gain of the sensor. Thus, when a pressure sensor outputs 10 mV per psi, the sensitivity is 10 mV/psi.

In the case of an ideal sensor, the value of the output signal is only affected by the input signal. However, no practical sensor is ideal. In the case of a pressure sensor, variations in temperature, humidity, vibration, or other conditions can affect the output signal. The degree to which one of these factors affects the output is called the *sensitivity* of the sensor to that factor.

Temperature affects pressure sensors the most. The effect of temperature on the operation of a pressure sensor can be characterized by the following equation:

$$S_o = aP + bT \tag{5-15}$$

where

S_o = the sensor output
P = the pressure signal
T = the temperature
a and b = constants

A good pressure sensor should be very sensitive to pressure and not very sensitive to temperature changes. This balance is obtained if the value of the constant b is small and the value of a is large.

Accuracy is a term used to specify the maximum overall error to be expected from a device, such as one that measures a process variable. Accuracy usually is expressed as the degree of *inaccuracy* and takes several forms:

- Measured variable, as the accuracy is ±1 psi in some pressure measurements. Thus, there is an uncertainty of ±1 psi in any value of pressure measured.
- Percentage of the instrument full-scale (FS) reading. Thus, an accuracy of ±0.5% FS in a 10-volt full-scale voltage meter would mean the inaccuracy or uncertainty in any measurement is ±0.05 volts.
- Percentage of instrument span, that is, percentage of the range of the instrument's measurement capability. Thus, for a device measuring ±2% of span for 20-50 psi range of pressure, the accuracy is (±0.02)(50-20) = ±0.6 psi.
- Percentage of the actual reading. Thus, for a ±5% of reading current meter, we would have an inaccuracy of ±1.0 milliamps (ma) for a reading of 20 mA of current flow.

Example 5-8 demonstrates several common error calculations for a pressure instrument.

EXAMPLE 5-8

Problem: A pressure instrument has a span of 0-100 psi. A measurement results in a value of 60 psi for pressure. Calculate the error if the accuracy is (a) ±0.5% FS, (b) ±1.00% of span, and (c) ±0.75% of reading. What is the possible pressure in each case?

Solution: Using the definitions just given we find the following:

Error = (±0.005)(100 psi) = ±0.5 psi. So, the actual pressure reading is in the range of 59.5 to 60.5 psi.

Error = (±0.01)(0-100)psi = ±1 psi. Thus, the actual pressure reading is in the range of 59 to 61 psi.

Error = (±0.0075)(60 psi) = ±0.45 psi. Thus, the pressure reading is in the range of 59.55 to 60.45 psi.

Potentiometric-type Sensor

The potentiometric-type sensor is one of the oldest types of electric pressure transducers. It converts pressure into a variable resistance. A mechanical device such as a diaphragm is used to move the wiper arm of a potentiometer as the input pressure changes. A direct current voltage (dc V) is applied to the top of the potentiometer (pot), and the voltage that is dropped from the wiper arm to the bottom of the pot is sent to an electronic unit, as shown in Figure 5-10. The output of the electronic unit is normally a 4-to-20 maDC current.

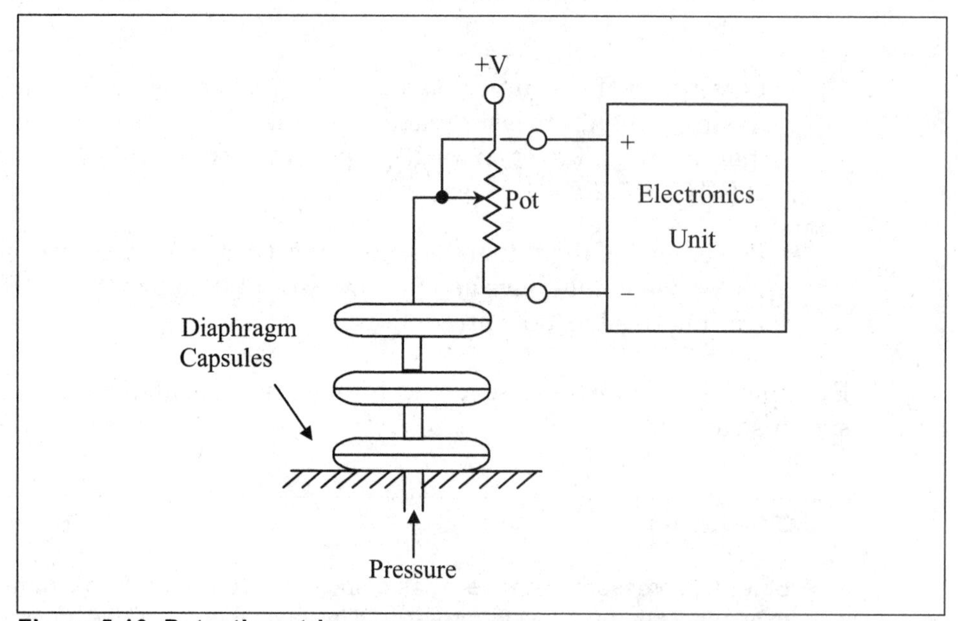

Figure 5-10. Potentiometric pressure sensor

Potentiometric pressure instruments normally cover a range of 5 psi to 10,000 psi. They are rugged instruments and can be operated over a wide range of temperatures. The potentiometer's (pot) large moving mass and low friction result in low frequency response and make them susceptible to vibration. Resolution is determined by the potentiometer's element. Wire-wound resistive elements have poor resolution while plastic elements have infinite resolution. Moreover, potentiometers are subject to wear because of the mechanical contact between the slider and the resistance element. Therefore, with potentiometers the instrument life is fairly short, and they tend to become noisier as the pot wears out.

Piezoelectric-type Sensor

A certain class of crystals, called piezoelectric, produce an electrical signal when they are mechanically deformed. The voltage level of the signal is proportional to the amount of deformation. Normally, the crystal is mechanically attached to a metal diaphragm. One side of the diaphragm is connected to the process fluid to sense pressure, and a mechanical linkage connects the diaphragm to the crystal.

The output voltage signal from the crystal is very small (normally in the microvolt range), so you must use a high-input impedance amplifier. The amplifier must be mounted within a few feet of the sensor to prevent signal loss. The crystals can tolerate temperatures up to 400°F, but they are affected by varying temperatures and must be temperature compensated.

Capacitance-type Sensor

Another example of an electronic unit connected to a pressure diaphragm is the variable capacitor pressure-sensing cell. A capacitor consists of two metal plates or conductors that are separated by an insulating material called a *dielectric*. In a capacitance-type pressure sensor, a differential pressure is applied to a diaphragm. This in turn causes a filling fluid to move between the isolating diaphragm and a sensing diaphragm. As a result, the sensing diaphragm moves toward one of the capacitor plates and away from the other, thus changing the capacitance of the device. Since capacitance is directly proportional to the distance between the plates, the pressure that is applied to the cell can be directly related to the change in capacitance. A pair of electrical leads is connected to an electronic circuit that measures the change in capacitance. This capacitance change is then converted into an electronic signal in the transmitter, which is calibrated in pressure units.

There are two common electrical methods for detecting the capacitance change. In one method, the change is detected by measuring the magnitude of an AC voltage across the plates when they are excited. In the other method, the sensing capacitor forms part of an oscillator, and the electronic circuit changes the frequency to tune the oscillator. These changes in frequency are then electronically converted into a pressure change.

Variable Inductance Sensor

Figure 5-11 shows another example of an electrical sensor used with a pressure-sensing diaphragm. Inductance, a fundamental property of electromagnetic circuits, is the ability of a conductor to produce induced voltage when the current in the circuit varies. A long wire has more

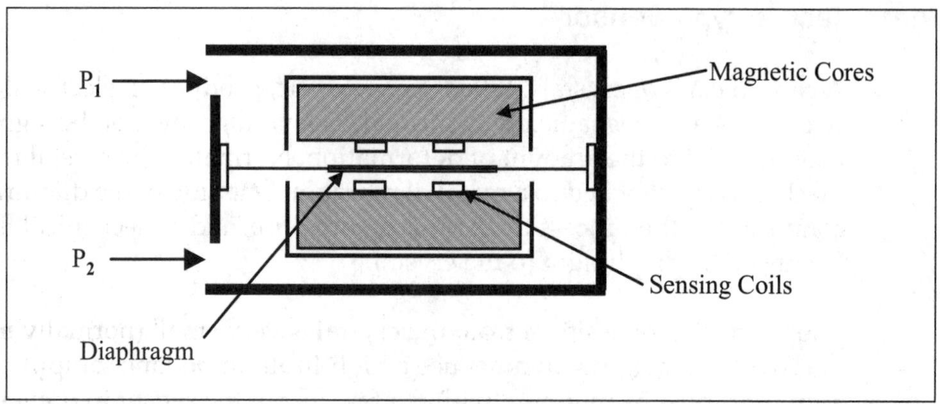

Figure 5-11. Variable inductance pressure sensor

inductance than a short wire since more conductor length is cut by magnetic flux, producing more induced voltage. Similarly, a coil of wire has more inductance than the equivalent length of straight wire because the coil concentrates magnetic flux. The core material around which the coil is wound also affects the inductance.

In other words, the inductance of a coil of wire or inductor depends on the number of turns and the magnetic properties of the material around the wire. The variable inductance device shown in Figure 5-11 uses two coils magnetically coupled through a core. The core changes properties as the applied pressure moves the diaphragm. The inductance is measured using a variety of electronic circuits. For example, you may use the pressure-sensing variable inductor as a component in an oscillator circuit.

Strain Gauge Pressure Sensors

The deformation of a material is called *strain,* and the mechanical force that produces the deformation is called *stress*. Strain is a unitless quantity, but it is common practice to express it as the ratio of two length units, such as in./in. or m/m.

A strain gauge is a device that changes resistance when stretched. A typical unbounded strain gauge is shown in Figure 5-12. It consists of multiple runs of very fine wire that are mounted to a stationary frame on one end and a movable armature on the other. The movable armature is generally connected to a pressure-sensing bellows or to a diaphragm. The multiple runs of wire amplify small movements along the direction of the wire. Very small pressure changes can be detected if there are a large number of wire runs.

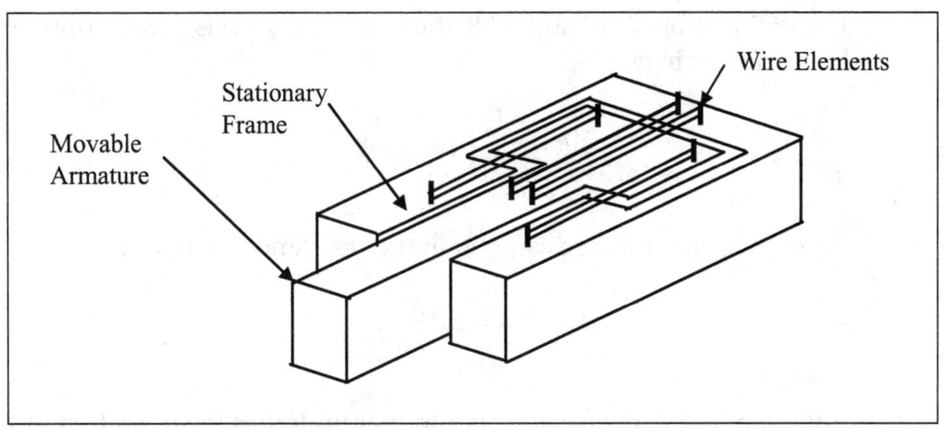

Figure 5-12. Unbonded strain gauge

In Chapter 3, we saw that the resistance of a metal wire is given by the following equation:

$$R_o = r\frac{l_o}{A_o} \tag{5-16}$$

where

R_o	=	the original wire resistance (Ω)
r	=	the resistivity of the wire in (Ω-m)
l_o	=	the wire length (m)
A_o	=	the starting cross-sectional area (m^2)

Suppose the wire is stretched. Then, we know that the wire will elongate by some amount Δl, so the new length is now $l = l_o + \Delta l$. It is also true that under this stress/strain condition, though the wire lengthens, its volume will remain constant. Because the volume unstretched is $V = l_o A_o$, it follows that the cross-sectional area must decrease by some area ΔA, so the volume (V) does not change. The new area will be $A_o - \Delta A$. Therefore,

$$V = l_o A_o = (l_o + \Delta l)(A_o - \Delta A) \tag{5-17}$$

Now, because both the length and the area have changed, the resistance of the wire will also change to

$$R = r\frac{l_o + \Delta l}{A_o - \Delta A} \tag{5-18}$$

Using Equations 5-17 and 5-18, the new resistance is given approximately by the following:

$$R \approx r\frac{l_o}{A_o}\left(1 + 2\frac{\Delta l}{l_o}\right) \tag{5-19}$$

From this equation, we can obtain the resistance change as

$$\Delta R \approx 2R\frac{\Delta l}{l_o} \tag{5-20}$$

Equation 5-20 is the basic equation that underlies the operation of wire strain gauges because it shows that the strain $(\Delta l/l_o)$ converts directly into a resistance change. Example 5-9 illustrates this concept.

EXAMPLE 5-9

Problem: Find the change in wire resistance for a strain gauge that has a nominal wire resistance of 100Ω, when it is subjected to a strain of 1000 μ m/m.

Solution: The change in strain gauge resistance can be found by using Equation 5-19:

$$\Delta R \approx 2R\frac{\Delta l}{l_o}$$

$$\Delta R = 2(100\Omega)(1000 \times 10^{-6}) = 0.2\Omega$$

Example 5-9 points out that in strain gauges the change in resistance is very small. This means that when using a strain gauge, sensitive electronic circuits must be used to obtain an accurate pressure measurement.

It is important to note that the signal produced by most electronic pressure-sensing elements cannot be sent more than a few feet. Therefore, most electronic pressure devices have a transmitter mounted on the sensor. The transmitter converts the measured signal into a 4-to-20-mA current signal, which is sent to a local indicator or to a central control area for indication, control, or recording.

Pressure Transmitter Applications

The most common application for a pressure transmitter is measuring liquid level in a process tank, as shown in Figure 5-13. Pressure transmitters determine level by using the principle that pressure is proportional to the height of the liquid multiplied by its specific gravity (SG). The pressure generated by the liquid is directly related to its height, and this pressure is independent of volume and the shape of the tank. Thus, 100 inches of water will produce 100 inH_2O of pressure since water has a specific gravity of one. If the specific gravity is different, the pressure changes proportionately. For example, if the specific gravity is 0.95, a liquid level of 100 inches produces a pressure of 95 inH_2O. If the specific gravity is 1.1, then a level of 100 inches produces 110 inH_2O of pressure.

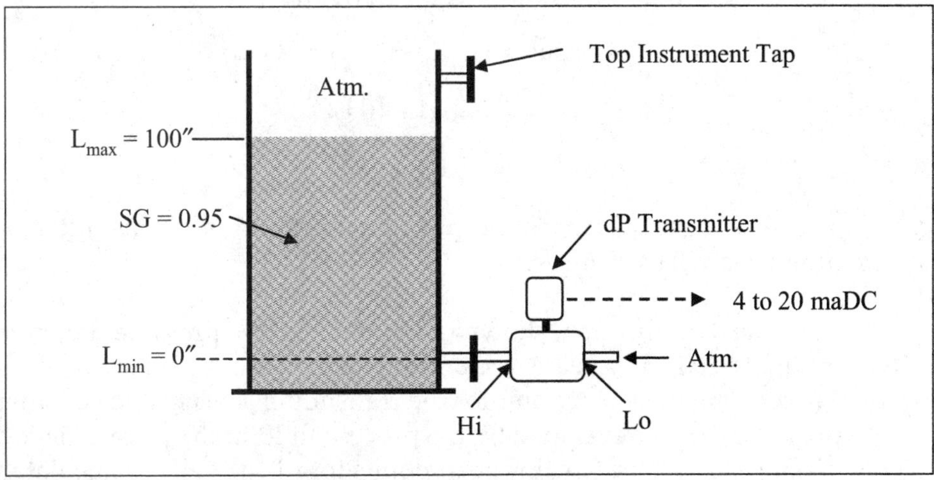

Figure 5-13. Open tank, dP cell horizontal to tap

Both gauge and differential pressure (dP) transmitters can be used to measure level. Gauge pressure transmitters are simply dP transmitters or cells that have their low-pressure port connected to the atmosphere. Either type of instrument is appropriate for use on an open tank or on a tank vented to the atmosphere, as shown in Figure 5-13. If the pressure inside a tank is either positive or negative, you must use a dP cell. In this case, you would connect the low-pressure port to the pressure above the fluid.

In the open-tank application shown in Figure 5-13, the pressure transmitter measures the pressure between the minimum fluid level (L_{min}) at the instrument tap and the maximum fluid level (L_{max}) in the tank using the high (Hi) pressure port on the pressure cell. The low (Lo) pressure side of the cell senses only the atmospheric pressure. Since atmospheric pressure is sensed by both sides of the dP cell, its effect is cancelled.

The transmitter output is a 4-to-20-milliamp (mA) current signal that represents the fluid level in the tank. The span points for the dP cell are calculated as follows:

$$4\text{ mA} = L_{min} \times SG$$

$$= 0\text{ inches} \times 0.95$$

$$= 0\text{ inH}_2\text{O}$$

$$20\text{ mA} = L_{max} \times SG$$

$$= 100\text{ inches} \times 0.95$$

$$= 95.0\text{ inH}_2\text{O}$$

So the span of the transmitter is given by the following:

$$\text{Span} = (L_{max} - L_{min}) \times SG$$

$$= (95.0 - 0)\text{ inH}_2\text{O}$$

$$= 95.0\text{ inH}_2\text{O}$$

The 4-to-20-maDC current output signal from the dP transmitter is calibrated from 0 to 95 inH_2O.

In the next application, shown in Figure 5-14, the pressure transmitter is mounted 20 inches below the bottom instrument tap on the process tank. To calculate the 4 mA point for the transmitter, we need to consider the pressure that is developed by the process fluid in the extra 20 inches of distance, d. In this case, the span points for the dP cell are calculated as follows:

$$4\text{ mA} = (L_{min} + d) \times SG$$

$$= (0 + 20)\text{ inches} \times 0.95$$

$$= 19\text{ inH}_2\text{O}$$

$$20\text{ mA} = (L_{max} + d) \times SG$$

$$= (100 + 20) \times SG$$

$$= 120\text{ inches} \times 0.95$$

$$= 114\text{ inH}_2\text{O}$$

So, the dP transmitter must be calibrated from 19 inH_2O to 114 inH_2O.

In the next application, shown in Figure 5-15, the tank is closed, and there is a "dry leg" between the instrument connection near the top of the tank

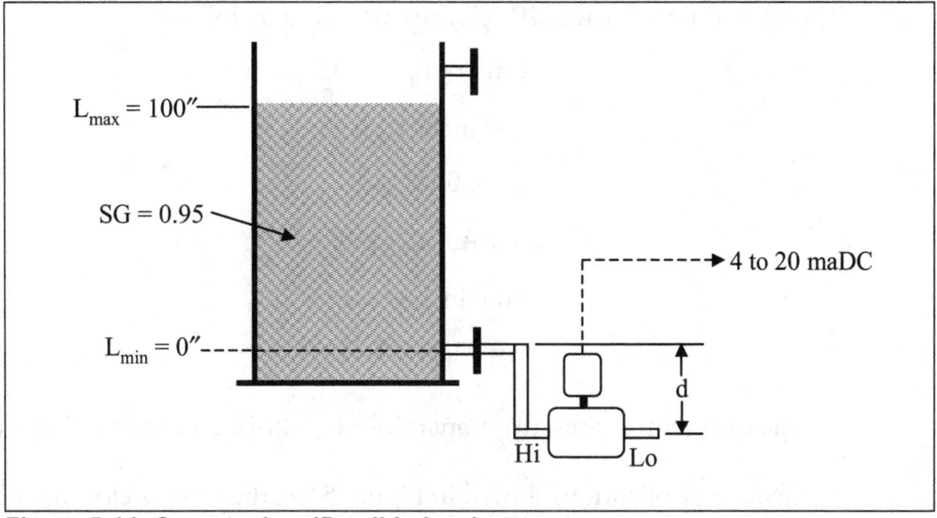

Figure 5-14. Open tank – dP cell below bottom tap

and the low (Lo) pressure port on the dP cell. A "dry leg" is defined as the length of instrument piping between the process and a pressure transmitter that is filled with air or another noncondensing gas.

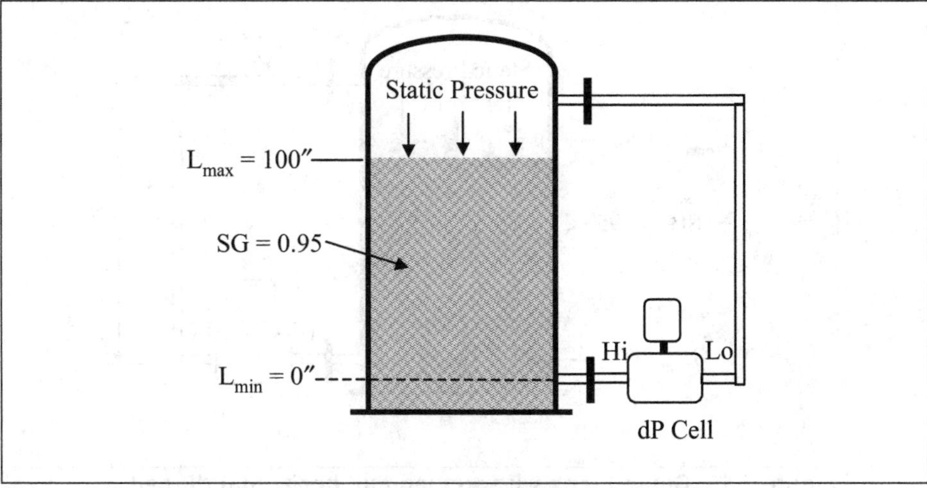

Figure 5-15. Closed tank with dry leg and horizontal dP cell

In the application shown in Figure 5-15, the high-pressure (HP) port senses the liquid level and the pressure above it. The low-pressure (LP) port senses only the pressure above the liquid. The impact of the static pressure in the closed tank is eliminated by using the differential pressure transmitter. The dP transmitter output is a 4-to-20-milliamp (mA) current signal that represents the fluid level in the tank.

The span points for the dP cell are calculated as follows:

$$4 \text{ mA} = L_{min} \times SG$$

$$= 0 \text{ inches} \times 0.95$$

$$= 0 \text{ inH}_2\text{O}$$

$$20 \text{ mA} = L_{max} \times SG$$

$$= 100 \text{ inches} \times 0.95$$

$$= 95.0 \text{ inH}_2\text{O}$$

So, the differential pressure transmitter is calibrated for 0 to 95.0 inH_2O.

In the next application, shown in Figure 5-16, the tank is closed, and there is a "wet leg" between the top instrument tap and the low-pressure (LP) side of the dP cell. A "wet leg" is defined as the length of instrument piping between the process and a pressure transmitter that is filled with a compatible fluid and kept at a constant height.

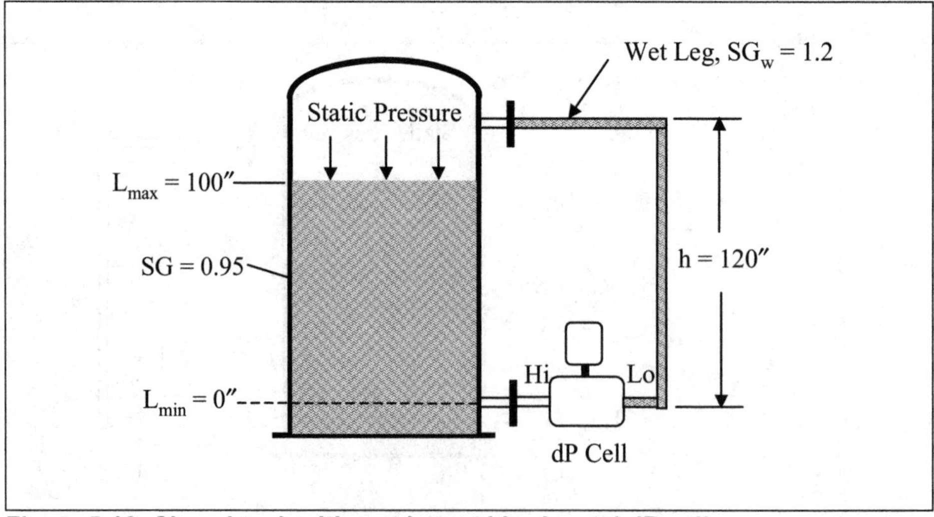

Figure 5-16. Closed tank with wet leg and horizontal dP cell

Since the reference leg is wet or filled, you must make a correction for the weight of the fluid in the wet leg. The specific gravity of the wet leg is 1.2 for the application shown in Figure 5-16. The fluid in the wet leg should be heavier than the process liquid. It is also important to maintain a constant height for the fill fluid in the wet leg.

The span points for the dP cell are calculated as follows:

$$4\text{ mA} = (L_{min} \text{x SG}) - (\text{h x } SG_w)$$

$$= (0'' \text{ x } 0.95) - (120'' \text{ x } 1.2)$$

$$= -144 \text{ inH}_2\text{O}$$

$$20\text{ mA} = (L_{max} \text{x SG}) - (\text{h x } SG_w)$$

$$= (100 \text{ inches x } 0.95) - (120'' \text{ x } 1.2)$$

$$= (95.0 - 144) \text{ inH}_2\text{O}$$

$$= -49 \text{ inH}_2\text{O}$$

So, the calibrated span for the dP transmitter is –144 to –49 inH_2O.

The final pressure transmitter application we will discuss is shown in Figure 5-17. It is a closed tank that has a wet leg and a dP transmitter below the bottom instrument tap. The dP cell is used to compensate for the static pressure in the process tank.

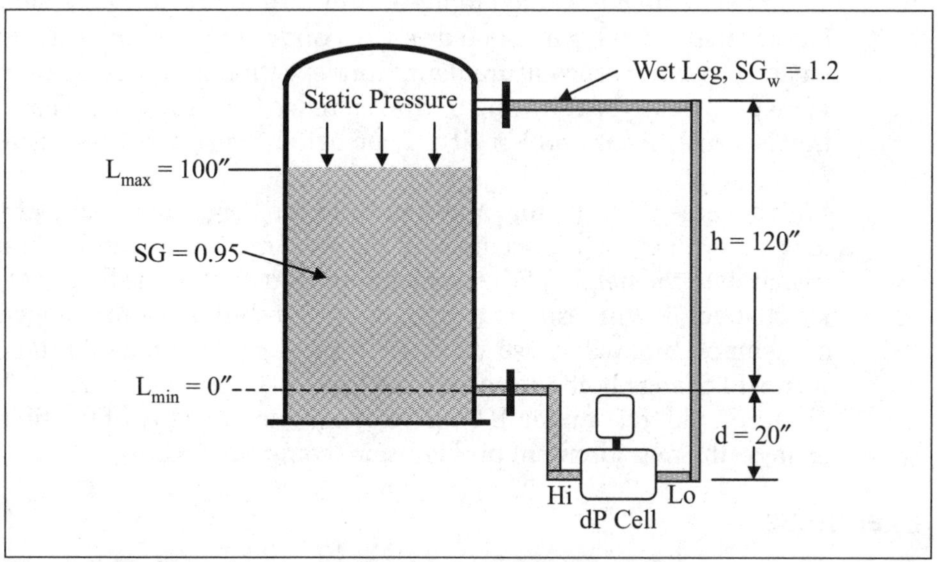

Figure 5-17. Closed tank with wet leg and dP cell below tap

Since the reference leg is wet or filled, we must make a correction for the weight of the fluid in the wet leg. We also need to correct for the fact that the dP cell is mounted 20 inches below the bottom instrument tap.

The span calibration points for the differential pressure instrument in Figure 5-17 are calculated as follows:

$$4\text{ mA} = (L_{min} + d) \times SG - (h + d) \times SG_w$$

$$= (0'' + 20'') \times 0.95 - (120'' + 20'') \times 1.2$$

$$= 19'' - 168''$$

$$= -149\text{ inH}_2\text{O}$$

$$20\text{ mA} = (L_{max} + d) \times SG - (h + d) \times SG_w$$

$$= (100'' + 20'') \times 0.95 - (120'' + 20'') \times 1.2$$

$$= 114'' - 168''$$

$$= -54\text{ inH}_2\text{O}$$

So, the calibrated span of the dP transmitter is –149 to –54 inH_2O.

In standard mount or dry leg applications, the accuracy of the level measurement is directly related to the accuracy of the pressure transmitter. However, the dry leg must be dry; any condensation in the leg will create an error in the level measurement. Condensation in most cases is prevented by doing heat tracing of the reference leg. However, if condensation is a significant problem, it may be better to use a wet leg installation.

With a wet leg installation, any change in the height or density of the wet leg fluid influences the accuracy of the level measurement. It is important to maintain the height of the wet leg fluid constant. If the height of the wet leg changes, it will result in a reference point shift. A large change in ambient temperature will cause a change in the density of the fill fluid. This in turn will change both the zero (4 mA) and maximum (20 mA) calibration points on the transmitter. If both the height and density of the fill fluid change, the measurement problems are compounded.

EXERCISES

5.1 Find force F_2, given the following data on the hydraulic press shown in Figure 5-4: A_1=1.0 in.^{2}, A_2=10 in.^{2}, and F_1=25 lbs.

5.2 A gas sample occupies a volume of 5 ft^3 at a pressure of 20 psi and a temperature of 70°F. What will be its volume if the pressure increases to 30 psi and the temperature remains at 70°F?

5.3 A sample of gas occupies a fixed volume at a pressure of 15 psi and a temperature of 30°C. Find the pressure of the gas if the temperature of the gas is increased to 40°C.

5.4 Find the gauge pressure if the absolute pressure reading is 50 psi and the local barometric pressure is 14.5 psi.

5.5 Find the absolute pressure if the reading of a pressure gauge is 21 psi and the local barometric pressure is 14.3 psi.

5.6 Find the displacement in inches for a mercury manometer with a specific gravity of 13.54, if the pressure applied is 1 psi.

5.7 Calculate the pressure detected by a U-tube manometer if the liquid in the manometer has a specific gravity of 2.95, and it is displaced 20 inches when pressure is applied.

5.8 Calculate the pressure detected by a mercury manometer if the mercury is displaced 4 inches. The specific gravity of the mercury in the manometer is 13.54.

5.9 List the sensing elements commonly used in pressure gauges.

5.10 List the two methods used in capacitance-type pressure sensors to detect the capacitance change produced by a pressure change.

5.11 A pressure instrument has a span of 0-100 inH_2O. A measurement results in a value of 50 inH_2O for pressure. Calculate the error if the accuracy is (a) ±2.0% FS, (b) ±1.0% of span, and (c) ±0.5% of reading. What is the possible pressure in each case?

5.12 Find the change in wire resistance for a strain gauge that has a nominal wire resistance of 150 Ω when it is subjected to a strain of 800 μm/m.

5.13 Calculate the span of a differential pressure transmitter that is mounted 10 inches below the bottom instrument tap on an open process tank that contains a liquid with a specific gravity of 1.1. Assume that the minimum level in the process tank is 0 inches, and the maximum level is 50 inches.

6

Level Measurement and Control

Introduction

This chapter discusses the basic principles of level measurement and control in industrial process applications. Sensing and controlling product levels in containers involves a wide range of materials, including liquids, powders, slurries, and granular bulk. In all level measurement a sensing device, element, or system interacts with material inside a container. A wide variety of physical principles are used to measure level. Common types include sight, pressure, electric, sonic, and radiation—each of which we discuss here. We will also explore level switches and a level control application.

Sight-type Instruments

There are three common sight-type level sensors: glass gauges, displacers, and tape floats. Glass gauges are the most widely used instruments for measuring the level in a process tank.

Glass Gauges

Two types of level glass gauges are used to measure liquid level: tubular and flat. The tubular type works in the same way as a manometer, that is, as the liquid level in a vessel rises or falls the liquid in the glass tube will also rise or fall. The gauges are made of glass, plastic, or a combination of the two materials. The material from which the transparent tubes are made must be able to withstand the pressure in the vessel, and they are generally limited to 450 psi at 400°F. Figure 6-1 shows two common applications of tubular sight glasses: an open or vented process vessel and

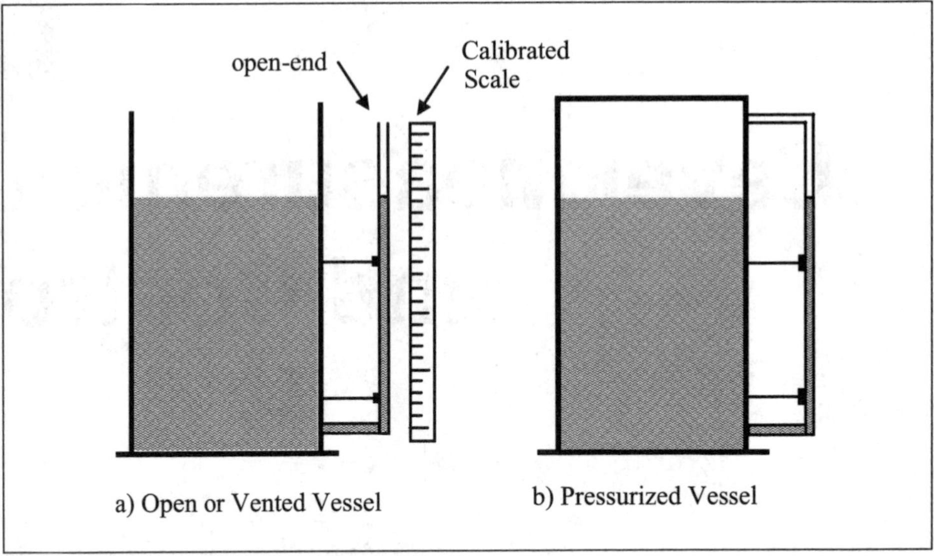

Figure 6-1. Tubular sight glass gauges

a pressurized vessel. For the pressurized tank, the upper end of the tube is connected to the tank. This creates an equilibrium pressure in both ends of the tube, and the liquid in the tube rises to the same level as the liquid in the vessel. A calibrated scale is normally mounted next to the sight gauge to indicate the level in the tank.

The pressure that the liquid exerts in the tank forces the liquid in the sight glass to rise to the same level as the liquid in the tank, as shown in Figure 6-1. You can calculate the pressure in the tank in psi using the Equation P = (0.03606)(SG)h, if the height of the liquid is given in inches. Example 6-1 illustrates this calculation.

EXAMPLE 6-1

Problem: Calculate the pressure in psi that is detected by a tubular sight glass gauge if the height of the liquid is 50 inches and the specific gravity of the liquid in the tank is 1.2.

Solution:

$$P = (0.03606)(SG)h$$

$$P = (0.03606)(1.2)(50 \text{ inches})$$

$$P = 2.16 \text{ psi}$$

The flat glass gauges are the most common type of sight gauges used in the process industries. They are used in a very wide range of pressure and temperature liquid services and in applications in which pressure and temperature are as high as 10,000 psi and 1000°F.

The bodies of flat sight gauges are either made of metal castings or forgings and a heavy glass or plastic front for viewing the level. A typical flat gauge design is shown in Figure 6-2.

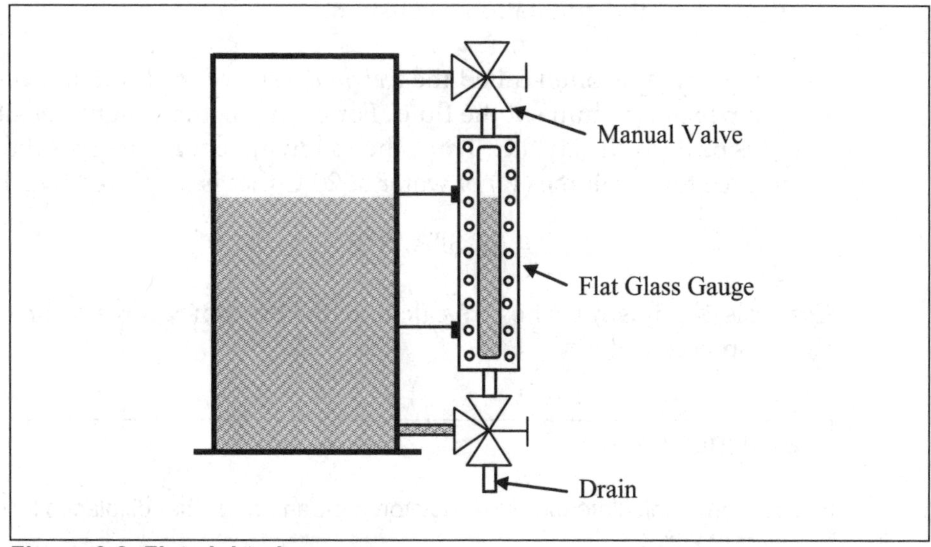

Figure 6-2. Flat sight glass gauge

There are two basic types of flat sight gauges: reflex and transparent. The reflex-type gauge produces a dark area where liquid is present and a light area where vapor is present. The reflex type gauge is normally chosen for liquids that are colorless, clear, and nonviscous. The transparent gauge is generally used when the liquid is colored, viscous, and corrosive.

Sight glass gauges are installed with manual shutoff valves at both ends. This isolates the gauges so maintenance can be performed, as shown in Figure 6-2. It is common practice to use ball-check valves for the shutoff valves because they will stop flow if the gauge glass breaks.

Displacers

Displacer level gauges operate on Archimedes' principle; they use the change in buoyant force acting on a partially submerged displacer. Archimedes' principle states that a body fully or partially immersed in a fluid is buoyed up by a force equal to the weight of the fluid displaced.

In equation form, Archimedes' principle for the buoyancy force, B, is given by the following:

$$B = W_d = \rho g V \tag{6-1}$$

where

W_d	=	the weight of the displaced fluid
ρ	=	the density of the displaced fluid
V	=	the volume of the displaced fluid
g	=	the gravitational constant.

The quantity ρg is often called the *weight density* of the fluid; it is the weight per unit volume of the fluid. For example, the weight density of water is 62.3 lbs/ft^3 at 20°C. Thus, the following equation gives the buoyancy force for a volume (V) of water at 20°C that is displaced by a body:

$$B = (62.3\text{lbs}/\text{ft}^3)V \tag{6-2}$$

Example 6-2 illustrates how to calculate the buoyancy force on an object that displaces water.

EXAMPLE 6-2

Problem: Calculate the buoyancy force on an object that displaces 2.0 ft^3 of water at 20°C.

Solution: Using Equation 6-2 to determine the buoyancy force, we obtain the following:

$$B = (62.3 \text{ lbs/ft}^3)\ V$$

$$B = (62.3 \text{ lbs/ft}^3)\ (2.0 \text{ ft}^3)$$

$$B = 124.6 \text{ lbs}$$

By measuring the buoyancy force produced by a displacer, you can determine a level value. If the cross-sectional area of the displacer and the density of the liquid are constant, then a unit change in level will result in a reproducible unit change in displacer weight.

The simplest level device of the displacer type consists of a displacer in a stilling well, which is suspended from a spring-weight scale, as shown in Figure 6-3.

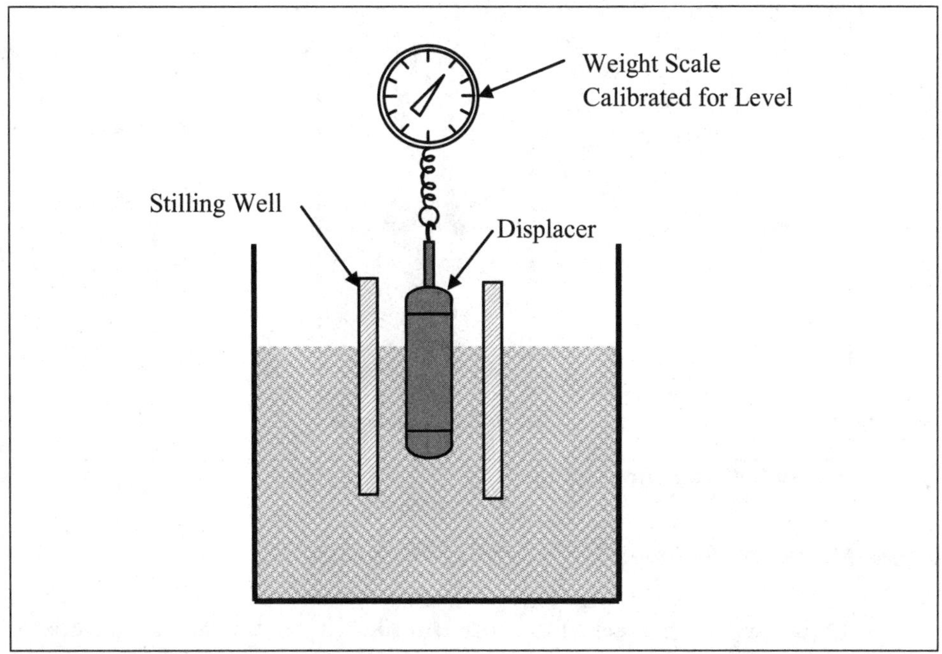

Figure 6-3. Displacer level gauge

The level scale can be calibrated from 0 to 100 percent, or in other level units.

Tape Float

Figure 6-4 shows one of the simplest, most direct methods for measuring level: the tape float. In the unit shown, a tape is connected to a float on one end and to a counterweight on the other to keep the tape under constant tension. The float motion makes the counterweight ride up and down a direct-reading gauge board. The gauge board is calibrated to indicate the liquid level in the tank.

Standard floats are normally cylindrical in shape for top-mounted designs and spherical or oblong for side-mounted designs. Small-diameter floats are used in higher-density materials; larger floats are used to detect liquid-liquid interface or for lower-density materials.

Pressure-type Instruments

Numerous sensors use basic pressure principles to measure level. The three common types that we will discuss here are differential pressure, bubblers, and diaphragm.

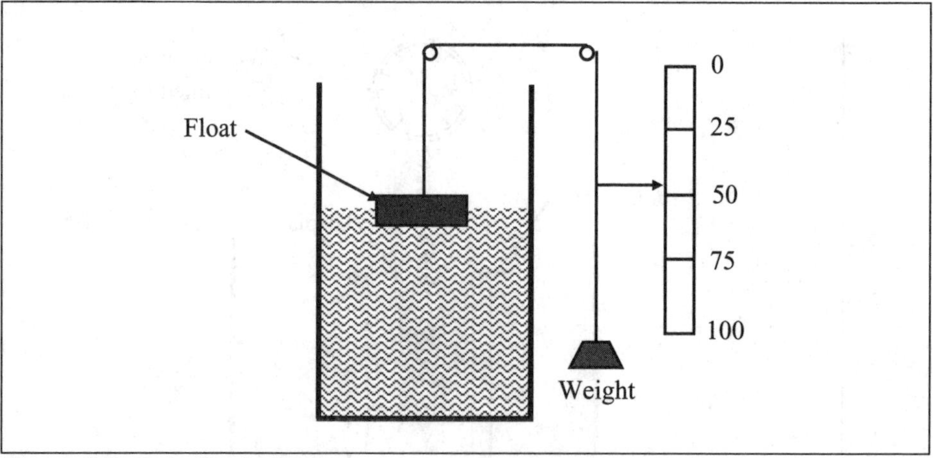

Figure 6-4. Tape float gauge

Level Measurement by Differential Pressure

Figure 6-5 shows a common open-tank differential pressure transmitter. In Chapter 5, we saw that there is a direct relationship between the hydrostatic pressure caused by a column of liquid, the specific gravity of the liquid, and the height of the vertical column of liquid. In most cases, the specific gravity of a fluid is constant, so pressure (P) is directly proportional to liquid level (h). This gives us

$$P = Kh \tag{6-3}$$

where K is a proportionality constant.

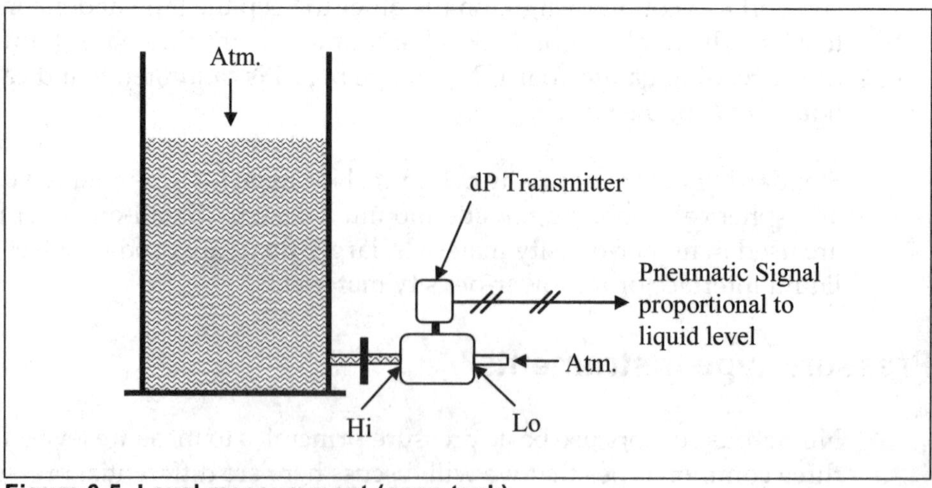

Figure 6-5. Level measurement (open tank)

In the open tank shown in Figure 6-5, the differential pressure (dP) transmitter converts the pressure of the liquid level into a pneumatic signal. Normally, a 3-to-15-psi pneumatic signal or 4-to-20-maDC current signal is used to represent the liquid level in the tank. The high side of the dP cell is connected to the bottom of the tank, and the low side of the dP cell is open to the atmosphere. Since the tank is open to the atmosphere, the pressure of the liquid on the high side of the dP transmitter is directly related to the level in the tank.

Figure 6-6 shows another application that uses a dP transmitter: measuring the level of a closed tank. In the example in Figure 6-6, if the pressure in the closed tank changes, an equal force is applied to both sides of the dP transmitter. Since the dP cell responds only to changes in differential pressure, a change in the static pressure on the liquid surface will not change the output of the transmitter. Thus, the dP cell responds only to changes in liquid level. In this example, a 4-to-20-maDC current signal is used to represent the level in the tank.

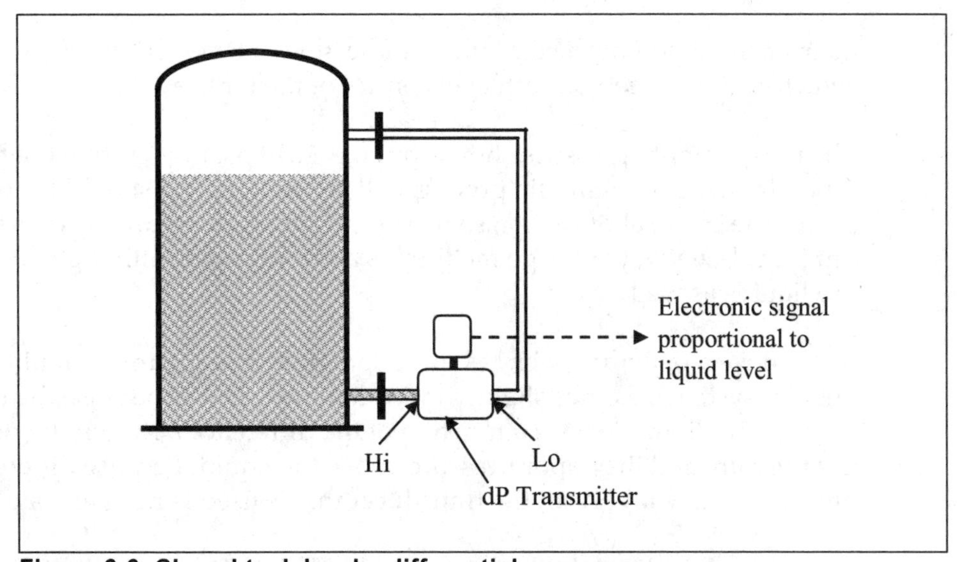

Figure 6-6. Closed tank level – differential pressure

Bubblers

Another example of a pressure-type level instrument is the bubbler system shown in Figure 6-7. When you use bubbler systems to measure liquid level, you install a dip tube in a tank with its open end a few inches from the bottom. A fluid is forced through the tube; when the fluid bubbles escape from the open end, the pressure in the tube equals the hydrostatic head of the liquid. As liquid level varies, the pressure in the dip tube changes correspondingly. This pressure is detected by a pressure-sensitive

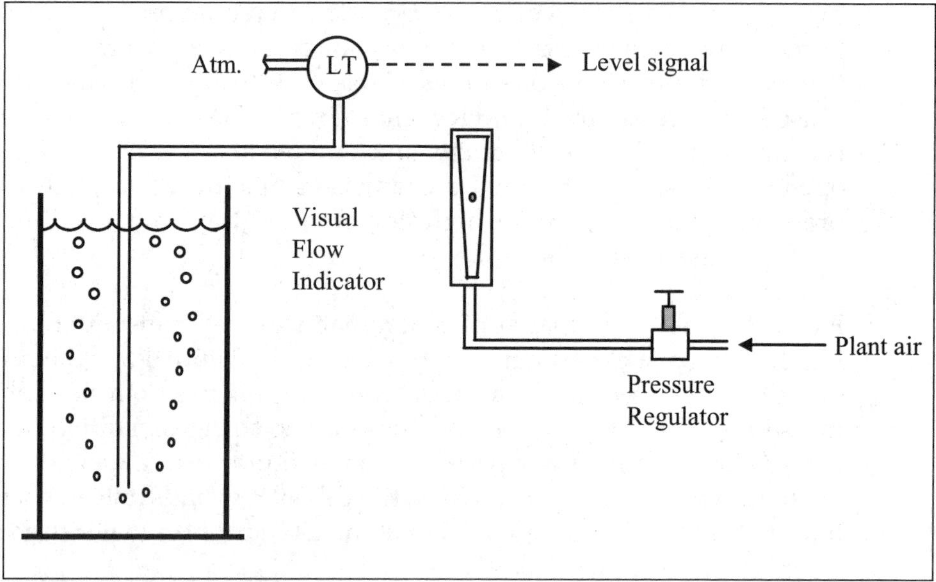

Figure 6-7. Air bubbler system

instrument, if you need continuous indication of level, or by pressure switches, if you need on/off level control of the tank level.

The purge supply pressure should be at least 10 psi higher than the highest hydrostatic pressure the process will encounter. You should keep the purge rate small, about 1 *scfh*, so no significant pressure drop occurs in the dip tube. Usually, the purge medium is air or nitrogen, although you can use liquids as well.

If you have a tank that operates under pressure or vacuum, installing a bubbler system becomes slightly more complex. This is because the measurement of liquid level is a function of the difference between the purge gas pressure and the vapor pressure above the liquid. Because differential pressure is now involved, the transducer that is used is normally a dP cell.

There are some disadvantages to using bubblers. The first is limited accuracy; another is that bubblers will introduce foreign matter into the process. Also, liquid purges can upset the material balance of the process, and gas purges can overload the vent system on vacuum processes. If the purge medium fails, not only do you lose the level indication on the tank, but the system is also exposed to process material that can cause plugging, corrosion, freezing, or safety hazards.

Diaphragm

Diaphragm detectors operate by the simple principle of detecting the pressure that the process material exerts against the diaphragm. Figure 6-8

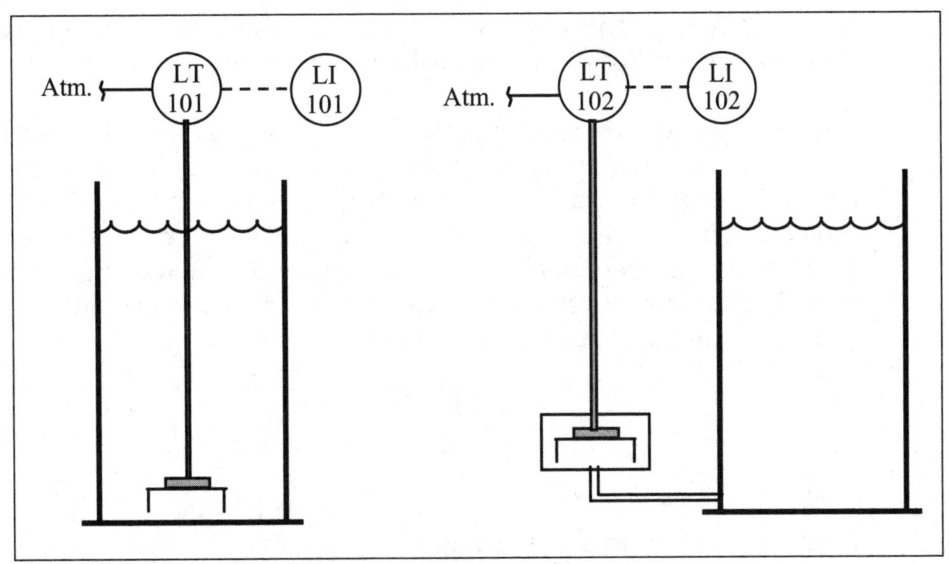

Figure 6-8. Diaphragm box level measurement

presents two versions of continuous level detection. The diaphragm box instrument shown on the right side of the figure isolates the captive air from the process fluid. The unit consists of an air-filled diaphragm that is connected to a pressure detector via air tubing. As the level rises above the diaphragm, the liquid head pressure compresses the captive air inside. A differential pressure element senses the air pressure, which is displayed as level.

The instrument shown on the left side of Figure 6-8 is submerged in the vessel. The static head of the liquid exerts an upward pressure on the diaphragm, which increases as the level rises.

Electrical-type Instruments

A wide variety of instruments and sensors use basic electrical principles to measure and detect level. The three common electrical-type level- measuring devices we will discuss here are capacitance probes, resistance tapes, and conductivity probes.

Capacitance Probes

A capacitor consists of two conductors separated by an insulator. The conductors are called *plates*, and the insulator is referred to as *the dielectric*. The basic nature of a capacitor is its ability to accept and store an electric charge. When a capacitor is connected to a battery, electrons will flow from the negative terminal of the battery to the capacitor, and the electrons on the opposite plate of the capacitor will flow to the positive terminal of the battery. This electron flow will continue until the voltage across the capacitor equals the applied voltage.

Capacitor size is measured in farads. A capacitor has the capacitance of one farad (1 F) if it stores a charge of one coulomb (1 C) when connected to a one-volt supply (1-V). Because this is a very large unit, one millionth of it (noted as a microfarad, μF) is commonly used. The electric size in farads of a capacitor is dependent on its physical dimensions and on the type of material (dielectric) between the capacitor plates. The equation for a parallel plate capacitor is given by the following:

$$C = \frac{KA}{d} \tag{6-4}$$

where

A	=	the area of the plates
d	=	the distance between plates
K	=	the dielectric constant, as listed in Table 6-1

For pure substances, the dielectric constant is a fundamental property of the material. The dielectric constant of any mixture of substances can be established experimentally.

Several laws for capacitive circuits are worth noting. One is that the capacitance of two or more capacitors connected in parallel is equal to the sum of the individual capacitances (i.e., $C_t = C_1 + C_2 + \ldots + C_n$). The second law is that for capacitors that are connected in series, the reciprocal of the total capacitance ($1/C_t$) equals the sum of the reciprocals of the individual capacitors (i.e., $1/C_t = 1/C_1 + 1/C_2 + \ldots + 1/C_n$).

A change in the characteristics of the material between the plates will cause a change in dielectric constant, which is often larger and more easily measured than changes in other properties. This makes the capacitance probes suitable for use to detect the level of material in vessels because changes in process level change the dielectric constant.

Table 6-1. Dielectric Constants

Material	Temperature °C	Dielectric Constant
Air	...	1.0
Oil	20	2-5
Water	25	78.5
Aqueous solutions	...	50-80
Glycol	25	37
Glass	...	3.7-10

As the temperature of the material increases, its dielectric constant tends to decrease. Temperature coefficients are on the order of 0.1%/°C. You can install automatic temperature compensator circuits to cancel the effect of temperature variations. Chemical and physical composition and structure changes affect the dielectric constant. When you intend to measure the dielectric constant of solids, keep in mind that variations in average particle size and in packing density will affect the dielectric constant. Current flow passing to the ground through material resistance tends to short out the capacitor. Shorting out the measured capacitance with a variable resistance can make the dielectric measurement very inaccurate if the resistance is low compared to the capacitive reactance.

Variations in process level cause changes in capacitance that you can measure by using an electronic circuit in the level instrument. As Figure 6-9 shows, the probe is insulated from the vessel and forms one plate of the capacitor; the metal vessel forms the other plate. The material between the two plates is the dielectric for the capacitor. As the liquid level in the tank rises, vapors that have a low dielectric constant are displaced by liquid that has the high dielectric value. An electronic instrument calibrated in units of level detects capacitance changes.

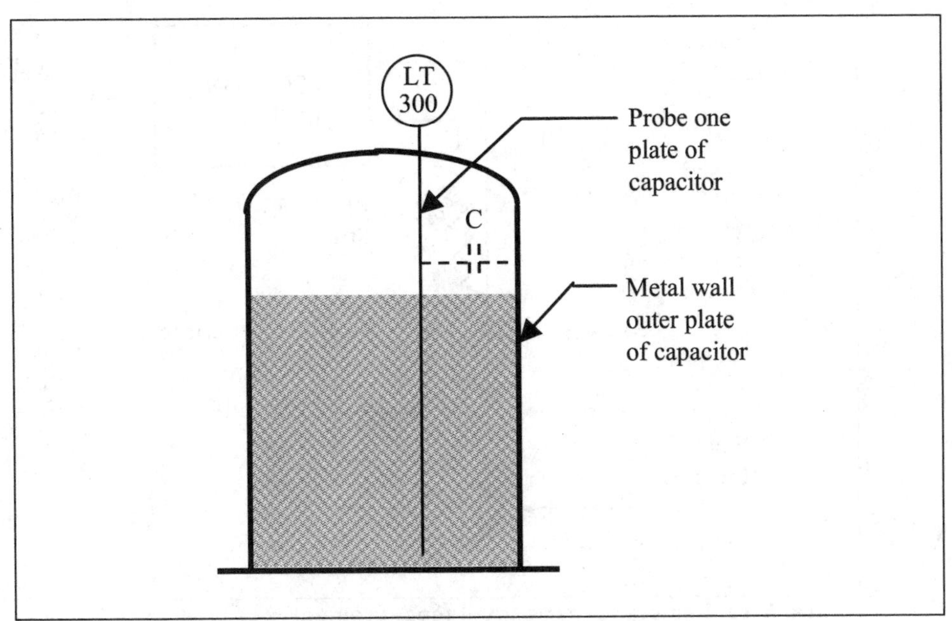

Figure 6-9. Capacitance level probe

To measure the level of conductive materials, use insulated (normally Teflon™-coated) probes. This measurement is largely unaffected by the fluid resistance. Therefore, this probe design is applicable to both conductive and nonconductive processes.

If the process material adheres to the probe when the level in the vessel is reduced, a layer of fluid will be left on the probe. When this layer is conductive, the wet portion of the probe will be coupled to ground. The instrument will not read the new level. Instead, it will register the level to which the probe is coated. Aside from changes in process material dielectric constant, this represents one of the most serious limitations of capacitance installations. It should be noted that if the probe coating is nonconductive, there will be much less pronounced interference with the accuracy of the measurement.

Resistance Tapes

Another electrical-type level instrument is the resistance tape. In these devices, resistive material is spirally wound around a steel tape, as shown in Figure 6-10. This type of probe is mounted vertically from top to bottom on a process tank. The pressure of the fluid in the tank causes the resistive tape to be short-circuited, thus changing the total resistance of the measuring tape. This resistance is measured by an electronic circuit and is directly related to the liquid level in the tank.

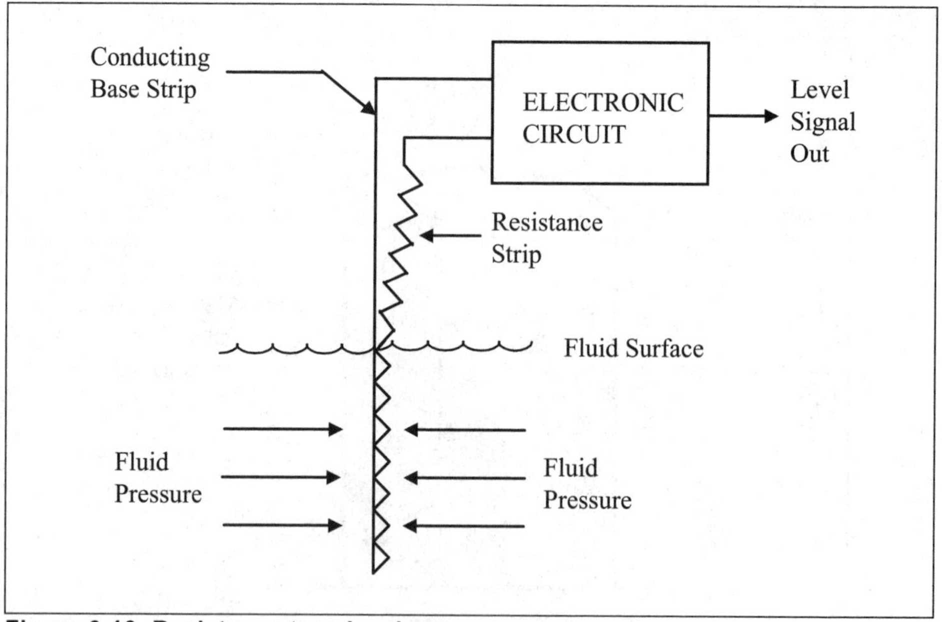

Figure 6-10. Resistance tape level measurement

Conductivity Probes

Conductivity probes operate on the principle that most liquids conduct electricity. Figure 6-11 illustrates how a typical conductivity level probe operates.

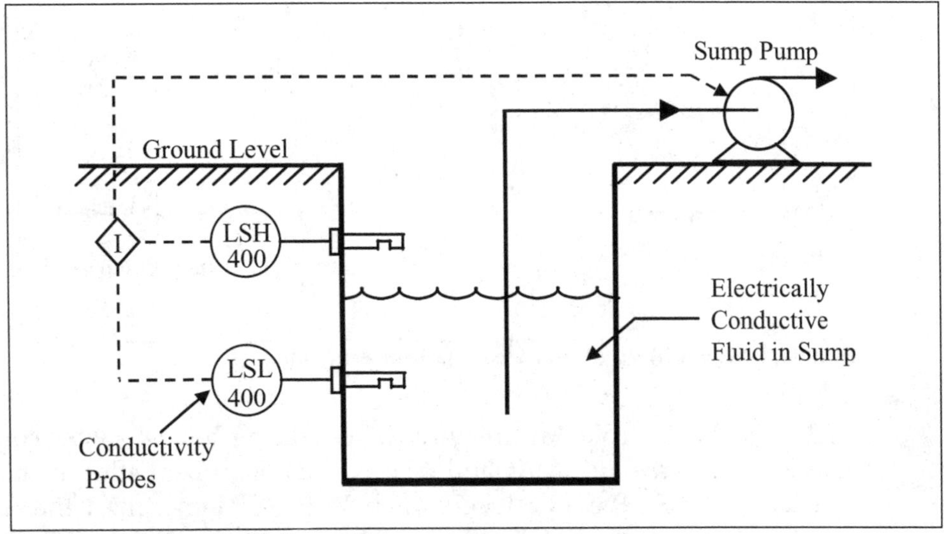

Figure 6-11. Conductivity probe

An electrode is shown above the liquid level on the left side of Figure 6-11. The circuit is therefore open, and no current is flowing through the level probe electrodes to energize it. When the liquid level rises, it establishes a conductive path at the electrode, closing the low-level switch. When the level probe is activated, it closes dry contacts in the instrument. This contact closure can be used to operate electric relays, pumps, solenoid valves, or other equipment.

These conductivity switches are available in a variety of configurations, and you can thus use them for on/off control of one component or to control several pieces of equipment. A typical application would be to control the liquid level in a sump. The relay ladder diagram in Figure 6-12 shows how you can use two conductivity probes in conjunction with an electromechanical control relay (CR1) to control the sump pump in Figure 6-11.

To start, let us assume that the sump is empty, so both level probes are open and the control relay is deenergized. Then, as the liquid level rises, the low-level switch (LSL-400) will close when the conducting liquid covers the probe. The control relay CR1 remains deenergized because the high-level switch is still open. When the level rises up to the high-level switch (LSH-400) it will close, energizing CR1. The first set of contacts,

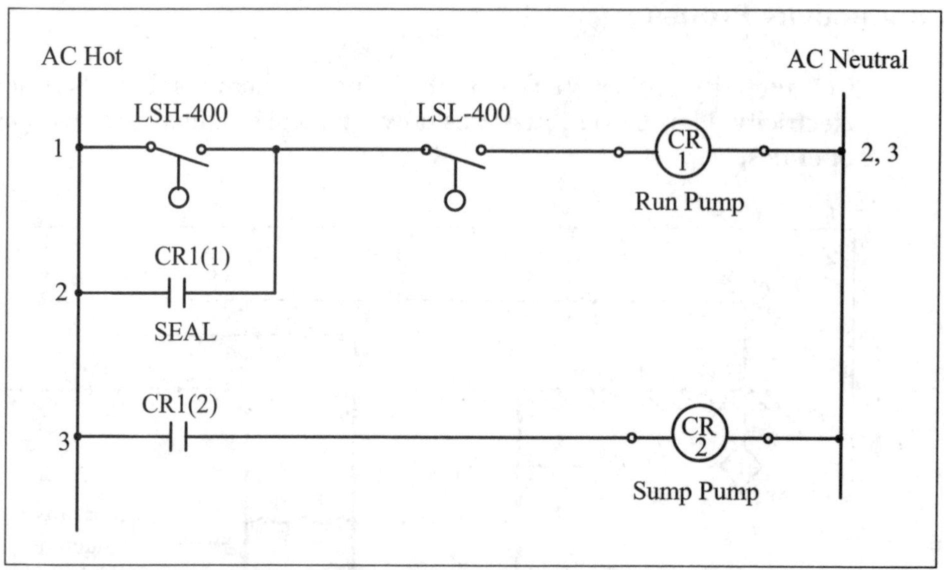

Figure 6-12. Sump level control – ladder diagram

CR1(1), of the control relay are wired across the high-level switch contacts. This is called a *seal-in circuit,* and it keeps CR1 energized after the liquid level drops below the high-level switch. When CR1 energizes, the second set of relay contacts CR1(2) and energizes sump pump relay CR2, turning on the sump pump. When the liquid level is pumped down below LSH-400, it will open this level switch, but control relay CR1 will remain ON because the seal-in contacts are still closed. When the liquid level falls below level switch LSL-400, the contacts open, turning off CR1 and the sump pump. The pump will remain off until the sump fills up to the high-level probe again. Then the control action will be repeated.

Sonic-type Instruments

The generation and detection of sound waves is another common method used to detect level. Sound waves are longitudinal mechanical waves. They can be generated in solids, liquid, and gases. The material particles transmitting sound waves oscillate in the direction of the propagation of the wave itself. There is a large range of frequencies within which longitudinal mechanical waves can be generated. Sound waves are confined to the sensation of hearing. This frequency ranges from about 20 cycles per second to about 20,000 cycles per second and is called the *audible* range. A longitudinal mechanical wave whose frequency is below the audible range is called an *infrasonic* wave, and a longitudinal mechanical wave whose frequency is above the audible range is called an *ultrasonic* wave.

Infrasonic waves of interest are usually generated by large sources, earthquake waves being an example. The high frequencies associated with ultrasonic waves can be produced by elastic vibrations of a quartz crystal induced by resonance with an applied alternating electric field. This is the common method used in ultrasonic level instruments. In this section the use of ultrasonic devices to detect level will be discussed.

Ultrasonic Level Measurement

Ultrasonic level sensors measure the time it takes sound waves to travel through material. Ultrasonic instruments operate at frequencies inaudible to the human ear and at extremely low power levels, normally a few thousandths of a watt. The velocity of a sound wave is a function of the type of wave being transmitted and the density of the medium in which it travels.

When a sound wave strikes a solid medium, such as a wall or a liquid surface, only a small amount of the sound energy penetrates the barrier; a large percentage of the wave is reflected. The reflected sound wave is called an *echo*.

Figure 6-13 is a block diagram of a typical ultrasonic level-measurement system. The sound waves are produced by the generator and transmitter circuit, and the transducer sends out the sound waves. These sound waves are reflected by the material or the level being measured. A transducer senses the reflected waves and converts the sound wave it received into an electrical signal. This signal is amplified by a receiver/amplifier circuit and sent to a wave-shaping circuit. A timing generator is used to synchronize the functions in the measurement system. The instrument measures the time that elapses between the transmitted burst and the echo signal. This elapsed time is proportional to the distance between the transducers and the object being sensed. You can easily calibrate the instrument to measure fluid or material level in a process vessel.

Radiation-type Instruments

The three common radiation-type level instruments—nuclear, microwave, and radar—are the subject of this section. These systems can be used to detect level on a wide variety of products, from liquids to bulk solids and slurries.

Nuclear Level Measurement

Nuclear radiation systems have the ability to "see" through tank walls, and thus they can be mounted on the outside of process equipment. This

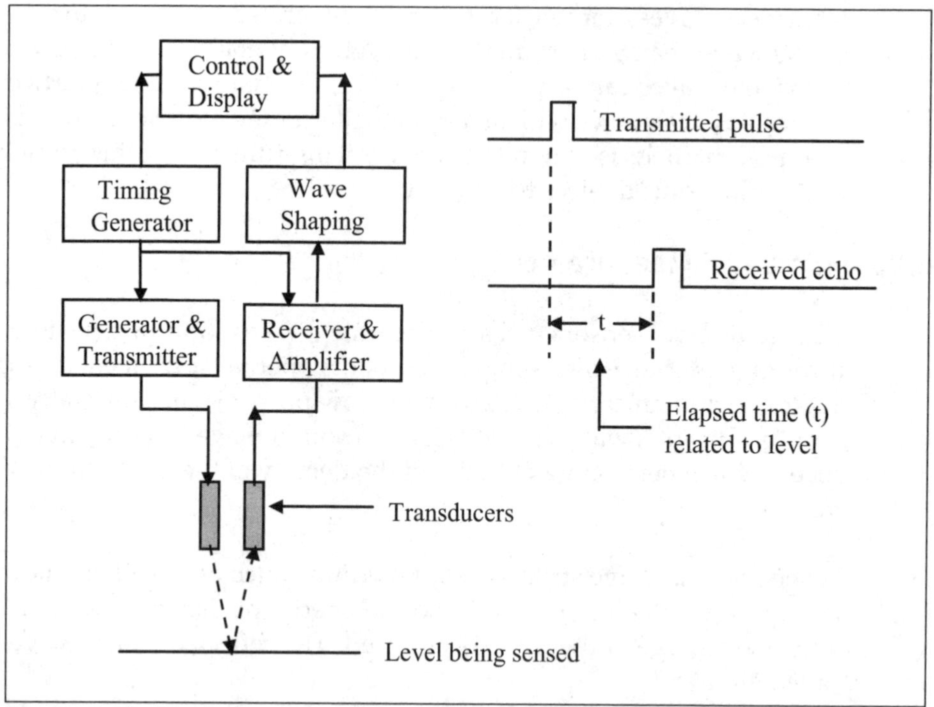

Figure 6-13. Block diagram of ultrasonic measurement system

reduces installation and repair costs. Two typical nuclear level instruments are shown in Figure 6-14. The nuclear level instrument shown in Figure 6-14a uses a single low-level gamma-ray source on one side of the process vessel and a radiation detector on the other side of the tank. You can measure level more accurately by placing several gamma sources at different heights on the tank, as shown in Figure 6-14b. The material in the tank has different transmissibility characteristics than air, so the instrument can provide an output signal that is proportional to the level of the material in the container.

Microwave Level Measurement

Microwave level measurement systems are another type of radiation detector commonly used in level-measurement applications. These systems consist of a transmitter and receiver as well as associated electronics equipment. The transmitter is a microwave oscillator and directional antenna. The receiver consists of an antenna, a high gain, a pulse-decoding circuit, and an output circuit.

In the transmitter, an AC voltage is converted into a +12Vdc supply. A pulse modulation circuit then pulses the dc voltage at about 1 kHz. The pulsed dc signal is sent to an oscillator and converted into a pulsed microwave signal. The transmitter antenna then radiates the signal toward the

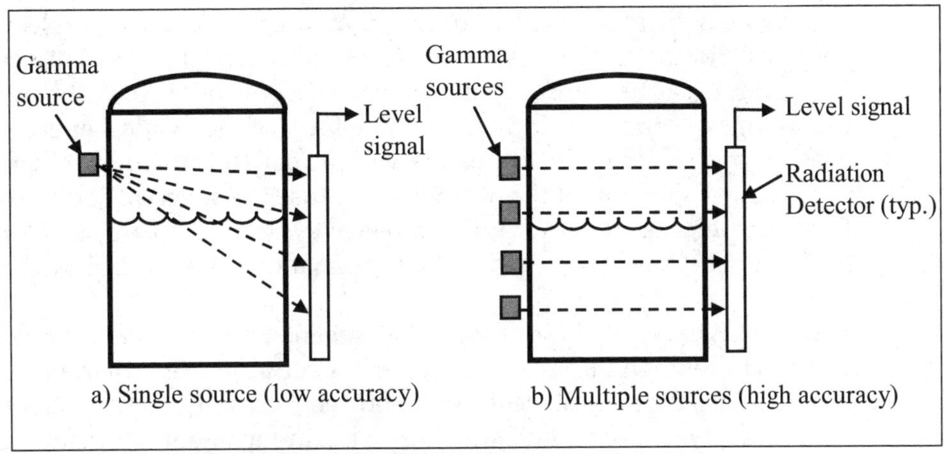

Figure 6-14. Nuclear level detection system

material in a process tank. In the receiver, the reflected microwave signal is received by a directional antenna and converted into a low-level pulsed dc signal.

Different materials have different effects on microwave signals. For example, low-level signals cannot penetrate metals but are reflected by them. Microwave signals are absorbed almost entirely by water and to varying degrees by water-based liquids or by products that have a high moisture content such as wood products, food grain, and the like.

Metal storage tanks or hoppers must have a detector window that is transparent to the microwave signals. If you store solid materials in closed metal vessels, you can construct the detector windows out of such materials as high-density polyethylene.

Signal transmission losses increase as conductivity rises and decrease as material dielectric constants rise. For example, air transmits microwaves with no loss. However, sea water attenuates microwaves very strongly. It is the dielectric constant and conductivity of material that determine whether or not the material is a good candidate for microwave level measurement.

Microwave systems are used for both point and continuous measurement of liquid, solid, and slurry levels. Microwave level instruments normally have accuracies in the range of ±1 to 2 percent.

Radar Level Instrument

Radar level instruments are the final type of radiation detector we will discuss. Radar instruments operate by transmitting a high-frequency (GHz) electromagnetic radiation and timing the transit time to the level surface and back. In this sense, they are similar to ultrasonic level instruments. However, radar has an advantage over ultrasonic because of the inherent limitations of sound. Sound waves are a form of mechanical energy that uses the molecules in the atmosphere to propagate. However, changes in the chemical makeup of the atmosphere cause the speed of sound to vary. For example, vapors are a typical cause of error in ultrasonic level measurements because they change the propagation speed and strength.

There are two types of radar level instruments: noncontact and guided wave (as shown in Figure 6-15). In the noncontact type shown in Figure 6-15a, the output electromagnetic energy of the radar antenna is very weak, typically about 1 mW. After the radar signal is transmitted into the air it begins to weaken and spread out very quickly. The signal reaches the liquid surface where it is reflected back to the instrument. The strength of the reflected signal is directly related to the dielectric constant of the liquid. Liquids with very low dielectric constants, like hydrocarbon fluids, reflect very little of the signal. On the return to the instrument, this weak signal loses more energy until the received signal may be less than 1 percent of its strength at initial transmission. Liquid turbulence and some foams can further complicate the measurement by scattering or absorbing the radar pulse.

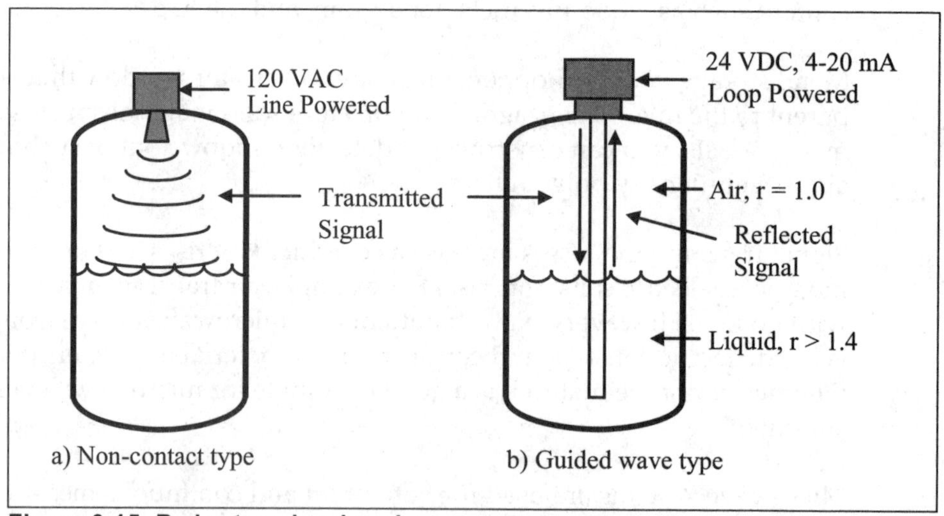

Figure 6-15. Radar-type level probes

The guided-wave radar unit shown in Figure 6-15b can overcome these problems. It is based on Time Domain Reflectometer (TDR) technology, which uses pulses of electromagnetic energy that are transmitted down the probe tube. When a radar pulse reaches a liquid surface that has a higher dielectric constant than the air in the process tank, the pulse is reflected back to the electronic unit. A high-speed electronic timing circuit precisely measures the transit time and calculates an accurate measurement of the liquid level in the tank.

The power output of the guided-wave radar probe is only 10 percent that of a conventional radar probe. This is possible because the waveguide offers a highly efficient path for the signal to travel down to reach the surface of the liquid and then bounce back to the receiver. Degradation of the signal is minimized. This means that you can accurately measure the level of liquids that have low dielectric constants of less than 1.7. Variations in media dielectric constant have no significant effect on performance.

Guided-wave radar level probes, like conventional radar probes, use transit time to measure the liquid level. The transit time for a signal reflected from a surface of hydrocarbon liquids with a dielectric constant of 2 to 3 is the same as the transit time for water with a dielectric constant of 80. Only the strength of the return signal is changed. Guided-wave radar level probes only measure transit time.

Since the speed of light is constant, no level movement is necessary to calibrate a guided wave device. In fact, you do not have to calibrate the instrument. You perform field configuration by entering process data into the instrument for the specific level application. You can configure several instruments in very quickly. All that is required is a 24 Vdc power source and the specifications for each process tank.

Level Switches

A wide variety of level switches are available for use in measurement and control applications. The most common types are inductive, thermal, float, rotating paddle, and ultrasonic level switches. These switches are used to indicate high and low levels in process tanks, storage bins, or silos. They are also used to control valves or pumps in order to maintain fluid level at a set value or to prevent tanks from being overfilled.

Inductive Switch

Inductive level switches are used on conductive liquids and solids. They are also used on the interface between conductive and nonconductive liquids. The inductive transducers are excited by an electrical source, causing them to radiate an alternating magnetic field. A built-in electronic circuit detects a change in inductive reactance when the magnetic field interacts with the conductive liquid. These switches are generally used in harsh environments because the probe is completely sealed and has no moving parts.

Thermal Switch

In a thermal-based level switch, a heated thermal resistor (thermistor) is used to detect the surface of a liquid. The level measurement a thermistor provides is based on the difference in thermal conductivity of air and liquid. Because the thermistor reacts to heat dispersion, it can be used with water- or oil-based liquids.

The thermistor has a negative temperature coefficient of resistance, which means that its resistance decreases as temperature rises. The internal temperature of any device depends upon the heat dispersion of its surrounding environment. Since heat dispersion is greater in a liquid than in a gas or air, the resistance of a thermistor level probe will change sharply whenever the probe enters or leaves a liquid. The operating range of a thermistor assures accurate and dependable operation in liquids up to 100°C.

Float Switch

Float switches are an inexpensive way to detect liquid level at a specific point. When you use float switches on process tanks, you must generally create a seal between the process and the switch. In most cases, magnetic coupling transfers the float motion to the switch or indicator mechanism. Figure 6-16 shows a typical magnetically activated level switch. In this configuration, a reed switch is positioned inside a sealed and nonmagnetic guide tube at a point where rising or falling liquid level should activate the switch. The float, which contains an annular magnet, rises or falls with liquid level and is guided by the tube.

In the example in Figure 6-16, the switch is normally closed and will open when the float and magnet are at the same level as the reed switch. You can use the switch opening, for example, to sound an alarm or to stop a pump. The switch can also be designed to close when activated by the magnet.

Rotating Paddle Switches

The rotating paddle level switches shown in Figure 6-17 are used to detect the presence or absence of solids in a process tank. A low-power synchronous motor keeps the paddle in motion at very low speed when no solids are present. Under such conditions, there is very low torque on the motor drive. When the level in the tank rises to the paddle, torque is applied to the motor drive and the paddle stops. The level instrument detects the torque and actuates a switch or set of switches. These switches can then be used to sound an alarm or to control the filling or emptying of the process tank or silo.

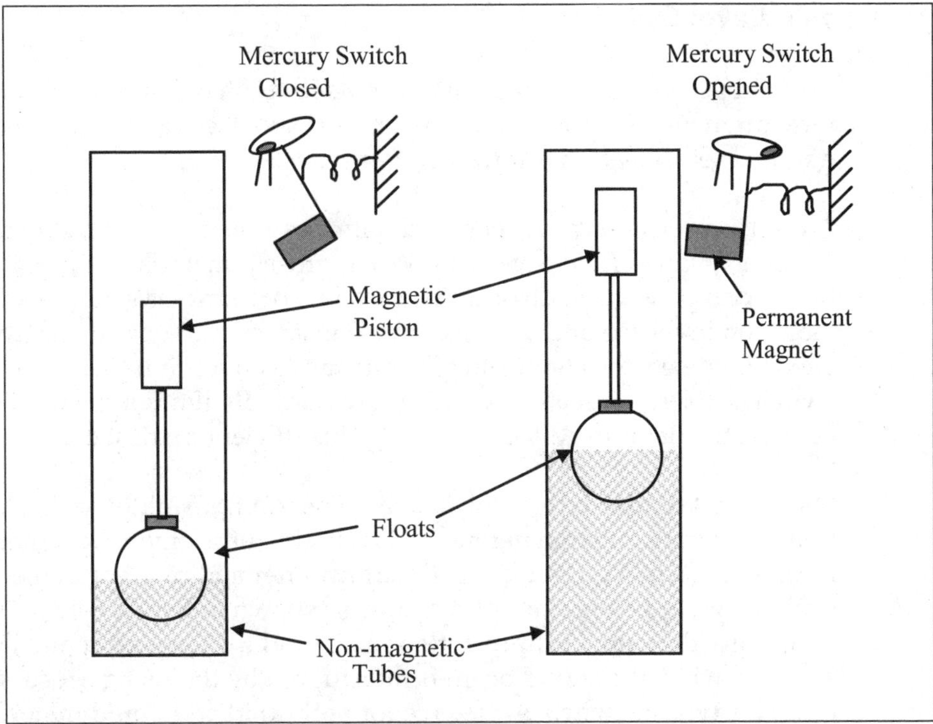

Figure 6-16. Magnetically activated level switch

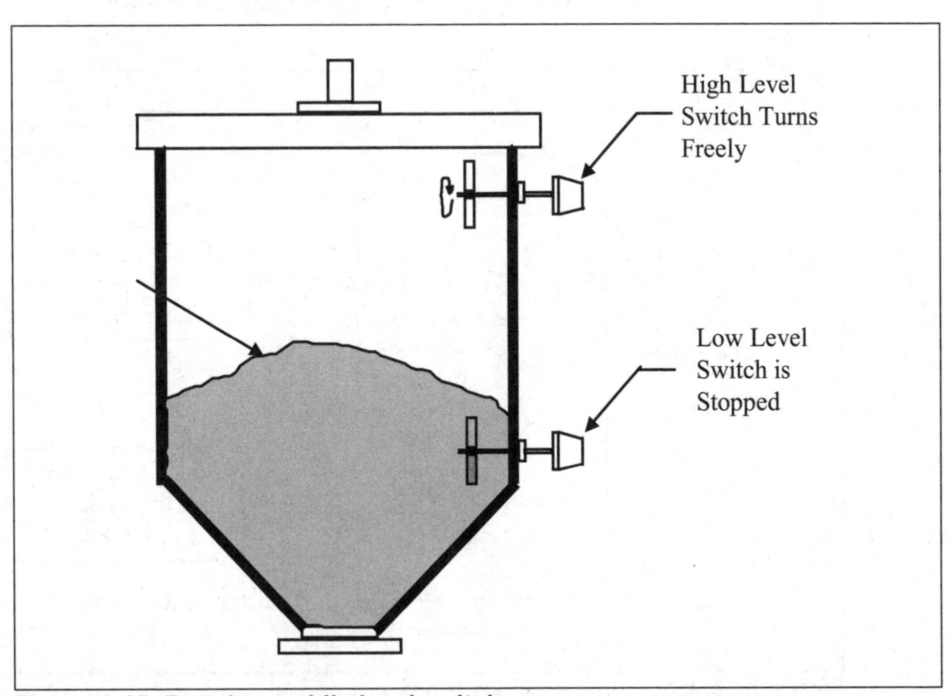

Figure 6-17. Rotating paddle level switches

Ultrasonic Level Switches

You can use ultrasonic level instruments for both continuous and point measurement. The point detectors or level switches can be grouped into damped sensors or on/off transmitter types.

A damped sensor-type level switch vibrates at its resonant frequency when no process fluid is present. When process material is present, vibration is dampened out. Most units use a piezoelectric crystal to obtain the vibration in the tip of the device. These instruments contain electronic circuits that detect the change in vibration and convert it into a dry-contact switch closure. These level switches are normally limited to liquid service because the damping effect of solids is insufficient in most cases.

On/off transmitter-type level switches contain transmitter and receiver units. The transmitter generates pulses in the ultrasonic range, which the receiver detects. You can locate the transmitter and receiver in the same probe or on opposite sides of the tank, as shown in Figure 6-18. In the latter design, the signal is transmitted in air, and the level switch will be actuated when the sound beam is interrupted by the rising process material. This type of switch is effective for both solid and liquid material applications. The sensor design in which the transmitter and receiver are mounted in the same unit is generally used only for liquids.

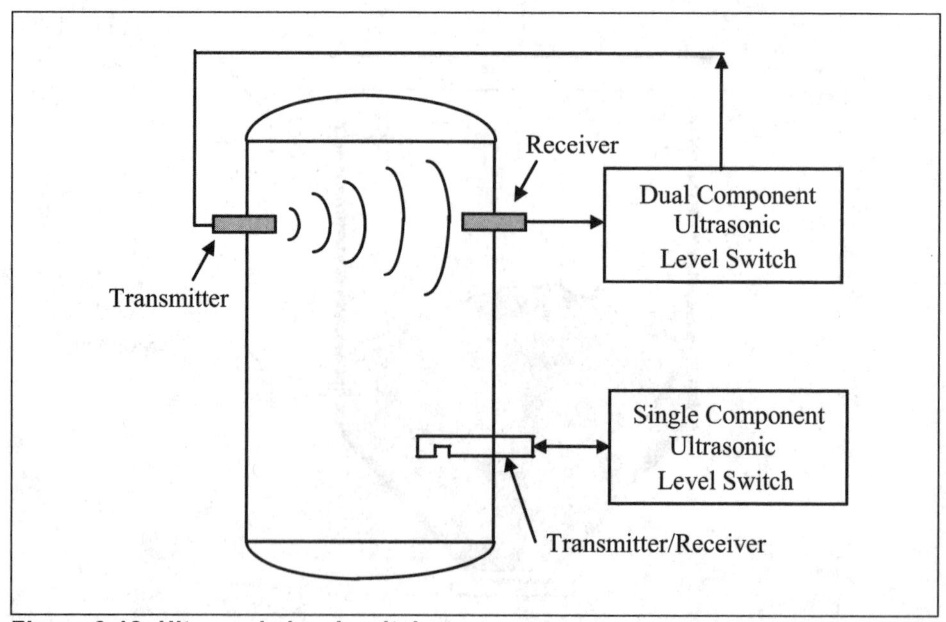

Figure 6-18. Ultrasonic level switches

EXERCISES

6.1 List three common sight-type level sensors.

6.2 Calculate the pressure in psi that is detected by a tubular sight glass gauge, if the height of the liquid is 100 inches and the specific gravity of the liquid in the tank is 0.95.

6.3 An object displaces 3 ft^3 of water at 20°C. Calculate the buoyancy force on the object.

6.4 Describe how a typical air bubbler level-measurement system operates.

6.5 List several disadvantages of bubbler level detection systems.

6.6 Explain how a typical capacitance probe operates, and list some factors that can cause the instrument to have calibration problems.

6.7 Describe how a typical ultrasonic level measurement system operates.

6.8 Explain how a nuclear level-detection system works.

6.9 Describe the basic principle of a guided-wave radar level probe.

6.10 List the level switches commonly used in process control.

6.11 Describe how a magnetic float-type level switch works.

6.12 Describe the basic principle behind the operation of a thermal level switch.

6.13 Explain how a rotating paddle switch operates in a typical level control application.

7 Temperature Measurement

Introduction

This chapter explores the more common temperature-measuring techniques and transducers used in process control, including filled-system thermometers, bimetallic thermometers, thermocouples, resistance temperature detectors (RTDs), thermistors, and integrated-circuit (IC) temperature sensors. We will discuss each transducer type in detail, but we will first consider the history of temperature measurement, temperature scales, and reference temperatures.

A Brief History of Temperature Measurement

The first known temperature-measuring device was invented by Galileo in about 1592. It consisted of an open container filled with colored alcohol and a long, narrow-throated glass tube with a hollow sphere at the upper end, which was suspended in the alcohol. When it was heated, the air in the sphere expanded and bubbled through the liquid. Cooling the sphere caused the liquid to move up the tube. Changes in the temperature of the sphere and the surrounding area could then be observed by the position of the liquid inside the tube. This "upside-down" thermometer was a poor indicator, however, since the level changed with atmospheric pressure, and the tube had no scale. Temperature measurement gained in accuracy with the development of the Florentine thermometer, which had a sealed construction and a graduated scale.

In the years to come, many thermometric scales were designed, all of which were based on two or more fixed points. However, no scale was universally recognized until the early 1700s, when Gabriel Fahrenheit, a

German instrument maker, designed, and made, accurate and repeatable mercury thermometers. For the fixed point on the low end of his temperature scale, Fahrenheit used a mixture of ice water and salt. This was the lowest temperature he could reproduce, and he labeled it "zero degrees." The high end of his scale was more imaginative; he chose the body temperature of a healthy person and called it 96 degrees.

The upper temperature of 96 degrees was selected instead of 100 degrees because at the time it was the custom to divide things into twelve parts. Fahrenheit, apparently to achieve greater resolution, divided the scale into twenty-four, then forty-eight, and eventually ninety-six parts. It was later decided to use the symbol °F for degrees of temperature in the Fahrenheit scale, in honor of the inventor. The Fahrenheit scale gained popularity primarily because of the repeatability and quality of the thermometers that Fahrenheit built.

Around 1742, Swedish astronomer Anders Celsius proposed that the melting point of ice and the boiling point of water be used for the two fixed temperature points. Celsius selected zero degrees for the boiling point of water and 100 degrees for the melting point of water. Later, the end points were reversed, and the centigrade scale was born. In 1948, the name was officially changed to the Celsius scale and the symbol °C was chosen to represent "degrees Celsius or centigrade" of temperature.

Temperature Scales

It has been experimentally determined that the lowest possible temperature is -273.15°C. The Kelvin temperature scale was chosen so that its zero is at -273.15°C, and the size of one Kelvin unit was the same as the Celsius degree. Kelvin temperature is given by the following formula:

$$T = T(°C) + 273.15 \tag{7-1}$$

Another scale, the Rankine scale (°R) is defined in the same way—with -273.15°C as its zero—and is simply the Fahrenheit equivalent of the Kelvin scale. It was named after an early pioneer in the field of thermodynamics, W. J. M. Rankine. The conversion equations for the other three modern temperature scales are as follows:

$$T(°C) = \frac{5}{9}(T(°F) - 32°) \tag{7-2}$$

$$T(°R) = T(°F) + 459.67 \tag{7-3}$$

$$T(°F) = \frac{9}{5}(T(°C) + 32°) \tag{7-4}$$

You can use these equations to convert from one temperature scale to another, as illustrated in Examples 7-1 and 7-2.

EXAMPLE 7-1

Problem: Express a temperature of 125°C in (a) degrees °F and (b) Kelvin.

Solution: (a) Convert to degrees Fahrenheit as follows:

$$T(°F) = \frac{9}{5}(T(°C) + 32°)$$

$$T(°F) = \frac{9}{5}(125°C) + 32°$$

$$T\,(°F) = (225 + 32)\,°F$$

$$T = 257°F$$

(b) Convert to Kelvin as follows:

$$T\,(K) = T(°C) + 273.15$$

$$T\,(K) = 125°C + 273.15$$

$$T = 398.15\,K$$

Reference Temperatures

We cannot build a temperature divider the way we can a voltage divider, nor can we add temperatures as we would add lengths to measure distance. Instead, we must rely on temperatures established by physical phenomena that are easily observed and consistent in nature.

The International Temperature Scale (ITS) is based on such phenomena. Revised in 1990, it establishes the seventeen reference temperatures shown in Table 7-1. The ITS-90, as this new version is called, is designed so that temperature values obtained on it do not deviate from the Kelvin thermodynamic temperature values by more than the uncertainties of the Kelvin values as they existed at the time the ITS-90 was adopted. Thermodynamic temperature is indicated by the symbol T and has the unit known as the Kelvin, symbol K. The size of the Kelvin is defined to be 1/273.16 of the triple point of water. A triple point is the equilibrium temperature at which the solid, liquid, and vapor phases coexist.

EXAMPLE 7-2

Problem: Express a temperature of 200°F in degrees Celsius and then degrees Kelvin.

Solution: First, we convert 200°F to degrees Celsius as follows:

$$T(°\mathrm{C}) = \frac{5}{9}(T(°\mathrm{F}) - 32°)$$

$$T = \frac{5}{9}(200 - 32)°\mathrm{C}$$

$$T = 93.33°\mathrm{C}$$

Now, we convert the temperature in degrees Celsius to Kelvin as follows:

$$T\,(K) = T(°\mathrm{C}) + 273.15$$

$$T\,(K) = 93.33°\mathrm{C} + 273.15$$

$$T = 366.48\ K$$

Table 7-1. Defining Fixed Points of the ITS-90

Description	K	°C
Vapor pressure (VP) point of helium	3 to 5	-270.15 to -268.15
Equilibrium hydrogen at triple point (TP)	13.8033	259.3467
Equilibrium hydrogen at VP point	≈17	≈-256.15
Equilibrium hydrogen at VP point	≈ 20.3	≈-252.85
Neon at TP	24.5561	248.5939
Oxygen at TP	54.3584	218.7916
Argon at TP	83.8058	189.3442
Mercury at TP	234.3156	38.8344
Water at TP	273.16	0.01
Gallium at melting point (MP)	302.9146	29.7646
Indium at freezing point (FP)	429.7485	156.5985
Tin at FP	505.078	231.928
Zinc at FP	692.677	419.527
Aluminum at FP	933.473	660.323
Silver at FP	1234.93	961.78
Gold at FP	1337.33	1064.18
Copper at FP	1357.77	1084.62

Since these fixed temperatures are our only reference, we must use instruments to interpolate between them. However, achieving accurate interpolation can require the use of some fairly exotic transducers, many of which are too complicated or expensive to use in process control applications.

Filled-System Thermometers

Many physical properties change with temperature, such as the volume of a liquid, the length of a metal rod, the electrical resistance of a wire, the pressure of a gas kept at constant volume, and the volume of a gas kept at constant pressure. Filled-system thermometers use the phenomenon of thermal expansion of matter to measure temperature change.

The filled thermal device consists of a primary element that takes the form of a reservoir or bulb, a flexible capillary tube, and a hollow Bourdon tube that actuates a signal-transmitting device and/or a local indicating temperature dial. A typical filled-system thermometer is shown in Figure 7-1. In this system, the filling fluid, either liquid or gas, expands as temperature increases. This causes the Bourdon tube to uncoil and indicate the temperature on a calibrated dial.

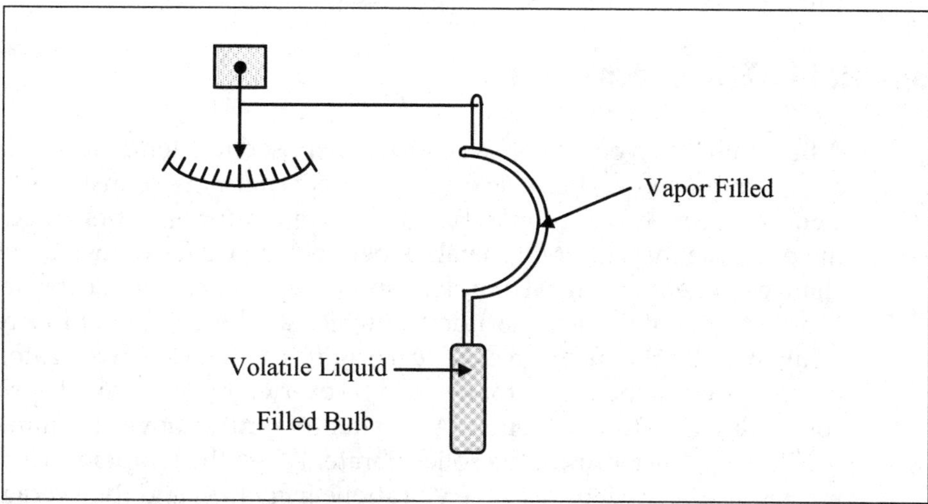

Figure 7-1. Filled bulb thermometer

The filling or transmitting medium is a vapor, a gas, mercury, or another liquid. The liquid-filled system is the most common because it requires a bulb with the smallest volume or permits a smaller instrument to be used.

The gas-filled system uses the perfect gas law, which states the following for an ideal gas:

$$T = kPV \tag{7-5}$$

where

T	=	temperature
k	=	constant
P	=	pressure
V	=	volume

If the volume of gas in the measuring instrument is kept constant, then the ratio of the gas pressure and temperature is constant, so that

$$\frac{P_1}{T_1} = \frac{P_2}{T_2} \tag{7-6}$$

The only restrictions on Equation 7-6 are that the temperature must be expressed in degrees Kelvin and the pressure must be in absolute units.

Example 7-3 shows how to calculate the temperature for a change in pressure of a fixed volume temperature detector.

Bimetallic Thermometers

A bimetallic strip curves or twists when exposed to a temperature change, as Figure 7-2 shows, because of the different thermal expansion coefficients of the metals used in it. Bimetallic temperature sensors are based on the principle that different metals experience thermal expansion with changes in temperature. To understand thermal expansion, consider a simple model of a solid, the atoms of which are held together in a regular array of forces that have an electrical origin. The forces between atoms can be compared to the forces that would be exerted by an array of springs connecting the atoms together. At any temperature above absolute zero (–273.15°C), the atoms of the solid vibrate. When the temperature is increased, the amplitude of the vibrations increases, and the average distance between atoms increases. This leads to an expansion of the whole body as the temperature is increased. The change in length that arises from a change in temperature (ΔT) is designated by ΔL. Through experimentation, we find that the change in length ΔL is proportional to the change in temperature ΔT and the original length L. Thus,

$$\Delta L = kL\Delta T \tag{7-7}$$

EXAMPLE 7-3

Problem: A gas in a fixed volume has a pressure of 30 psi at a temperature of 20°C. What is the temperature in °C if the pressure in the detector has increased to 35 psi?

Solution: First, convert the temperature of 20°C to the absolute scale of Kelvin using Equation 7-1:

$$T = T\,(°C) + 273.15$$

$$T = 20 + 273.15 = 293.15°$$

Now, use Equation 7-6 to find the new temperature that the detector is measuring:

$$T_1 = \frac{P_2}{P_1} T_1 = \left(\frac{35 \text{ psi}}{30 \text{ psi}}\right) 293.15° = 342°$$

Finally, convert this value for T_2 to °C:

$$T_2 = (342° - 273.15°) = 68.85°C$$

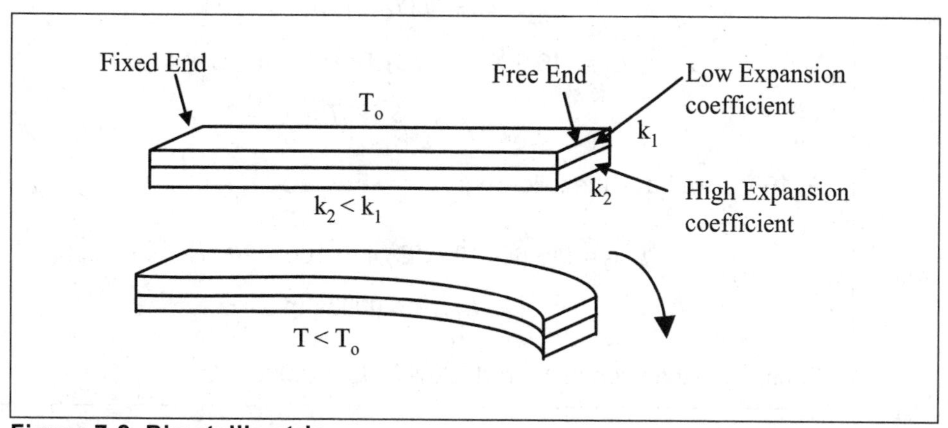

Figure 7-2. Bimetallic strip

where k is called the *coefficient of linear expansion*. This coefficient has different values for different materials. Table 7-2 lists the experimentally determined values for the average coefficient of linear expansion of several common solids in the temperature range of 0 to 100°C.

Table 7-2. Thermal Expansion Coefficient

Material	Expansion Coefficient (*k*)
Aluminum	25×10^{-6}°C
Copper	16.6×10^{-6}°C
Steel	6.7×10^{-6}°C
Beryllium/copper	9.3×10^{-6}°C

Example 7-4 illustrates how to calculate the expansion of a metal rod with a temperature increase.

EXAMPLE 7-4

Problem: How much will a 4 m-long copper rod expand when the temperature is changed from 0 to 100°C?

Solution: First, find the length of the rod at 0°C and at 100°C, then find the change in length. Using Equation 7-7 at 0°C,

$$\Delta L = kL\Delta T$$

or

$$L_o = kL(T_o - T_{20}) + L$$

$$L_o = (16.6 \text{ X } 10^{-6}/°\text{C})\ 4 \text{ m } (0 - 20)°\text{C} + 4 \text{ m}$$

$$L_o = 3.9987\text{m}$$

At 100°C we have the following:

$$L_{100} = (16.6 \text{ X } 10^{-6}/°\text{C})\ 4 \text{ m}(100 - 20)°\text{C} + 4 \text{ m}$$

$$L_{100} = 4.00531 \text{ m}$$

Thus, the expansion in the rod is $L_{100} - L_o = 0.00661$ m

Figure 7-3 shows a typical bimetallic dial thermometer using a spiral wound element. The spiral element provides a larger bimetallic element in a smaller area, and it can measure smaller changes in temperature. It is a low-cost instrument but has the disadvantages of relative inaccuracy and a relatively slow response time. It is normally used in temperature measurement applications that do not require high accuracy.

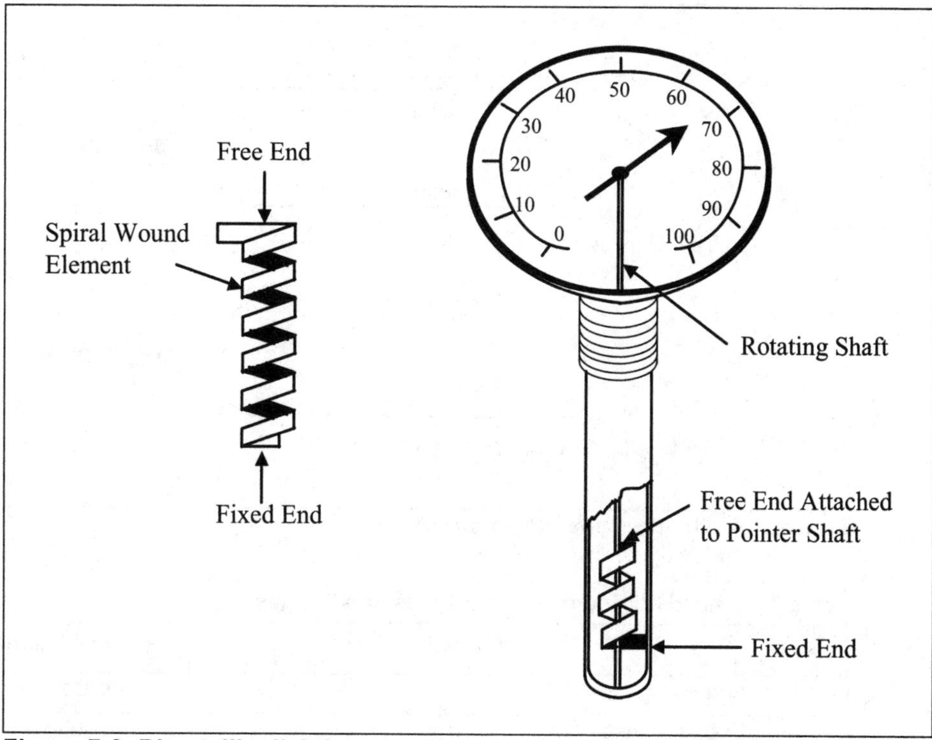

Figure 7-3. Bimetallic dial thermometer

Thermocouples

Thermoelectric Circuit

When two wires composed of dissimilar metals are joined at both ends and one of the ends is heated, a continuous current flows in the "thermoelectric" circuit. Thomas Seebeck made this discovery in 1821. This thermoelectric circuit is shown in Figure 7-4(a). If this circuit is broken at the center, as shown in Figure 7-4(b), the new open-circuit voltage (known as "the Seebeck voltage") is a function of the junction temperature and the compositions of the two metals.

All dissimilar metals exhibit this effect, and this configuration of two dissimilar metals joined together is called a *thermocouple*, which is abbreviated TC. The most common TCs and their normal temperature ranges are listed in Table 7-3.

For small changes in temperature, the Seebeck voltage is linearly proportional to temperature:

$$V_{ab} = \alpha(T_1 - T_2) \tag{7-8}$$

where α, the Seebeck coefficient, is the constant of proportionality.

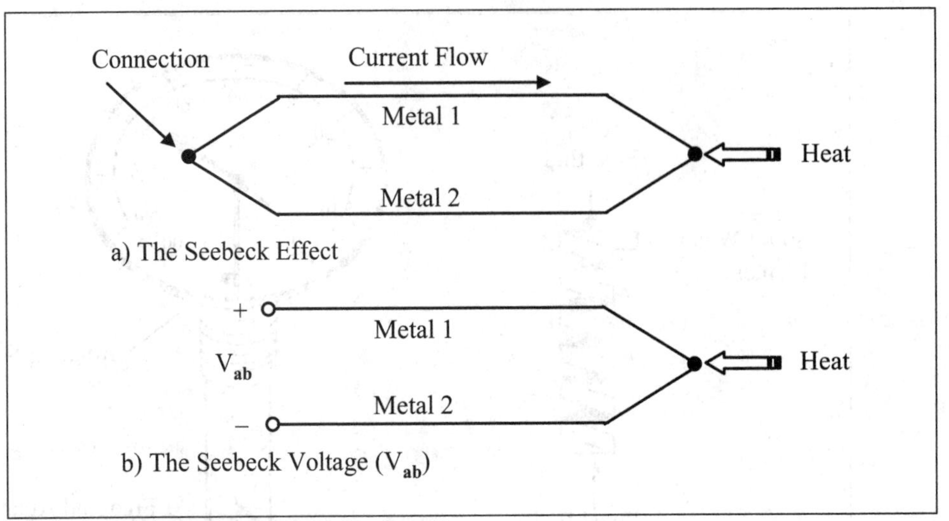

Figure 7-4. The thermoelectric circuit

Table 7-3. Standard Thermocouple Types and Ranges

Type	Material	Normal Range, °C
J	Iron-constantan	–190 to 760°C
T	Copper-constantan	–200 to 37°C
K	Chromel-alumel	–190 to 1260°C
E	Chromel-constantan	–100 to 1260°C
S	90% platinum + 10% rhodium-platinum	0 to 1482°C
R	87% platinum + 13% rhodium-platinum	0 to 1482°C

Example 7-5 shows how to calculate the Seebeck voltage for a thermocouple.

EXAMPLE 7-5

Problem: Find the Seebeck voltage for a thermocouple with α= 40 μv/°C if the junction temperatures are 40°C and 80°C.

Solution: Using Equation 7-8, the Seebeck voltage can be found as follows:

$$V_{ab} = \alpha\,(T_1 - T_2)$$

$$V_{ab} = 40\ \mu v/°C\ (80°C - 40°C) = 1.6\ mv$$

Thermocouple Tables

To take advantage of the voltage that thermocouples produce, scientists have developed comprehensive tables of voltage versus temperature for many types of thermocouples. These tables (presented in Appendix B) give the voltage that results for a particular type of thermocouple when the reference junctions are at 0°C and the measurement junction is at a given temperature. For example, according to Table B-4 in Appendix B, for a type K thermocouple at 200°C with a 0°C reference, the voltage produced is as follows:

$$V(200°C) = 8.13 \text{ mV} \qquad \text{(type K, 0°C)}$$

Conversely, if a voltage of 29.14 mV is measured with a type K thermocouple with a 0°C reference, we find from Table B-4 that

$$T(29.14 \text{ mV}) = 700°C \qquad \text{(type K, 0°C)}$$

However, in most cases the TC voltage does not fall exactly on a table value. When this happens, you must interpolate between the table values that bracket the desired value. You can find an approximate value of temperature by using the following interpolation equation:

$$T_m = T_l + \frac{(T_h - T_l)}{(V_h - V_l)}(V_m - V_l) \qquad (7\text{-}9)$$

In this equation, the measured voltage V_m lays between a higher voltage V_h and lower voltage V_l given in the tables. The temperatures that correspond to these voltages are T_h and T_l. Example 7-6 illustrates this concept.

EXAMPLE 7-6

Problem: A voltage of 6.22 mV is measured with a type J thermocouple at a 0°C reference. Find the temperature of the measurement junction.

Solution: From Table B-1 in Appendix B we see that V_m = 6.22 mv lies between V_l = 6.08 mv and V_h = 6.36 mv, with corresponding temperatures of T_l = 115°C and T_h = 120°C, respectively. Therefore, using Equation 7-10, we can find the junction temperature as follows:

$$T_m = 115°C + \frac{(120°C - 115°C)}{(6.36mV - 6.08mV)}(6.22mV - 6.08mV)$$

$$T_m = 117.5°C$$

Measuring Thermocouple Voltage

You cannot measure the Seebeck voltage directly because you must first connect a voltmeter to the thermocouple, and the voltmeter leads create a new thermoelectric circuit.

Consider a digital voltmeter (DVM) that is connected across a copper-constantan (type T) thermocouple. The voltage output is shown in Figure 7-5.

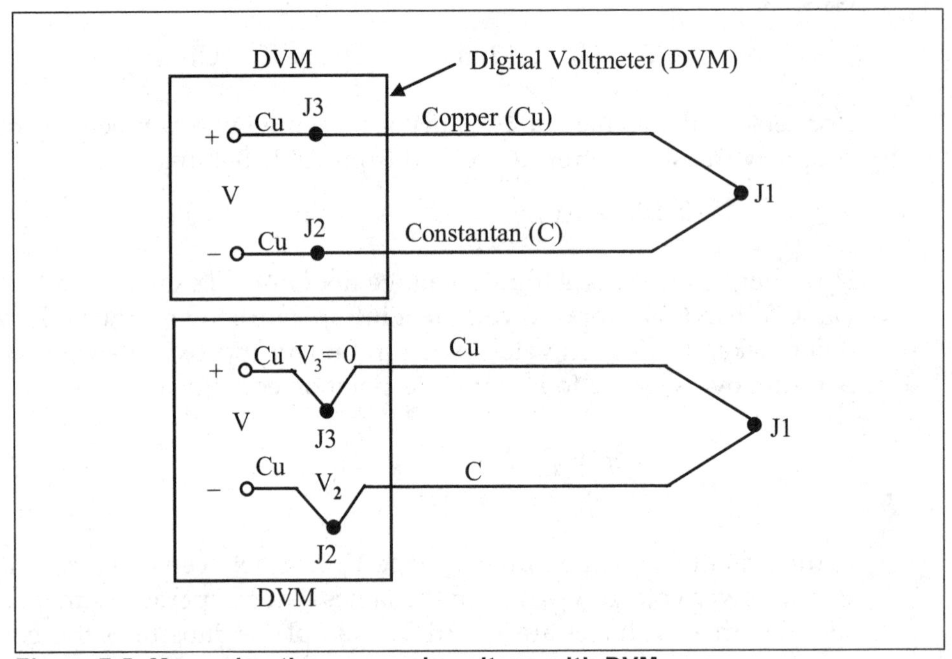

Figure 7-5. Measuring thermocouple voltage with DVM

We would like the voltmeter to read only V_1, but to measure the output of junction J1 we have connected the voltmeter, which created two more metallic junctions: J2 and J3. Since J3 is a copper-to-copper junction, it creates no thermal voltage ($V_2 = 0$). However, J2 is a copper-to-constantan junction, which will add a voltage V_2 in opposition to V_1. The resultant voltmeter reading V will be proportional to the temperature difference between J1 and J2. This indicates that we cannot find the temperature at J1 unless we first find the temperature of J2.

One way to determine the temperature of J2 is to physically put the junction into an ice bath. This forces its temperature to be 0°C and establishes J2 as the "reference junction," as shown in Figure 7-6. Since both voltmeter terminal junctions are now copper-copper, they create no thermal voltage, and the reading "V" on the voltmeter is proportional to the temperature difference between J1 and J2.

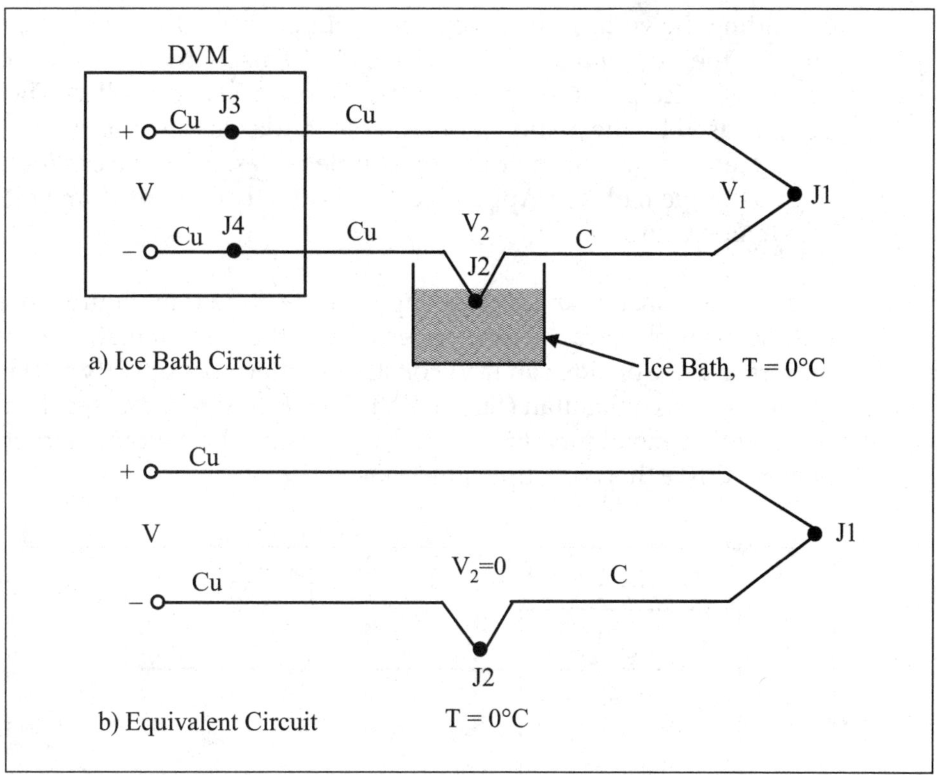

Figure 7-6. External reference junction

Now the voltmeter reading is as follows:

$$V = V_1 - V_2 = \alpha\,(T_{J1} + T_{J2})$$

If we specify T_{J1} in degrees Celsius,

$$T_{J1}\,(°C) + 273.15 = T_{J1}(K)$$

then V becomes the following:

$$V = V_1 - V_2$$

$$V = \alpha\,(T_{J1} + 273.15) - (T_{J2} + 273.15)$$

$$V = \alpha\,(T_1 - T_2)$$

$$V = \alpha\,(T_{J1} - 0)$$

$$V = \alpha\, T_{J1}$$

We presented this somewhat involved discussion to emphasize that the ice-bath junction output, V_2, is not zero volts. It is a function of absolute temperature.

By adding the voltage of the ice-point reference junction, we have now referenced the TC voltage reading (V) to 0°C. This method is very accurate because the ice-point temperature can be precisely controlled. The ice point is used by the National Bureau of Standards (NBS) as the fundamental reference point for their thermocouple tables. Now we can look at the thermocouple tables in Appendix B and directly convert from voltage V into temperature.

The copper-constantan thermocouple circuit shown in Figure 7-6 is a unique example because the copper wire is the same metal as the voltmeter terminals. Consider an iron-constantan thermocouple (type J) instead of the copper-constantan (Figure 7-7). The iron wire increases the number of dissimilar metal junctions in the circuit, since both voltmeter terminals become *Cu-Fe* thermocouple junctions.

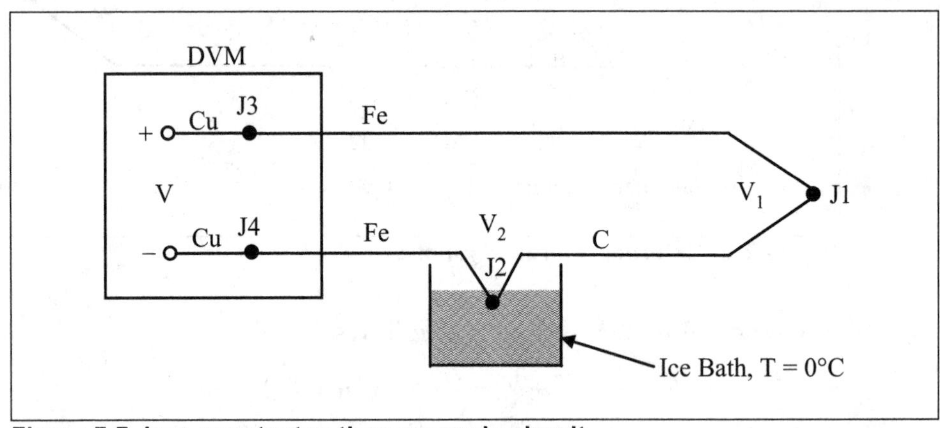

Figure 7-7. Iron-constantan thermocouple circuit

This circuit will provide moderately accurate measurements as long as the voltmeter (positive) and (negative) terminals (J3 and J4) are the same temperature. The thermoelectric effects of J3 and J4 act in opposition,

$$V_1 = V$$

if,

$$V_3 = V_4$$

that is, if

$$T_{J3} = T_{J4}$$

If both front-panel terminals are not at the same temperature, an error will result. To gain a more precise measurement, you should extend the copper voltmeter leads so the copper-to-iron junctions are made on an "isothermal" (same temperature) block, as shown in Figure 7-8.

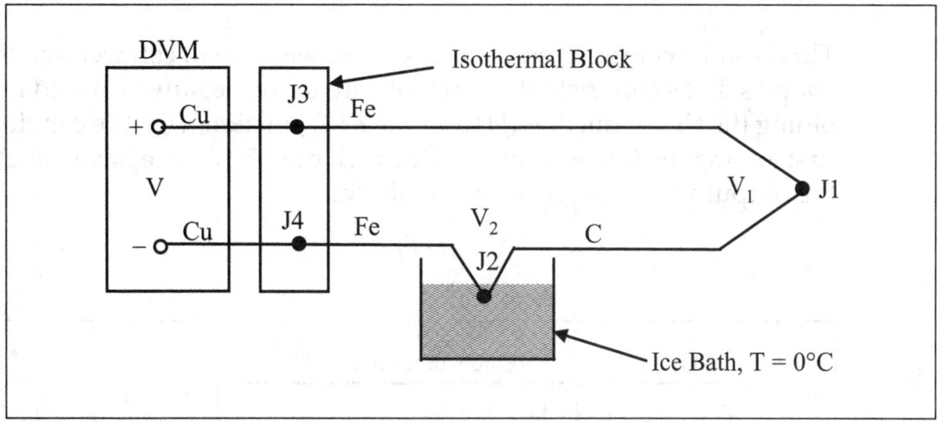

Figure 7-8. Using an isothermal block in a thermocouple circuit

The isothermal block is not only a good electrical insulator. It is also a good heat conductor, and this helps to hold J3 and J4 at the same temperature. The absolute block temperature is unimportant because the two *Cu-Fe* junctions act in opposition. Thus,

$$V = \alpha\,(T_1 - T_{ref})$$

The circuit in Figure 7-8 will provide accurate readings, but it is desirable to eliminate the ice bath if possible. One way to do this is to replace the ice bath with another isothermal block, as shown in Figure 7-9.

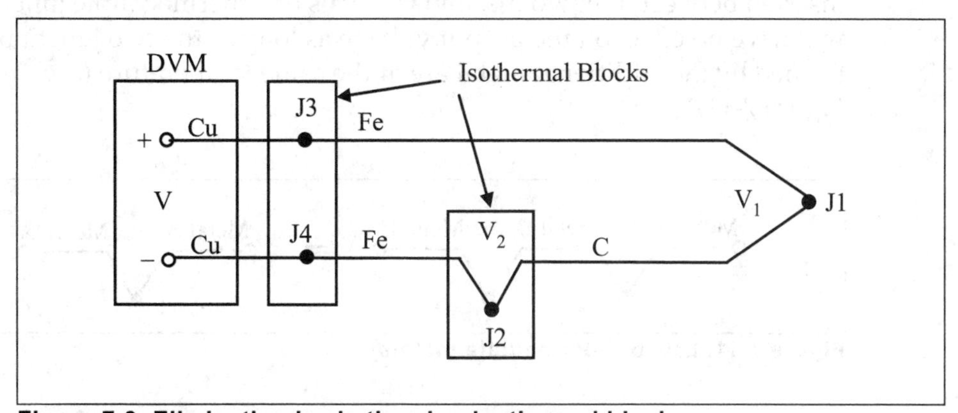

Figure 7-9. Eliminating ice bath using isothermal block

The new block is at reference temperature T_{ref}, and because J3 and J4 are still at the same temperature we can again show that

$$V = \alpha\,(T_1 - T_{ref})$$

This is still a complicated circuit because we have to connect two thermocouples. Let's eliminate the extra *Fe* wire in the negative (–) lead by combining the *Cu-Fe* junction (J4) and the *Fe-C* junction (J_{ref}). We can do this by first joining the two isothermal blocks (Figure 7-10). We have not changed the output voltage V. It is still as follows:

$$V = \alpha\,(T_1 - T_{ref})$$

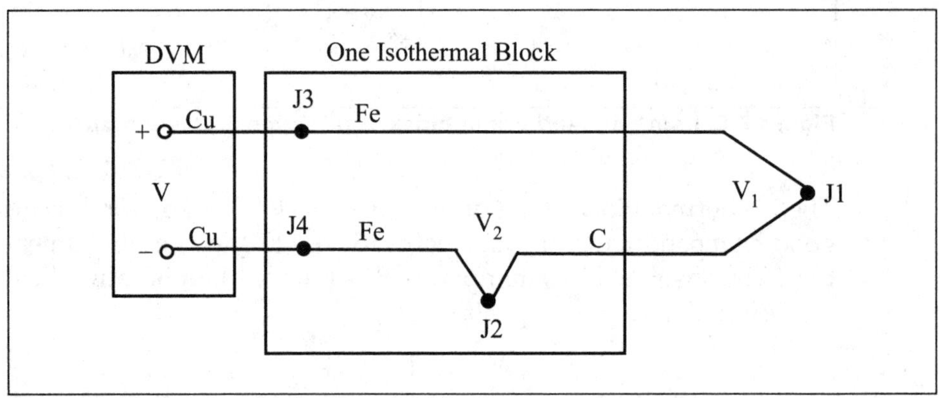

Figure 7-10. Joining the isothermal blocks

Now we can use the law of intermediate metals to eliminate the extra junction. This empirical "law" states that a third metal (in this case, iron) inserted between the two dissimilar metals of a thermocouple junction will have no effect on the output voltage as long as the two junctions formed by the additional metal are at the same temperature (see Figure 7-11).

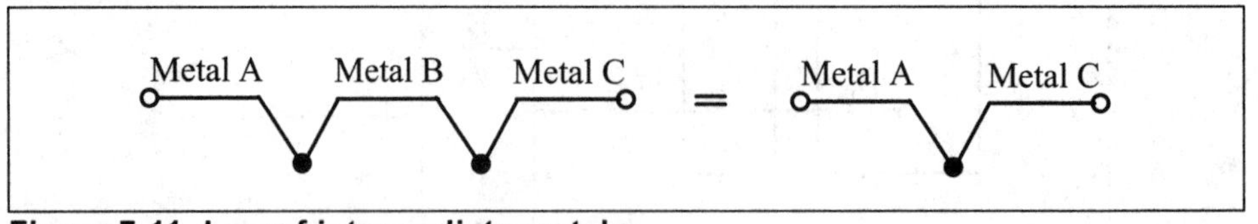

Figure 7-11. Law of intermediate metals

This is a useful result, as it eliminates the need for the iron wire in the negative lead. This can be seen in Figure 7-12. There again, $V = \alpha\,(T_1 - T_{ref})$, where α is the Seebeck coefficient for a *Fe-C* thermocouple. Junctions J2 and J3 take the place of the ice bath. These two junctions now become the "reference junction."

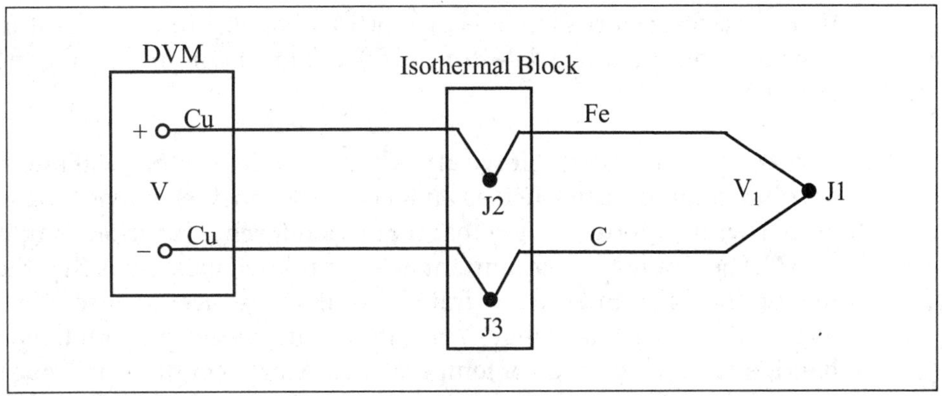

Figure 7-12. Equivalent thermocouple circuit

Now we can proceed to the next logical step: directly measure the temperature of the isothermal block (the reference junction) and use that information to compute the unknown temperature, T_{J1}. The thermistor is a temperature-measuring device whose resistance R_t is a function of temperature. It provides us with a method for measuring the temperature of the reference junction (Figure 7-13). We assume that the junctions J2 and J3 and the thermistor are all at the same temperature because of the design of the isothermal block. Using a digital voltmeter under computer control, we simply do the following:

1. Measure R_t to find T_{ref}, and convert T_{ref} to its equivalent reference junction voltage, V_{ref}.
2. Measure V and subtract V_{ref} to find V_1, and convert V_1 to temperature T_{J1}.

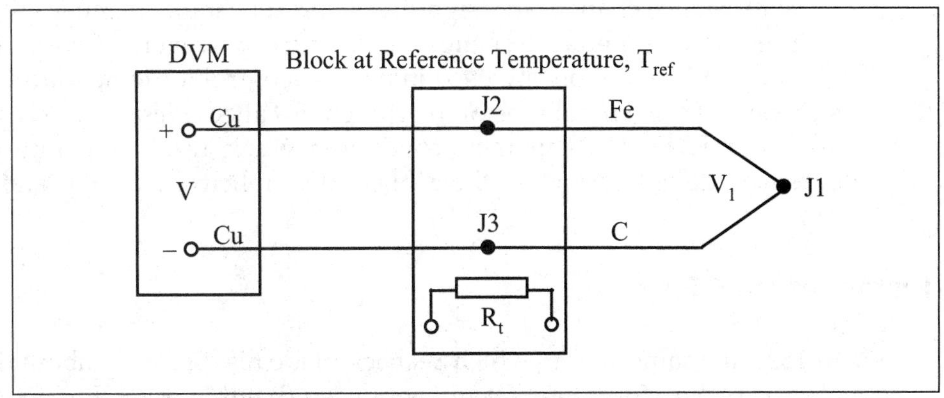

Figure 7-13. Standard thermocouple measurement circuit

This procedure is known as "software compensation" because it relies on a computer's software to compensate for the effect of the reference junction. The isothermal terminal block temperature sensor can be any device that has a characteristic that is proportional to absolute temperature: a resistance temperature detector (RTD), a thermistor, or an integrated-circuit sensor.

The logical question is, "If we already have a device that will measure absolute temperature (such as an RTD or thermistor), why do we even bother with a thermocouple that requires reference junction compensation?" The single most important answer to this question is that the thermistor, the RTD, and the integrated-circuit transducer are useful only over a limited temperature range. You can use thermocouples, on the other hand, over a wide range of temperatures. Moreover, they are much more rugged than thermistors (as evidenced by the fact that thermocouples are often welded to metal process equipment or clamped under a screw on the equipment). They can be manufactured easily, either by soldering or welding.

In short, thermocouples are the most versatile temperature transducers available. Furthermore, a computer-based temperature-monitoring system can perform the entire task of reference compensation and software voltage-to-temperature conversion. As a result, using a thermocouple in process control becomes as easy as connecting a pair of wires. The one disadvantage of the computer-based approach is that the computer requires a small amount of extra time to calculate the reference junction temperature, and this introduces a small dead time into a control loop.

Resistance Temperature Detectors

In principle, any material could be used to measure temperature if its electrical resistance changes in a significant and repeatable manner when the surrounding temperature changes. In practice, however, only certain metals and semiconductors are used in process control for temperature measurement. This general type of instrument is called a *resistance temperature detector* or RTD. RTDs are the second most widely used temperature measurement device because of their inherent simplicity, accuracy, and stability.

History of the RTD

In 1821, the same year in which Seebeck made his discovery about thermoelectricity, Sir Humphrey Davy announced that the resistivity of metals showed a marked dependence on temperature. Fifty years later, Sir William Siemens recommended platinum as the element to be used in a resistance thermometer. His choice proved most correct, since platinum is used to this day as the primary element in all high-accuracy resistance thermometers. In fact, the platinum resistance temperature detector, or PRTD, is used today as an interpolation standard from the oxygen point (-182.96°C) to the antimony point (630.74°C). Platinum is especially suited to this purpose because it can withstand high temperatures while maintaining excellent stability and good linearity.

C. Meyers proposed the classical RTD construction using platinum in 1932. He wound a helical coil of platinum on a crossed mica web and mounted the assembly inside a glass tube. This construction minimized strain on the wire while maximizing resistance. Although this construction produced a very stable element, the thermal contact between the platinum and the measured point was quite poor. This resulted in a slow thermal response time. The fragility of the structure limits its use today primarily to that of a laboratory standard.

In a more rugged construction technique, the platinum wire is wound on a glass or ceramic bobbin, as illustrated in Figure 7-14. The winding reduces the effective enclosed area of the coil, which minimizes a magnetic pickup and its related noise. Once the wire is wound onto the bobbin, the assembly is then sealed with a coating of molten glass. The sealing process ensures that the RTD will maintain its integrity under extreme vibration, but it also limits the expansion of the platinum metal at high temperatures. Unless the coefficients of expansion of the platinum and the bobbin match perfectly, stress will be placed on the wire as the temperature changes. This will result in a strain-induced resistance change, which may cause permanent change in the resistance capacity of the wire.

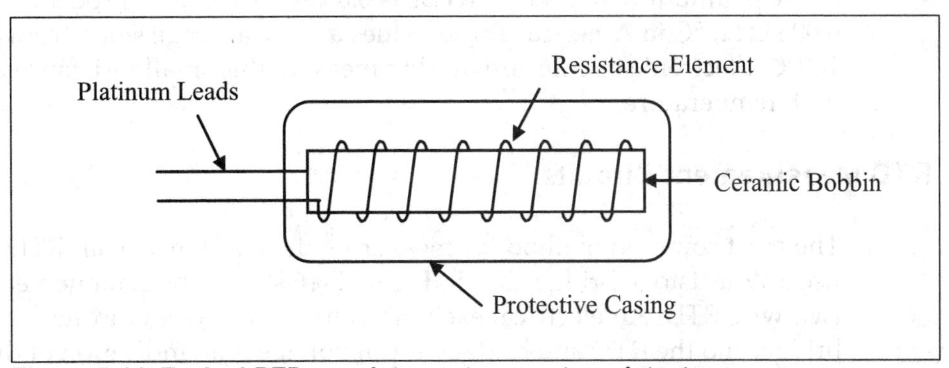

Figure 7-14. Typical RTD – resistance temperature detectors

Metal-Film RTDs

In the newest construction technique, a platinum or metal-glass slurry film is deposited or screened onto a small flat ceramic substrate. It is then etched with a laser-trimming system and sealed (see Figure 7-15). The laser makes adjustment cuts to achieve the desired base resistance. The

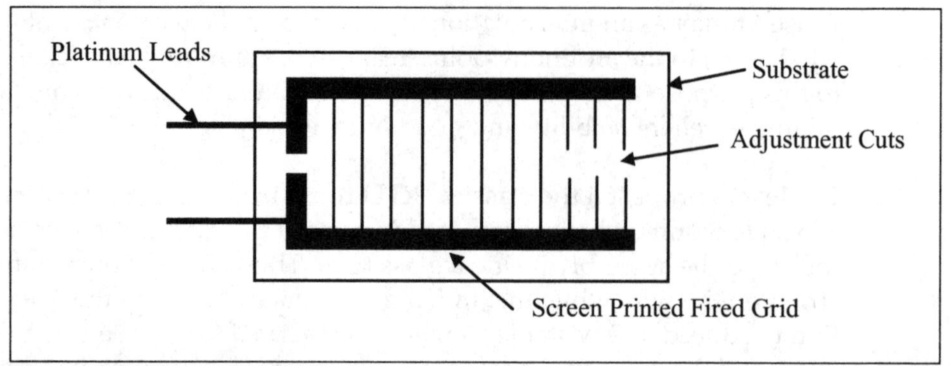

Figure 7-15. Metal-film RTD

metal-film RTD substantially reduces assembly time and has the further advantage of providing increased resistance for a given size. Because of the technology with which the device is manufactured, the device size itself is small, normally about one-half inch by one-half inch. This means it can respond quickly to rapid changes in temperature. Film RTDs are currently less stable than wound RTDs, but they are becoming more popular because of their decided advantages in size and low production cost.

The common values of resistance for platinum RTDs range from 10 Ω to several thousand ohms. The single most common value is 100 Ω at 0°C. The resistance only changes about 0.4 percent per one degree change in centigrade temperature. This change in resistance with temperature is called the *temperature coefficient* (α). The standard temperature coefficient of the platinum wire used in RTDs is 0.00385 Ω/°C in Europe and 0.00392 Ω/°C in America. These values are the average slope from 0 to 100°C. Electronic circuits are used to measure this small resistance change with temperature.

RTD Measurement Circuits

The most common method for measuring the resistance of an RTD is to use a Wheatstone bridge circuit. Figure 7-16 shows the arrangement for a two-wire RTD. An electrical excitation current is passed through the bridge, and the RTD and bridge output voltage is an indication of the RTD resistance. The circuit uses a very stable excitation power source, three high-precision resistors that have a very low temperature coefficient, and a high-input impedance amplifier to measure the resistance change of the RTD with changes in temperature.

The RTD is generally located on process equipment or piping, and the measurement circuit can be located hundreds of feet from a control room at the process plant. Since both the slope and the absolute value of resis-

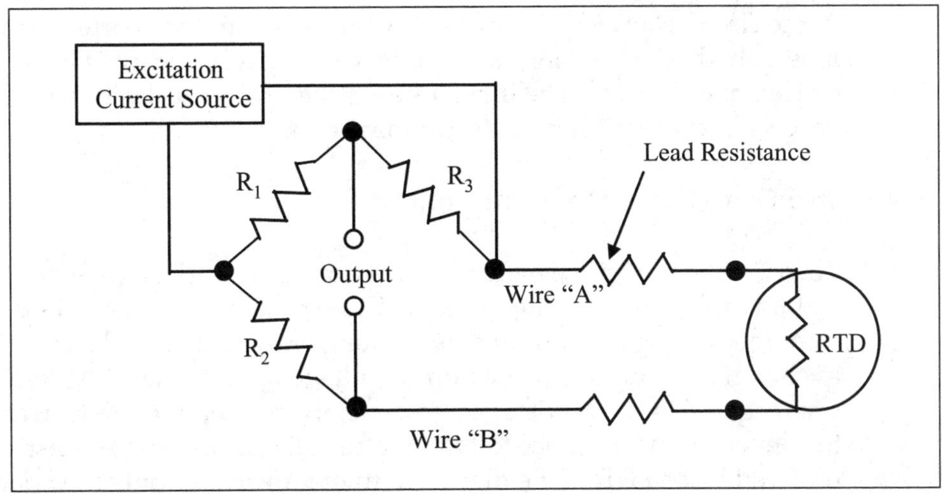

Figure 7-16. Two-wire RTD bridge circuit

tance of a typical RTD are small numbers, the length of the wire from the RTD to the Wheatstone bridge circuit can be significant. This is especially true when we consider that the measurement wires that lead to the sensor may be several ohms or even tens of ohms. A small lead resistance can introduce significant error into the measurement of output temperature. For example, a 10 Ω lead wire on a field-mounted RTD implies a 10 Ω/0.385Ω/°C = 26°C error in a measurement. Even the temperature coefficient of the lead wire can contribute a measurable error.

The standard method for avoiding this problem has been to use a three-wire connection to the Wheatstone bridge measurement circuit, as shown in Figure 7-17.

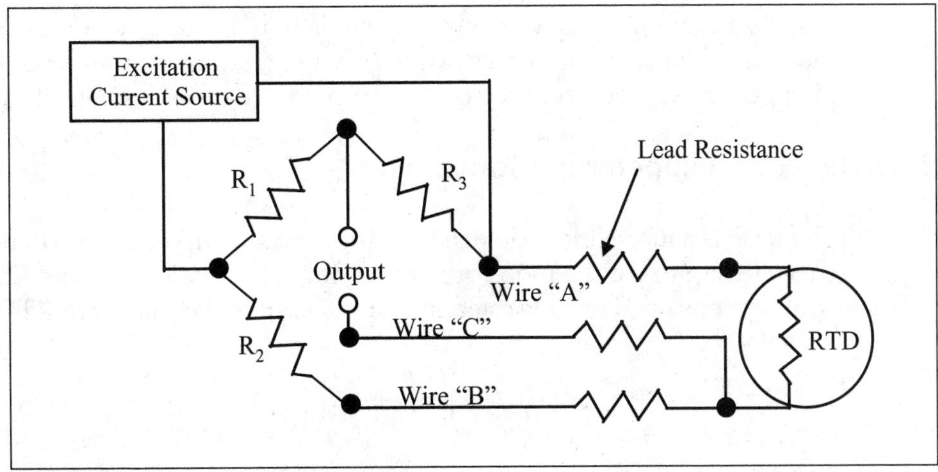

Figure 7-17. Three-wire RTD bridge

In the circuit shown in Figure 7-17, if wires A and B are perfectly matched in length, their impedance effects will cancel because each is in an opposite leg of the bridge. The third wire, C, acts as a sense lead and carries a very small current (in the microampere range).

Four-Wire Resistance Measurement

The Wheatstone bridge method of measuring the resistance of an RTD has certain problems associated with it. These problems are solved by the technique of using a current source along with a remotely located DVM, as shown in Figure 7-18. The output voltage read by the DVM is directly proportional to RTD resistance, so you only need one conversion equation to convert from resistance to temperature. The three bridge resistors are replaced by one RTD. The digital voltmeter measures only the voltage dropped across the RTD and is insensitive to the length of the lead wires.

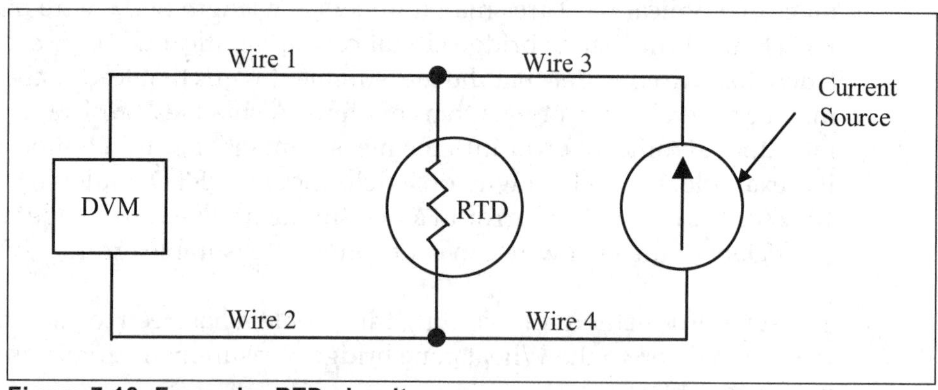

Figure 7-18. Four-wire RTD circuit

One disadvantage of a four-wire system is that, obviously, it requires one more wire than a three-wire bridge. However, this is a small price to pay given the increased accuracy of the temperature measurement it provides.

Resistance to Temperature Conversion

The RTD is a more linear device than the thermocouple, but it still requires curve fitting to yield a more precise reading. The Callendar-Van Dusen equation has been used for years to approximate the platinum RTD curve:

$$R = R_o + \alpha R_o \left[T - \delta \left(\frac{T}{100} - 1 \right) \left(\frac{T}{100} \right) - \beta \left(\frac{T}{100} - 1 \right) \left(\frac{T^3}{100} \right) \right] \quad (7\text{-}10)$$

where

R = resistance at temperature T

R_o = resistance at $T = 0°C$
α and δ = constants
β = 0 if $T > 0$, or β is 0.1 (typical) if $T < 0$

You determine the exact values for coefficients α, β, and δ by testing the RTD at four temperatures and then solving the resulting equations. Typical values for platinum RTDs are as follows:

$$\alpha = 0.00392$$

$$\delta = 1.49$$

$$\beta = 0 \text{ if } T > 0°C, \text{ and } \beta = 0.1 \text{ for } T < 0°C$$

Example 7-7 illustrates a typical calculation to obtain the resistance ratio for a platinum RTD.

EXAMPLE 7-7

Problem: Calculate the resistance ratio for a platinum RTD with $\alpha = 0.00392$ and $\delta = 1.49$ when $T = 100°C$.

Solution: Since T is greater than 0°C, $\beta = 0$, the Callendar-Van Dusen equation reduces to the following:

$$R = R_o + \alpha R_o \left[T - \delta \left(\frac{T}{100} - 1 \right) \left(\frac{T}{100} \right) \right]$$

$$R = R_o + (0.00392) R_o \left[100 - 1.49 \left(\frac{100}{100} - 1 \right) \left(\frac{100}{100} \right) \right]$$

$$R = R_o + (0.00392)(100) R_o$$

So,

$$\frac{R}{R_o} = 1.392$$

Thermistors

Like the RTD, the thermistor is also a temperature-sensitive resistor. The name *thermistors* is derived from the term "*therm*ally sensitive res*istors*," since the resistance of the thermistor varies as a function of temperature.

While the thermocouple is the most versatile temperature transducer and the RTD is the most linear, "most sensitive" are the words that best describe thermistors. The thermistor exhibits by far the largest value change with temperature of the three major categories of sensors.

A thermistor's high resistance change per degree change in temperature provides excellent accuracy and resolution. A standard 2,000-ohm thermistor with a temperature coefficient of 3.9%/°C at 25°C will have a resistance change of 78 ohms per °C change in temperature. A 2000 Ω platinum RTD would have a change of only 7.2 ohms under the same conditions. So, a standard thermistor is over ten times more sensitive than a RTD. This allows the thermistor circuit to detect minute changes in temperature that could not be observed with an RTD or thermocouple circuit. A thermistor connected to a bridge circuit can readily indicate a temperature change of as little as 0.0005°C.

The cost of this increased sensitivity is loss of linearity, as the curves in Figure 7-19 show. The thermistor is an extremely nonlinear device that is highly dependent on process parameters. Consequently, manufacturers have not standardized thermistor curves to the same extent as they have RTD and thermocouple curves.

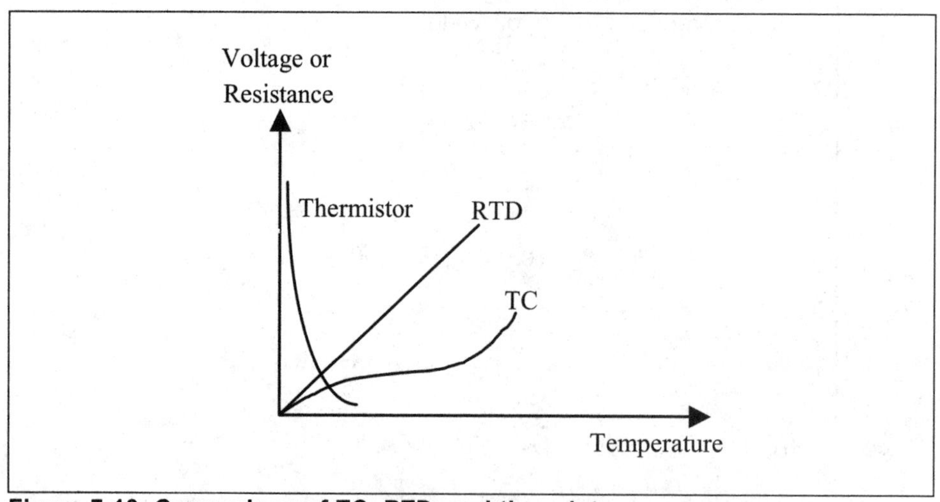

Figure 7-19. Comparison of TC, RTD, and thermistor

You can approximate an individual thermistor curve very closely by using the Steinhart-Hart equation:

$$\frac{1}{T} = A + B \ln R + C(\ln R)^3 \tag{7-11}$$

where

T	=	temperature (K)
R	=	resistance (Ω) of the thermistor
A, B, and C	=	curve-fitting constants

You can find the constants A, B, and C by selecting three data points on the published data curve and solving the three simultaneous equations. When you choose data points that span no more than 10°C within the nominal center of the thermistor's temperature range, this equation approaches a remarkable +0.01°C curve fit.

Example 7-8 illustrates a typical calculation to obtain the temperature for a thermistor with a known resistance.

A great deal of effort has gone into developing thermistors that approach a linear characteristic. These are typically three- or four-lead devices that require the use of external matching resistors to linearize the characteristic curve. Modern data acquisition systems with built-in microprocessors have made this type of hardware linearization unnecessary.

The high resistivity of the thermistor affords it a distinct measurement advantage. The four-wire resistance measurement is not required as it is with RTDs. For example, a common thermistor value is 5,000 Ω at 25°C. With a typical temperature coefficient of 4%/°C, a measurement lead resistance of 10 Ω produces only a 0.05°C error. This is a factor of five hundred times less than the equivalent RTD error.

Because thermistors are semiconductors, they are more susceptible to permanent decalibration at high temperatures than are RTDs or thermocouples. The use of thermistors is generally limited to a few hundred degrees Celsius, and manufacturers warn that extended exposures, even well below maximum operating limits, will cause the thermistor to drift out of its specified tolerance.

Thermistors can be manufactured very small, which means they will respond quickly to temperature changes. It also means that their small thermal mass makes them susceptible to self-heating errors. Thermistors are more fragile than RTDs or thermocouples, and you must mount them carefully to avoid crushing or bond separation.

Integrated-Circuit Temperature Sensors

Integrated-circuit temperature transducers are available in both voltage and current-output configurations (Figure 7-20). Both supply an output

EXAMPLE 7-8

Problem: A typical thermistor has the following coefficients for the Steinhart-Hart equation:

$$A = 1.1252 \times 10^{-3}/K$$

$$B = 2.3478 \times 10^{-4}/K$$

$$C = 8.5262 \times 10^{-8}/K$$

Calculate the temperature when the resistance is 4000 Ω.

Solution: Using Equation 7-11, we obtain the following:

$$\frac{1}{T} = A + B \ln R + C(\ln R)^3$$

$$\frac{1}{T} = 1.1252x10^{-3}/K + (2.3478x10^{-4}/K)(\ln 4000) + (8.5262x10^{-8}/K)(\ln 4000)^3$$

$$\frac{1}{T} = 1.1252x10^{-3}/K + 1.9471x10^{-3}/K + 0.0486x10^{-3}/K$$

$$\frac{1}{T} = 3.1209x10^{-3}/K$$

$$T = 320.4\ K$$

Now convert from Kelvin to Celsius:

$$T = (320.4 - 273.15)°C$$

$$T = 47.25°C$$

that is linearly proportional to absolute temperature. Typical values are one microampere of current per one-degree temperature change in Kelvin (1 μA/K) and ten millivolts per one-degree change in Kelvin (10 mv/K).

Except for the fact that these devices provide an output that is very linear with temperature, they share all the disadvantages of thermistors. They are semiconductor devices and thus have a limited temperature range.

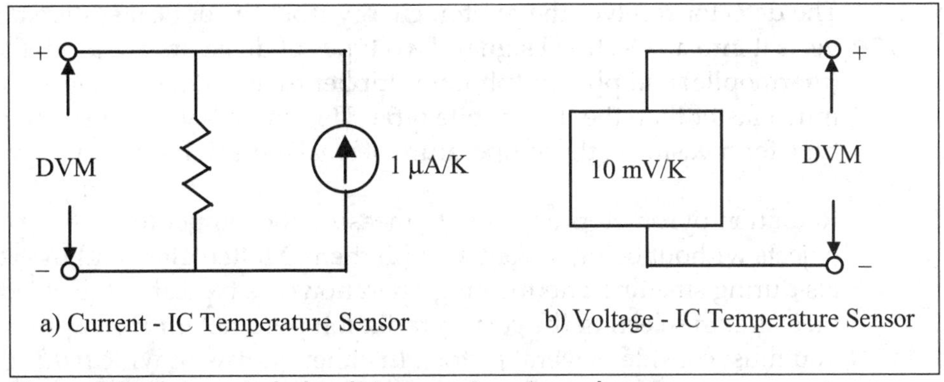

Figure 7-20. Integrated-circuit temperature transducers

Integrated-circuit temperature sensors are normally only used in applications that have a limited temperature range. One typical application is in temperature data acquisition systems where they are used for thermocouple compensation.

Radiation Pyrometers

A pyrometer is any temperature-measuring device that includes a sensor and a readout. However, in this section we will discuss only radiation-type pyrometers. A radiation pyrometer is a noncontact temperature sensor that infers the temperature of an object by detecting its naturally emitted thermal radiation. An optical system collects the visible and infrared energy from an object and focuses it on a detector, as shown in Figure 7-21. The detector converts the collected energy into an electrical signal to drive a temperature display or control unit.

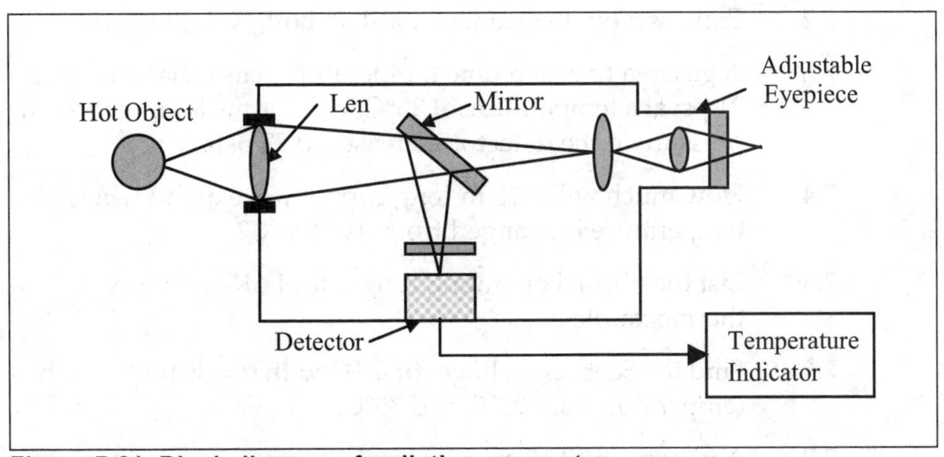

Figure 7-21. Block diagram of radiation pyrometer

The detector receives the photon energy from the optical system and converts it into an electrical signal. Two types of detectors are used: thermal (thermopile) and photon (photomultiplier tubes). Photon detectors are much faster than the thermopile type. This enables you to use the photon type for measuring the temperature of small objects moving at high speed.

Radiation pyrometers are used to measure the temperature of very hot objects without being in contact with them. Molten glass and molten metals during smelting and forming operations are typical of the objects they measure. In selecting the correct radiation pyrometer for an application you must consider several factors. In either narrow or wide fields of view, the cross-sectional area can vary greatly. It can be rectangular, circular, and slot shaped, depending on the kind of apertures used in the instrument. In some instruments, a telescopic eye magnifies the radiant energy so much smaller objects at longer distances can be measured. Hot objects as small as 1/16 inch in diameter can be measured with some instruments.

The construction of the instrument components, such as the lens and curved mirrors, control the sight path. The materials of construction determine the optical characteristics of the device. For example, glass does not transmit light well beyond 2.5 microns. It is therefore suitable only for high-temperature applications where high-energy outputs are present. Other common optical materials are quartz (transmitting well to 4 microns) and crystalline calcium fluoride (transmitting well up to about 10 microns). Band pass filters are used in some instruments to cut off unwanted light at certain wavelengths.

EXERCISES

7.1 Express a temperature of 133°C in both °F and Kelvin.

7.2 Express a temperature of 400°F in both °C and Kelvin.

7.3 A gas in a fixed-volume temperature sensor has a pressure of 21 psi at a temperature of 25°C. What is the temperature in °C if the pressure in the detector increases to 22 psi?

7.4 How much will a 10 m-long copper rod expand when the temperature is changed from 0 to 100°C?

7.5 List the normal operating ranges for J-, K-, and S-type thermocouples.

7.6 Find the Seebeck voltage for a TC with $\alpha = 36\ \mu v/°C$ if the junction temperatures are 25°C and 75°C.

7.7 A voltage of 10.10 mv is measured across a type K thermocouple at a 0°C reference. Find the temperature at the measurement junction.

7.8 A voltage of 19.50 mv is measured across a type J thermocouple at a 0°C reference. Find the temperature at the measurement junction.

7.9 Explain the purpose and function of the isothermal blocks in a digital voltmeter temperature-measuring circuit.

7.10 Calculate the resistance ratio for a platinum RTD with $\alpha = 0.00392$ and $\delta = 1.49$ when $T = 80°C$.

7.11 Calculate the temperature measured by a thermistor when the resistance is 2,800 Ω. Assume that the thermistor has the following coefficients for the Steinhart-Hart equation: $A = 1.1252 \times 10^{-3}/K$, $B = 2.3478 \times 10^{-4}/K$, and $C = 3.5262 \times 10^{-8}/K$.

7.12 List a typical temperature measurement application for a radiation pyrometer.

8

Analytical Measurement and Control

Introduction

This chapter discusses the basic principles of chemical analytical measurement and control, with an emphasis on the following areas: conductivity, pH, density, humidity, turbidity, and gas analysis. We introduce the basic principles of electromagnetic (EM) radiation and describe several common phototransducers that use EM radiation to measure analytical variables.

Conductivity Measurement

Measuring conductivity means determining a solution's ability to conduct electric current. This ability is referred to as *specific conductance*—or, simply, *conductivity*—and is expressed in "mhos," which is the reciprocal of ohm (the unit used to express resistance).

Aqueous solutions of acids, bases, or salts are known as electrolytes; they are conductors of electricity. Conductivity measurements are generally made to detect electrolytic contaminants around water and waste-treatment areas. The degree of electrical conductivity of such solutions is affected by three factors: the nature of the electrolyte, the concentration of the solution, and the temperature of the solution. A measurement of the conductivity at a fixed temperature can be a measurement of the solution's concentration, which can be expressed in percentage terms by weight, parts per million, or other applicable units.

If you know the conductivity values of various concentrations of an electrolyte, you can then determine the concentration by passing current through a solution of known dimensions and measuring its electrical resistivity or conductivity.

The primary element in an electrical conductivity system is the *conductivity cell* (see Figure 8-1). Such cells consist of a pair of electrodes whose areas and spacing are precisely fixed, a suitable insulating member to confine the conductive paths, and suitable fittings for supporting and protecting the cell. The conductance between two electrodes varies as follows:

$$C \propto \frac{A}{L} \tag{8-1}$$

where

C	=	conductance (mho)
A	=	the area of electrodes (cm^2)
L	=	the distance between electrodes (cm)

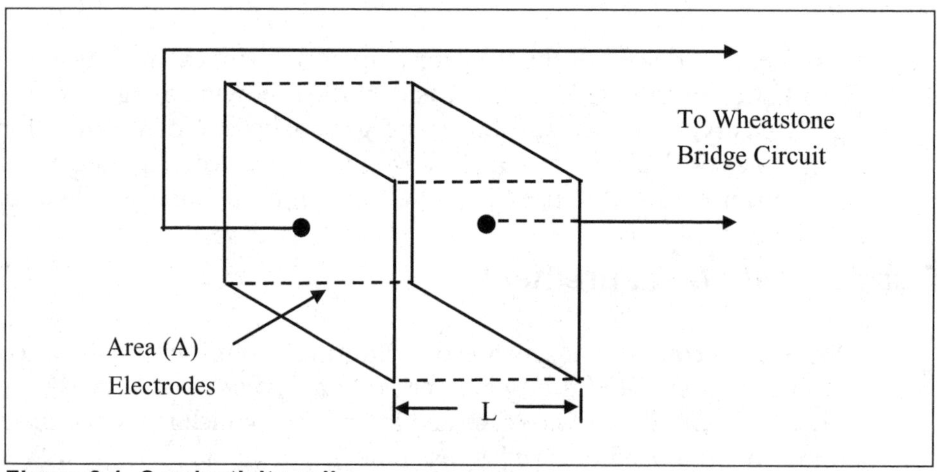

Figure 8-1. Conductivity cell

To establish a common basis for comparing conductivities of different solutions, a standard conductivity cell is considered for which the following is true:

$$A = 1\ \text{cm}^2$$

$$L = 1\ \text{cm}$$

This volume of solution is completely insulated from its surroundings. Unless this insulated condition is met, no simple relationship exists between conductance and A/L. This is because there are additional conducting paths available between electrodes. This conductance of the solution in a unit cell is called its *specific conductivity* k (sometimes called *conductance*) and has units of mhos per centimeter (mho/cm). In common practice, the specific conductance in mhos per centimeter is referred to as the solution's "conductivity" and is given in units of mhos only. For example, an electrolyte with a specific conductance of 0.03 mho/cm is said to have a conductivity of 0.03 mho. The two terms are synonymous as they are commonly used.

Conductivity instruments normally use some variation of an AC Wheatstone bridge circuit to measure solution conductivity. Alternating current is used to eliminate electrode polarization effects. Using platinum electrodes can further reduce these effects. Conductivity analyzers generally use an AC frequency of 50 to 600 Hz, and the normal range of resistance measured is from 100 to 60,000 Ohms.

The instrument can be calibrated for either resistance or conductance because conductance is the reciprocal of resistance,

$$C = \frac{1}{R} \tag{8-2}$$

where

C	=	conductance in mhos
R	=	resistance in ohms

Example 8-1 shows how to calculate conductance based on the resistance of a typical solution.

EXAMPLE 8-1

Problem: A solution has a measured resistance of 50,000 ohms. Calculate its conductance.

Solution: Since conductance is the reciprocal of resistance, we have

$$C = \frac{1}{R}$$

$$C = \frac{1}{50K\Omega} = 20\mu mhos$$

Hydrogen-Ion Concentration (pH) Measurement

The symbol pH represents the acidity or alkalinity of a solution. It is a measure of a key ingredient of aqueous solutions of all acids and bases: the hydrogen-ion concentration.

Earlier techniques for measuring the hydrogen-ion concentration used paper indicators. When the indicator was added to the sample, it would produce a certain color change in relation to the value of the pH. The result could then be compared with a standard to evaluate the hydrogen-ion concentration.

Such a method does not lend itself well to automatic measurement, nor can it be used with liquids that are normally colored. For these reasons, a method of measurement was developed that was based on the potential created by a set of special electrodes in the solution. This method has become the industrial standard used to measure pH and to control processes involving pH. However, to understand this method you must know what pH is and be familiar with the fundamentals of a solution's properties.

Basic Theory of pH

Stable chemical compounds are electrically neutral. When they are mixed with water, many of them break up into two or more charged particles. The charged particles formed in the dissociation are called *ions*. The amount by which a compound will dissociate varies from one compound to another as well as with the temperature of the solution. At a specified temperature, a fixed relationship exists between the concentration of the charged particles and the neutral undissociated compound. This relationship is called the *dissociation constant* or *ionization constant*:

$$K = \frac{[M+][A-]}{[MA]} \tag{8-3}$$

where

K = the dissociation constant
$[M+]$ = the concentration of positive ions
$[A-]$ = the concentration of negative ions
$[MA]$ = the concentration of undissociated ions

An acid such as hydrochloric (HCl) breaks up completely into positively charged hydrogen ions and negatively charged chloride ions ($HCl \rightarrow H^+ + Cl^-$). For such acids, the dissociation constant is practically infinity, hence, they are called "strong" acids. On the other hand, an acid such as acetic ($HC_2H_3O_2$ or C H_3COOH) breaks up as follows: ($CH_3COOH \leftrightarrow H^+ + CH_3COO^-$). For this acid, very few hydrogen ions show up in the solution—less than one in every one hundred undissociated molecules. It therefore has a low dissociation constant and is called a "weak" acid. Thus, the strength of an acid solution depends on the number of hydrogen ions available. That number depends, in turn, on the weight of the compound and, in water, the dissociation constant of the particular compound.

When the free hydroxyl ions (OH^-) predominate, the solution is alkaline or basic. For example, sodium hydroxide (NaOH) in water completely dissociates as follows: ($NaOH \rightarrow Na^+ + OH^-$). It is therefore a strong base. On the other hand, ammonium hydroxide (NH_4OH) weakly dissociates into $NH^+{}_4$ ions and OH^- ions and thus is a weak base.

Pure water dissociates into H^+ and OH^- ions, as follows: ($HOH \rightarrow H^+ + OH^-$). However, it is considered extremely "weak" because very little of the HOH breaks up into H^+ and OH^- ions. So few water molecules dissociate vis-à-vis those that are undissociated that the value of (HOH) can be considered equal to one or 100 percent. The ionization constant of water has been determined to have a value of 10^{-14} at 25°C. The product of the activities $(H^+)(OH^-)$ is then 10^{-14}.

If the concentrations of hydrogen ions and hydroxyl ions are the same, they must be 10^{+7} and 10^{-7}, respectively. Regardless of what other compounds are dissolved in the water, the product of the concentrations of H^+ ions and OH^- ions is always 10^{-14}. Therefore, if you add a strong acid to water, many hydrogen ions are added and will reduce the hydroxyl ions accordingly. For example, if you add HCl at 25°C until the H^+ concentration becomes 10^{-2}, the OH^- concentration must become 10^{-12}.

It is awkward to work in terms of small fractional concentrations such as $1/10^7, 1/10^{12}, 1/10^2$. For that reason, in 1909 Sorenson proposed that, for convenience, the expression "pH" be adopted for hydrogen-ion concentration to represent degree of acidity or activity of hydrogen ions. This term was derived from the phrase *the power of hydrogen*. Sorenson defined the pH of a solution as the negative of the logarithm of the hydrogen-ion concentration, or

$$pH = -\log_{10} [H^+] \tag{8-4}$$

He also defined it as the base 10 log of the reciprocal of the hydrogen-ion concentration:

$$pH = \log 1/[H^+]$$

This means that if the hydrogen-ion concentration is $1/10^x$, the pH is said to be x. In pure water, where the concentration of the hydrogen ion is $1/10^7$, the pH is therefore 7.

If we know the OH ion concentration but not the H^+ ion concentration of an aqueous solution, we can still calculate the pH by using the following relationship:

$$[OH^-][H^+] = 10^{-14} \tag{8-5}$$

For example, if $[OH^-] = 10^{-2}$, then $[H^+] = 10^{-14} / 10^{-2} = 10^{-12}$, and the pH = –(–12) = 12. The acidity or basicity of a solution can be expressed either in terms of its H^+ ion concentration, its OH^- ion concentration, or its pH, as shown in Table 8-1.

Table 8-1. Acidity or Basicity of Aqueous Solutions

Solution	$[H^+]$	$[OH^-]$	pH	Nature of Solution
0.01 *M* HCl	10^{-2}	10^{-12}	2	Acidic
10^{-5} *M* HCl	10^{-5}	10^{-9}	5	Acidic
H_2O (pure water)	10^{-7}	10^{-7}	7	Neutral
10^{-5} *M* NaOH	10^{-9}	10^{-5}	9	Basic
0.01 *M* NaOH	10^{-12}	10^{-2}	12	Basic

Example 8-2 shows a typical calculation to obtain hydrogen ion concentration and pH values based on a given OH ion concentration.

EXAMPLE 8-2

Problem: The $[OH^-]$ ion concentration of an aqueous solution is 10^{-8}. What are the values of the H^+ ion concentration and pH?

Solution: First, use Equation 8-5 — $[OH^-][H^+] = 10^{-14}$ — to determine $[H^+]$. Since $[OH^-] = 10^{-8}$, we obtain $[H^+] = 10^{-14}/10^{-8} = 10^{-6}$. Then, we determine the pH value using Equation 8-4:

$$pH = -\log_{10} 10^{-6} = -(-6) = 6$$

pH Electrode Systems

Industrial electrode systems for pH determinations consist of two separate electrodes. The first is the active or measuring electrode, which produces a voltage that is proportional to the hydrogen-ion concentration. The second is a reference electrode, which serves as a source of constant voltage against which the output of the measuring electrode is compared.

A number of measuring electrodes have been developed for pH applications, but the glass electrode has evolved as the universal standard for industrial process purposes. Typically, it consists of an envelope of special glass that is designed to be sensitive only to hydrogen ions. It contains a neutral solution of constant pH (called *buffer solution*) and a conductor, immersed in the internal solution, which makes contact with the electrode lead.

The electrode operates on the principle that a potential is observed between two solutions of different hydrogen-ion concentration when a thin glass wall separates them. The solution within the electrode has a constant concentration of hydrogen ions. Whenever the hydrogen-ion concentration of the solution being measured differs from that of the neutral solution within the electrode, a potential difference (or voltage) is developed across the electrode. If the solution being measured has a pH of 7.0, the potential difference is 0. When the pH of the measured solution is greater than 7.0, a positive potential exists across the glass tip. When the pH is less than 7.0, a negative potential exists.

The glass electrode responds predictably within the 0 to 14 pH range. It develops 59.2 mV per pH unit at 25°C, which is consistent with the Nernst equation:

$$E = \frac{RT}{F} \ln \frac{[H^+]outside}{[H^+]inside} \tag{8-6}$$

where

E	=	potential (v)
R	=	the gas law constant
F	=	Faraday's number, a constant
T	=	temperature (K)
$[H^+]$	=	the hydrogen-ion concentration

Many types of glass electrodes are available, and which one you choose to use will usually depend on the temperature range and physical characteristics of your process. The reference electrode is used to complete the circuit so the potential across the glass electrode can be measured.

Because of the temperature coefficient of the glass electrode, you must compensate the system for temperature if it is to continue to read pH correctly. This is done manually at a point where the temperatures do not vary widely. Otherwise, it is done automatically by means of a temperature element, which is located in the vicinity of the electrodes and connected into the circuit. Therefore, the industrial pH assembly can consist of as many as three units: the glass measurement electrode, the reference electrode, and a temperature element mounted in holders of differing designs.

pH Measurement Applications

Applications for pH measurement and control can be found in waste-treatment facilities, pulp and paper plants, petroleum refineries, power generation plants, and across the chemical industry. In other words, continuous pH analyzers can be found in almost every industry that uses water in its processes.

Figure 8-2 shows an example of pH control—a P&ID of the manufacturing of disodium phosphate using flow and pH control. The automatic control system shown in the figure produces a high-purity salt and prevents unnecessary waste of both reagents (i.e., soda ash and phosphoric acid).

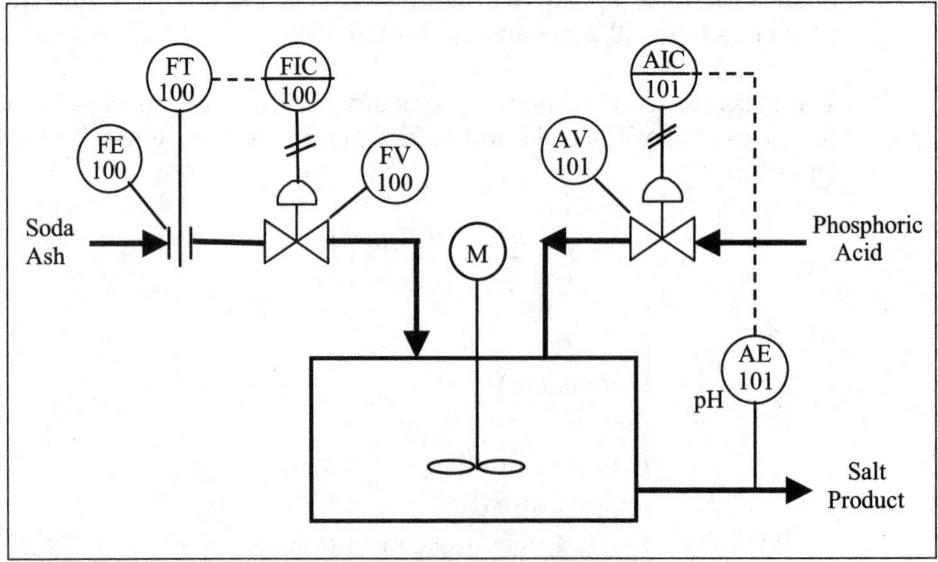

Figure 8-2. P&ID for production of disodium phosphate

Density and Specific Gravity Measurement

Control of the more common variables, such as flow, temperature, and pressure, is the basic criterion for process control. However, there are cases where measuring density or specific gravity (SG) is the best way to determine and control the concentration of a process solution. For a fluid, density is defined as the mass per unit volume and is usually expressed in units of grams per cubic centimeter (g/cm^3) or pounds per cubic foot (lb/ft^3). The specific gravity of a fluid is the ratio of the density of the fluid to the density of water at 60°F (15.5°C).

Hydrometer

The simple hand hydrometer consists of a weighted float that has a small-diameter stem proportioned such that more or less of the scale is submerged according to the specific gravity. These hydrometers are widely used for "spot" or off-line intermittent specific gravity measurements of process liquids. A typical use for a hydrometer is checking the state of discharge for a lead-acid cell by measuring the specific gravity of the electrolyte. Concentrated sulfuric acid is 1.835 times as heavy as water (the reference with a specific gravity of 1) for the same volume. Therefore, the specific gravity of the acid equals 1.835.

In a fully charged automotive cell, the mixture of sulfuric acid and water results in a specific gravity of 1.280 at a room temperature of 70°F. As the cell discharges, more water is formed, lowering the specific gravity. When the gravity falls to about 1.150, the cell is completely discharged. Specific gravity readings are taken with a battery hydrometer, as shown in Figure 8-3. Note that the calibrated float that has the SG marks will rest higher in an electrolyte of higher specific gravity.

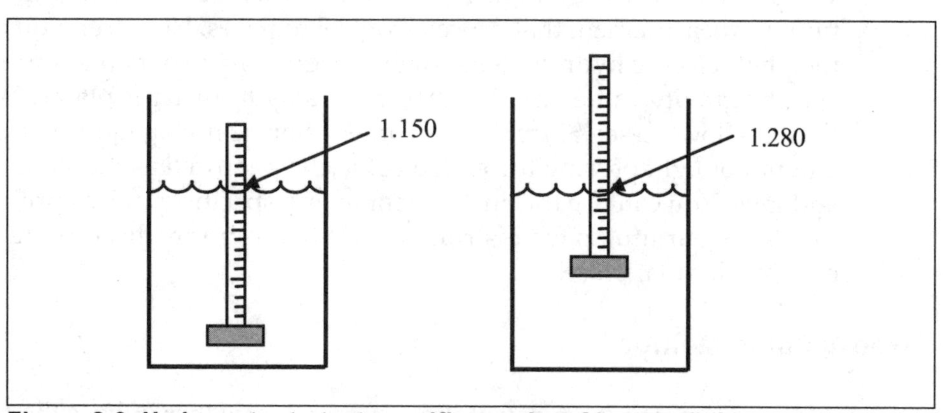

Figure 8-3. Hydrometer to test specific gravity of battery fluid

The importance of specific gravity is evidenced by the fact that the open-circuit voltage of the lead-acid cell is given approximately by the following equation:

$$V = (SG + 0.84) \text{ volts} \tag{8-7}$$

For $SG = 1.280$, the voltage is $1.280 + 0.84 = 2.12$ V. This value is for a fully charged single-cell battery.

Example 8-3 shows how to calculate the no-load voltage of a battery.

EXAMPLE 8-3

Problem: The specific gravity of a six-cell lead-acid battery is measured as 1.22. Calculate the no-load voltage of the battery.

Solution: Use Equation 8-7 to calculate the voltage of a single cell.

$$V = SG + 0.84 \text{ (volts)}$$

$$V = 1.22 + 0.84\text{V}$$

$$V = 2.06\text{V}$$

Since the battery has six cells, the no-load voltage of the battery is as follows:

$$\text{V} = 6 \times (2.06)\text{V}$$

$$\text{V} = 12.36\text{V}$$

The problem with the simple hydrometer is that it cannot perform the continuous measurement that process control requires. However, you can use the photoelectric hydrometer shown in Figure 8-4 to obtain a continuous specific gravity value. In this instrument, a hydrometer is placed in a continuous-flow vessel. Since the instrument stem is not opaque, it affects the amount of light passing through a slit to the photocell as the stem rises and falls. You can use this instrument in any specific gravity application that is not harmful to the instrument and that is normally accurate to two or three decimal places.

Fixed-Volume Method

A common continuous density-measuring device that utilizes the fixed-volume density principle is the so-called displacement meter, which is schematically illustrated in Figure 8-5. In this device, liquid flows continuously through the displacer chamber, with the buoyant body, or displacer, completely immersed. The buoyant force is exerted on the displacer. It is dependent on the weight of the displaced liquid and, in turn, is a function of the volume and specific gravity.

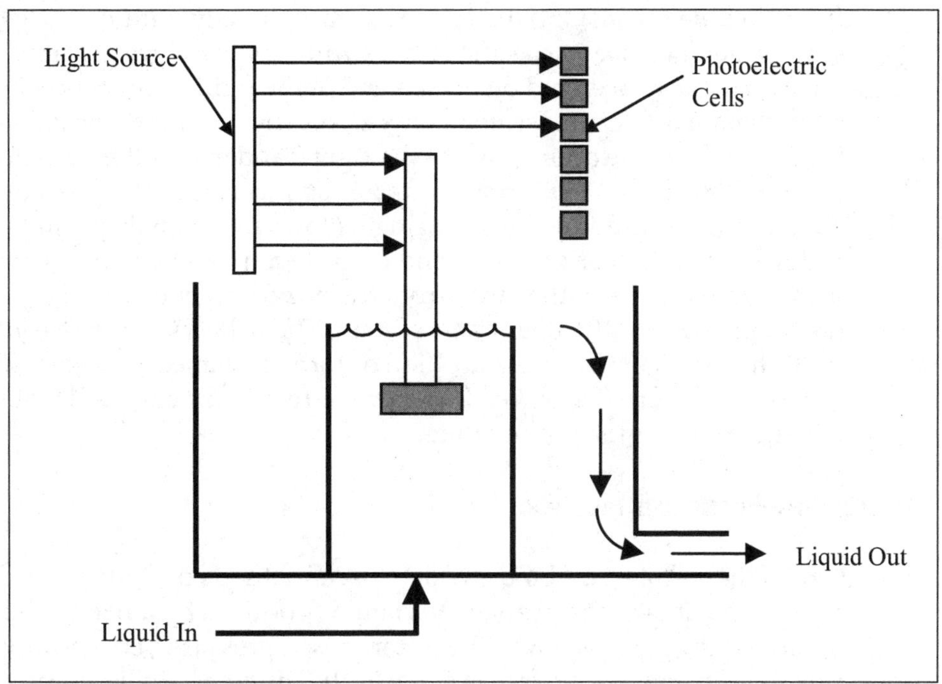

Figure 8-4. Photoelectric hydrometer

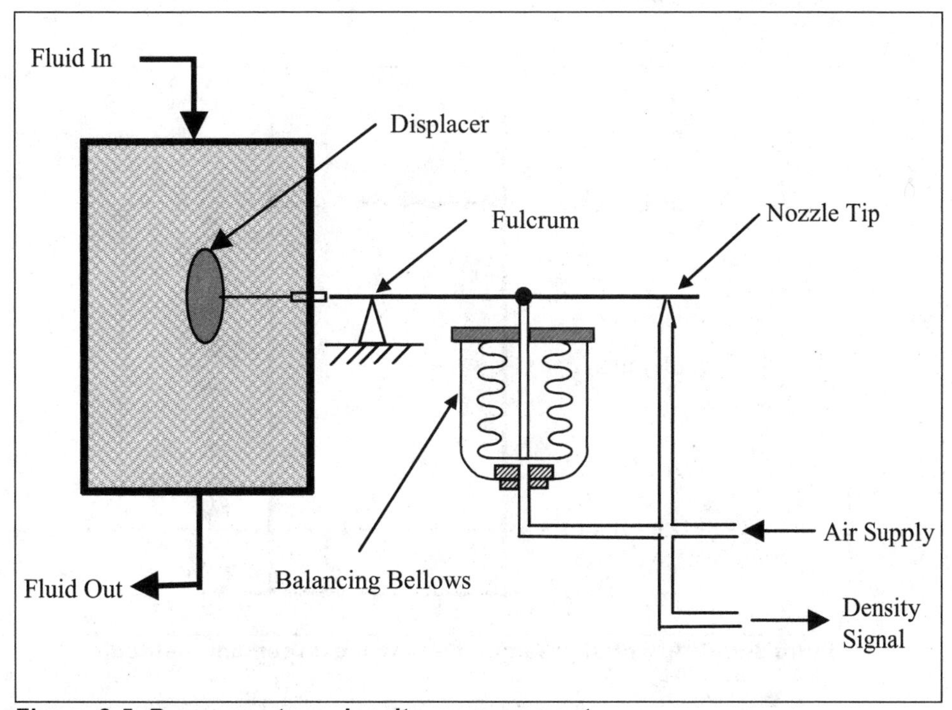

Figure 8-5. Buoyancy-type density measurement

If the volume is constant, the force will vary directly with the specific gravity. An increase in specific gravity will produce a greater upward force on the displacer and on the left end of the rigid beam. Since the beam moves about a fulcrum located between the displacer and the balancing bellows, this upward force causes the right-hand end of the beam to move closer to the nozzle. This, in turn, creates an increase in the pressure at the nozzle that is sensed by the balancing bellows, which will expand. As the bellows expand, the right-hand end of the beam moves away from the nozzle, moving the baffle away from the nozzle tip and causing a reduction in pressure in the pneumatic system. The bellows will move just enough to reestablish a new position or torque balance with somewhat different pressure. This is read on a pressure instrument that is calibrated in density or specific gravity units.

Differential-Pressure Method

One of the simplest and most widely used methods of continuous density measurement uses the pressure variation produced by a fixed height of liquid. As Figure 8-6 shows, the difference in pressure between any two elevations below the surface is equal to the difference in liquid pressure (head) between these elevations, regardless of the variation in level above the higher elevation.

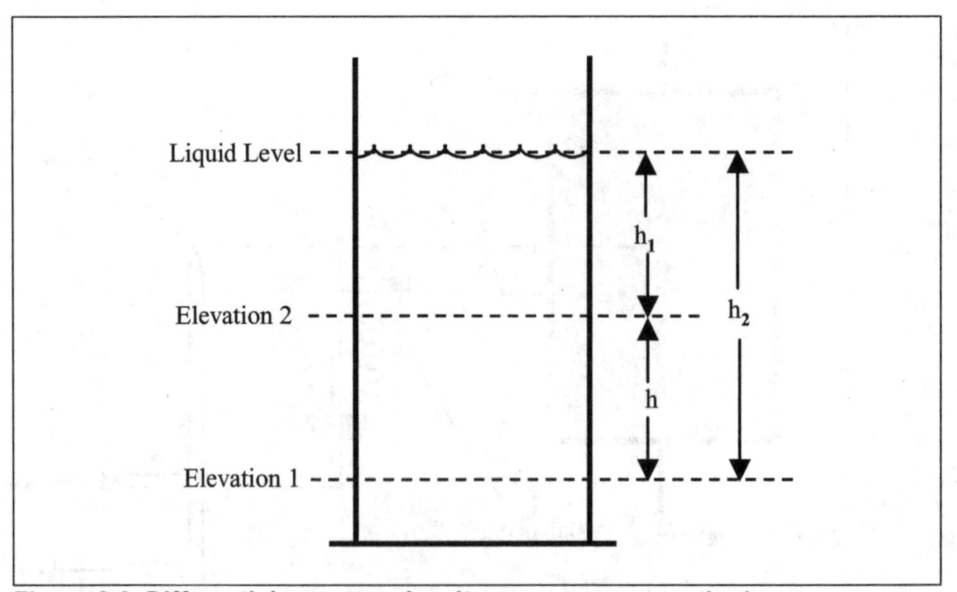

Figure 8-6. Differential pressure density measurement method

This difference in elevations is represented by dimension *h*. This dimension must be multiplied by the specific gravity of the liquid to obtain the difference in head in inches of water, which is the standard unit for measurement calibration. To measure the change in head that results from a change in specific gravity from minimum (SG_{min}) to maximum (SG_{max}), you must calculate the difference between hSG_{min} and hSG_{max}:

$$dP = h\,(SG_{max} - SG_{min}) \tag{8-8}$$

where

dP	=	differential pressure (inches of water)
h	=	the difference in elevations (inches)
SG_{min}	=	minimum specific gravity
SG_{max}	=	maximum specific gravity

It is common practice to measure only the span of actual density changes. This is done by elevating the instrument "zero" to the minimum pressure head to be encountered, allowing the entire instrument working range to be devoted to the differential caused by density changes. For example, if $SG_{min} = 1.0$ and $h = 100$ in., the range of the measuring instrument must be elevated $h \times SG_{min}$, or 100 in. of water. For $SG_{min} = 0.9$ and $h = 100$ in., the elevation would be 90 in. of water. Thus, the two principal relationships to be considered in this type of measuring device are as follows:

$$\text{Span} = h\,(SG_{max} - SG_{min}) \tag{8-9}$$

$$\text{Elevation} = h \times SG_{min} \tag{8-10}$$

The calculation of the span for a typical differential-pressure type density instrument is illustrated in Example 8-4.

EXAMPLE 8-4

Problem: Calculate the span in inches of a differential-pressure type density instrument if the minimum specific gravity is 0.90, the maximum specific gravity is 1.10, and the difference in liquid elevation is 20 inches.

Solution: Using Equation 8-9, we obtain the following:

$$\text{Span} = h\,(SG_{max} - SG_{min})$$

$$\text{Span} = 20 \text{ inches}(1.10 - 0.90)$$

$$\text{Span} = 4 \text{ inches}$$

Figure 8-7 shows an instrument that uses a differential-pressure method to determine density. This application consists of a process tank and two bubbler tubes that are installed in the fluid of the tank, so that the end of

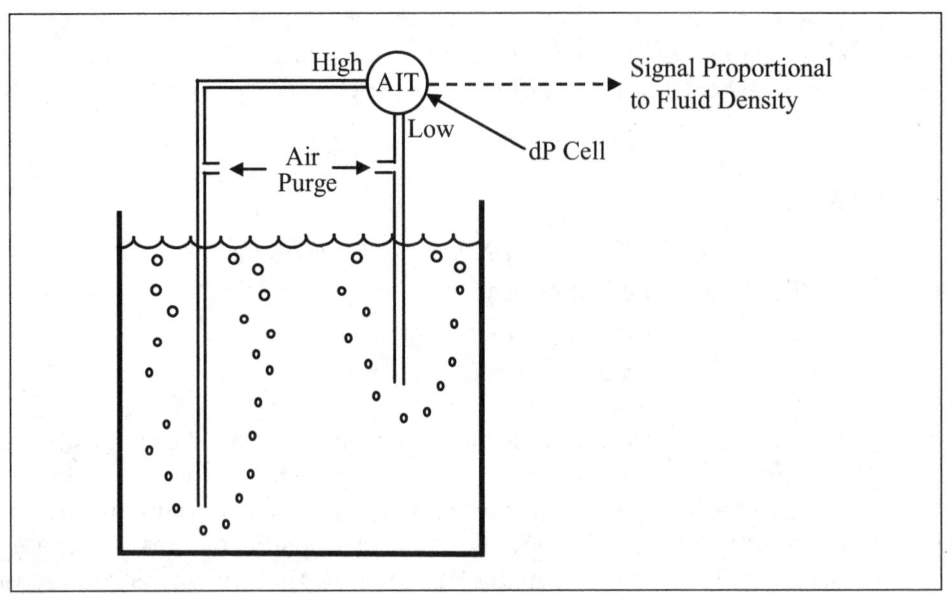

Figure 8-7. Air purge dP density measurement method

one tube is lower than the end of the other. The pressure required to bubble air into the fluid is equal to the pressure of the fluid at the ends of the bubble tubes.

Since the outlet of one tube is lower than that of the other, the difference in pressure will be the same as the weight of a constant-height column of liquid. Therefore, the differential-pressure measurement is equivalent to the weight of a constant volume of the liquid and can be represented directly as density.

Nuclear Method

You can also use nuclear devices to measure density. They operate on the principle that the absorption of gamma radiation increases with the density of the material being measured. A representative radiation-type density-measuring element is depicted in Figure 8-8. It consists of a constant gamma-ray radiation source (which can be radium, cesium, or cobalt), which is mounted on a wall of the pipe, and a radiation detector, which is mounted on the opposite side. Gamma rays are emitted from the source, through the pipe, and into the detector. Materials flowing through the pipeline and between the source and detector absorb radioactive energy in proportion to their densities. The radiation detector measures the radioactive energy that is not absorbed by the process material. The amount that is measured varies inversely with the density of the process stream. The radiation detector unit converts this energy into an electrical signal, which is transmitted to an electronics module.

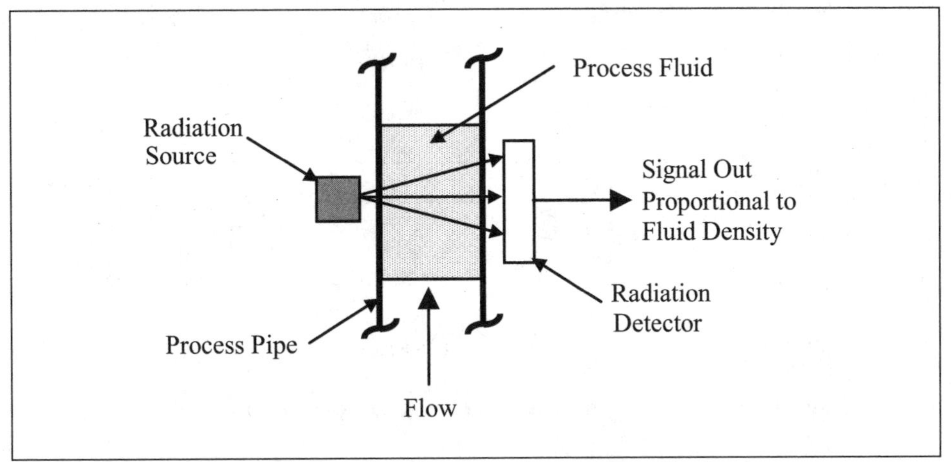

Figure 8-8. Nuclear radiation density measurement method

Vibration Method

The most widely used method for determining process fluid density is based on the fact that the natural oscillating frequency of a fluid changes with changes in density. The common transducers based on this principle are (1) the vibrating U-tube transducer, (2) the vibrating cylinder transducer, (3) the vibrating vane transducer, (4) the vibrating single-tube transducer, and (5) the vibrating twin-tube transducer.

In the *vibrating U-tube* transducer, the amplitude of the vibration changes when the density of the process fluid in the fixed-volume tube changes. You can measure this change with the proper mechanical and electronic equipment calibrated to read in density units.

The *vibrating cylinder* transducer consists of a thin-walled cylinder located concentrically inside the sensor housing. The process fluid flows through the thin cylinder, and the entire mass is electrically excited into a resonance frequency. This resonance frequency changes as the mass of the fluid changes. The instrument can be used for both gases and liquids. Its primary advantage is that it can measure the density of both gas and gas-liquid combinations.

The *vibrating vane* transducer oscillates at its natural frequency when inserted into the process fluid. When the density of the fluid changes, additional energy is required to maintain the natural frequency of the sensor. More energy is required if the density of the fluid increases and less if the density decreases.

The *vibrating single-tube* transducer is shown in Figure 8-9. This sensor has two cantilevered masses, which are quite similar to end-to-end tuning

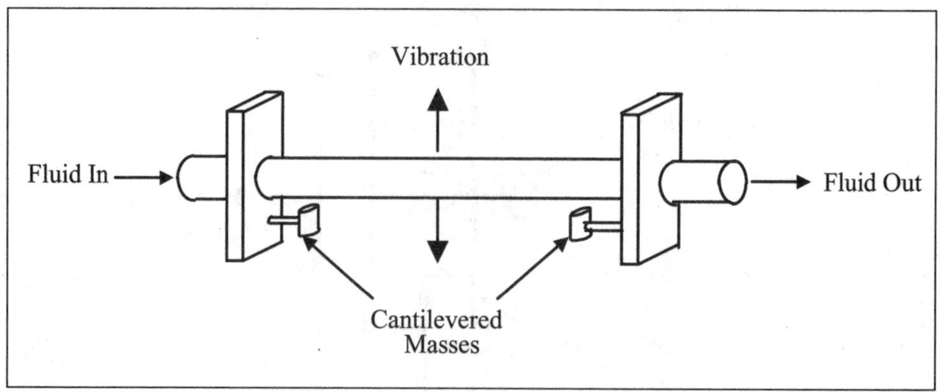

Figure 8-9. Vibrating single-tube density sensor

forks. The active part of the device consists of a flow-through tube for the process fluid, where the cantilevered masses are mounted on the same structure under the flow tube. This structure is driven into oscillation at a frequency that is determined by the combined mass of the device and the process fluid at any given instant. A change in the density of the flowing fluid will change the frequency of the device, and this change is an indirect measure of the density or specific gravity of the fluid.

The *vibrating twin-tube* transducer has two parallel flow tubes. The process fluid flows through the tubes and converges into a single flow at each end. This device is similar to a tuning fork since the tubes are driven at their natural frequency. This oscillation varies inversely with changes in fluid density. Two drive coils and their associated electronic circuits keep the tubes in resonance.

Humidity Measurement

Humidity is an expression of the amount of moisture in a gas or gases, whether isolated or as part of the atmosphere. There are two basic quantitative measures of humidity: absolute and relative. *Absolute humidity* is the amount of water vapor present in each cubic foot or other unit volume, and is expressed in various units such as dew point, grains of water per pound of air, or pounds of water per million standard cubic feet. *Relative humidity* describes the ability of air to moisten or dry materials. It compares the actual amount of water vapor that is present with the maximum amount of water vapor the air could hold at that temperature. For example, air that is considered saturated at 50°F (100% relative humidity) would be considered quite dry if heated to 100°F (19% relative humidity). The graph in Figure 8-10 shows the maximum amount of moisture that can be held by air at various temperatures.

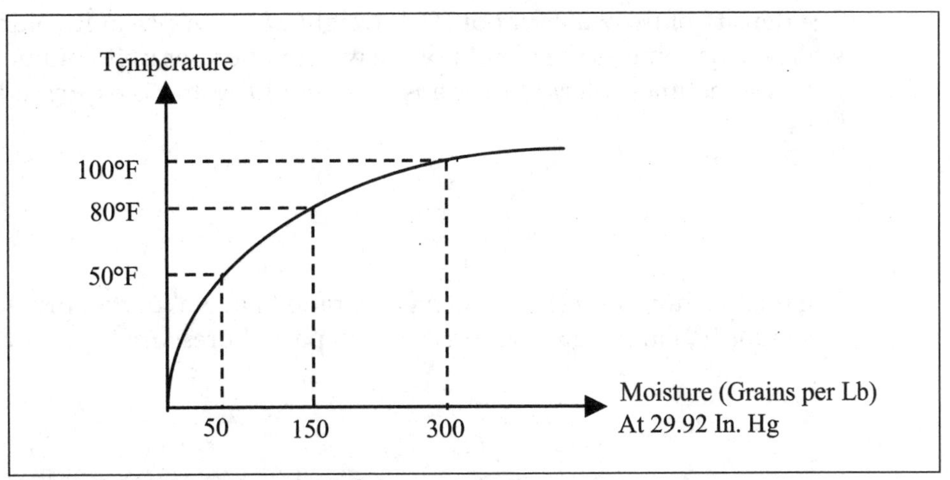

Figure 8-10. Maximum air moisture content

You can calculate humidity by assuming that the atmosphere to be measured is an ideal gas. The ideal gas law states that the partial pressure of one of the constituents of a mixture is the pressure that would exist if only that constituent occupied the volume of the mixture at the same temperature. This is expressed mathematically by the following:

$$W_v = \frac{P_v V}{R_v T} \tag{8-11}$$

and

$$W_g = \frac{P_g V}{R_g T} \tag{8-12}$$

where

W	=	the weight of the constituent (v = water vapor and g = gas)
P	=	the absolute partial pressure of the constituent
V	=	volume
T	=	absolute temperature
R	=	a gas constant for a given gas

The absolute humidity (H_a) (weight of water vapor per unit weight of dry gas) is therefore as follows:

$$H_a = \frac{W_v}{W_g} = \frac{R_g P_v}{R_v P_g} \tag{8-13}$$

The density of the water vapor (D_v) (weight of vapor per unit volume, which is the same as the weight of the water vapor per unit volume of dry gas) is sometimes referred to as absolute humidity and is expressed as follows:

$$D_v = \frac{P_v}{R_v T} \tag{8-14}$$

The relative humidity (H_t), which is the ratio of actual partial pressure of the vapor (P_v) in the gas to the saturation partial pressure (P_{sat}), is then

$$H_v = \frac{P_v}{P_{sat}} \cong \frac{D_v}{P_{sat}} \tag{8-15}$$

You can find the value of saturation pressure in the steam tables of engineering handbooks.

Dew point is defined as the temperature at which the air or a gas becomes saturated. If the mixture is cooled at constant pressure to the dew point, the vapor will begin condensing.

Dry-bulb and wet-bulb temperatures are also used to determine the humidity of a gas or air mixture. You measure the dry-bulb temperature of a gas mixture by using an ordinary thermal measuring element. The wet-bulb temperature is measured by a thermal element covered by a wick and fully wetted by water vapor. The difference between dry-bulb temperatures and wet-bulb temperatures is sometimes called the *wet-bulb depression*. This is caused by the cooling effect on the bulb that results when the water on the wick evaporates.

Humidity affects materials in diverse ways, so you can measure the water present, the dew point, and other variables with a wide variety of instruments that employ quite different methods and principles. Humidity measurements are considered to be inferred because they depend on differences between two thermometers, the expansion or contraction of different materials, the temperature at which the water vapor will condense, or the temperature at which certain salt solutions are in equilibrium.

Relative Humidity Measurement

A well-established empirical method for measuring relative humidity is based on psychrometry. It involves the reading of two thermometers, one bulb directly exposed to the atmosphere and the other covered by a continuously wet wick. Actually, the second bulb measures the thermodynamic equilibrium temperature reached between the cooling that is effected by evaporation of water and the heating effected by convection. The device used for making this measurement is called a *psychrometer*.

Wet- and dry-bulb psychrometers for the continuous industrial measurement of humidity come in a wide variety of configurations. A typical system is shown schematically in Figure 8-11. It consists of a temperature recorder connected to two temperature bulbs, "wet" and "dry." You locate the dry bulb in the open and a wick or a porous, ceramic-covered wet bulb in the moving airstreams. A reservoir of supply air ensures that the wick has the proper wetness. You can convert the wet-bulb/dry-bulb temperature readings to suitable relative humidity values by using appropriate scales or charts.

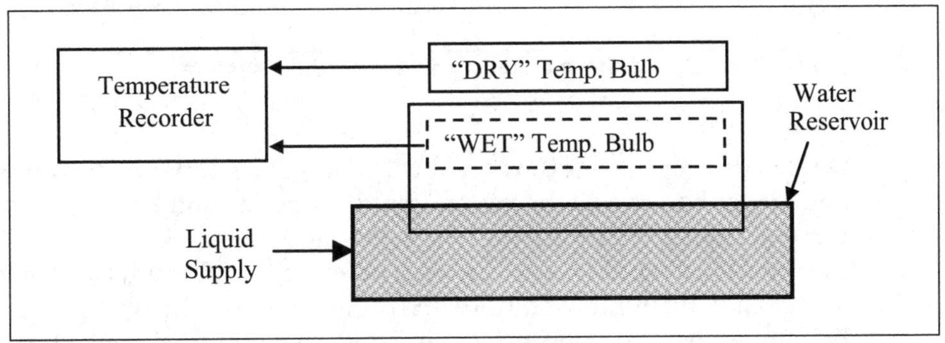

Figure 8-11. Wet and dry recording psychrometer

Dew-Point Measurement

As we noted earlier, dew point is the temperature at which a mixture of air and water vapor is saturated. The classical method for determining dew point is to slowly cool a polished surface until condensation takes place. The temperature of the surface when the first droplet appears is considered the dew point. You can use this method to determine the absolute humidity or partial pressure of the vapor.

One widely used approach for continuously measuring dew point is based on the temperature of vapor equilibrium. You measure the temperature at which a saturated solution of a hygroscopic salt (lithium chloride) achieves vapor equilibrium with the atmosphere. You use electric heating to reach the temperature of the salt solution, since it is much higher than the temperature of pure water.

In terms of the structure of the dew-measuring system (Figure 8-12), a tube containing a temperature-measuring element is wrapped with a glass fiber that has been wetted with a saturated lithium chloride salt solution.

Two conductors are wrapped around the assembly in contact with the wick and supplied with low-voltage (25 V) alternating current. Current flow through the salt solution generates heat, which raises the temperature. When the temperature of vapor equilibrium is reached, water evapo-

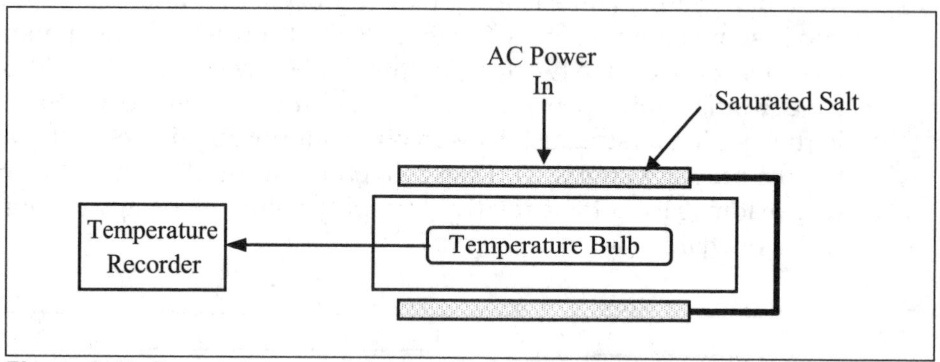

Figure 8-12. Dew point measuring and recording system

rates, reducing current flow and heat input. The temperature cannot go any higher because all the water would evaporate and heat input would cease. It cannot fall because all the salt would then go into solution, and too much heat would be generated. Therefore, equilibrium is reached with a portion of the lithium chloride in solution and conductive, and the remainder dry and nonconductive. Thus, heat input is balanced with heat loss. The thermometer bulb, when placed inside the metal tube, will measure temperature or dew point. This is also a measure of absolute humidity and can be expressed in grains of moisture per pound of dry air, percentage of water vapor by volume, and other units.

Since there is no provision for cooling with this method, it can be used only for conditions where the equilibrium temperature of lithium chloride is above ambient temperature. This corresponds to a minimum of approximately 12 percent to 15 percent relative humidity over ordinary temperature ranges. The method can be used up to saturation, or 100 percent relative humidity. Ambient temperatures may vary from 200°F to –30°F. You must measure dew points at higher temperatures on a cooled sample.

Principles of Electromagnetic Radiation

To study the analytical and optical instruments used in the process industry, you must have an understanding of electromagnetic (EM) radiation. In the following section we discuss EM radiation. We then look at several common transducers that use EM radiation to measure analytical variables.

Electromagnetic Spectrum

EM radiation is a form of energy in classical physics that consists of electrical and magnetic waves. As illustrated in Figure 8-13, *light* is the visible portion of the electromagnetic spectrum. Light is defined as the EM radiation that can affect the eye. All these waves are electromagnetic in nature and have the same speed, *c*, in free space. They differ in wavelength, and thus frequency only, which means that the sources that give rise to them differ from the instruments that are used to make measurements with them.

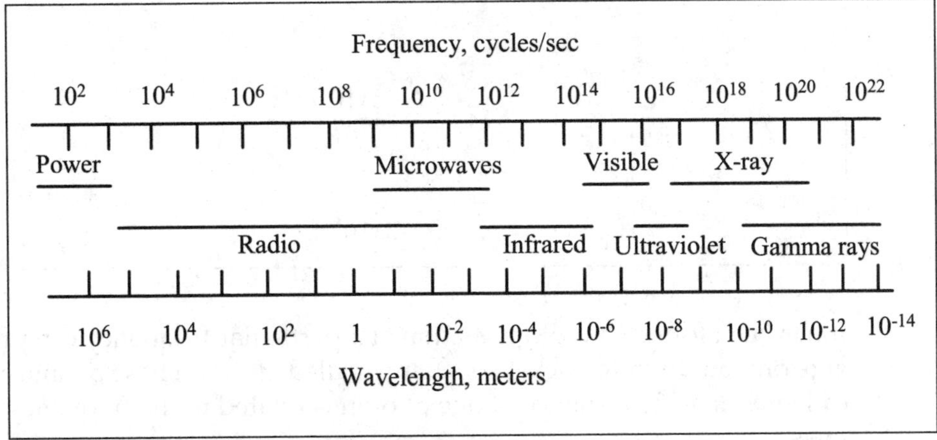

Figure 8-13. Electromagnetic spectrum

The electromagnetic spectrum has no definite upper or lower limit. The regions labeled in Figure 8-13 represent frequency intervals within which there exist a common body of experimental technique. All regions in the electromagnetic spectrum overlap. For example, we can produce EM radiation of wavelength 10^{-3} meter either by using microwave techniques that employ electronic oscillators or by infrared techniques, which use incandescent sources.

It has been found that the wavelength and the frequency of the EM radiation that propagate through a vacuum are related by the following equation:

$$c = \lambda f \tag{8-16}$$

where

c	=	the speed of EM radiation in a vacuum or 3×10^8 m/s
λ	=	wavelength in meters
f	=	the frequency in cycles per second or hertz (Hz)

Example 8-5 illustrates how to determine the wavelength of EM radiation for a given frequency.

EXAMPLE 8-5

Problem: Given an EM radiation that has a frequency of 10^8 Hz, find its wavelength.

Solution: Use Equation 8-16 to calculate the wavelength:

$$c = \lambda f$$

or

$$\lambda = \frac{c}{f} = \frac{3x10^8 \, m/s}{10^3 s^{-1}}$$

$$\lambda = 3x10^5 \, m$$

It has been found that EM radiation at a particular frequency can propagate only in discrete packets of energy called *quanta*. These quanta are called *photons*. The energy of one photon is related to the frequency as follows:

$$E_p = hf = \frac{hc}{\lambda} \tag{8-17}$$

where

E_p	=	the photon energy in joules
h	=	Planck's constant with a value of 6.63 x 10^{-34} joules
f	=	frequency in Hz
λ	=	the wavelength in meters

The energy of one photon is very small compared to the electrical energy we normally experience. This can be illustrated by Example 8-6.

Since electromagnetic radiation is energy in motion, the power of a light source is given by joules per second or watts. The intensity of a light source is given by the following:

$$I = \frac{P}{A} \tag{8-18}$$

EXAMPLE 8-6

Problem: A microwave source emits a pulse of EM radiation at 2×10^9 Hz with energy (E) of 1 joule. Find the photon energy and calculate the number of photons in the pulse.

Solution: First, use Equation 8-17 to calculate photon energy (E_p) as follows:

$$E_p = hf$$

$$E_p = (6.63 \times 10^{-34} \text{ joule-s})(2 \times 10^9 \text{ s}^{-1})$$

$$E_p = 1.326 \times 10^{-24} \text{ joule}$$

Next, we find the number of photons as follows:

$$N = E/E_p = 1 \text{ joule}/(1.326 \times 10^{-24} \text{ joule/photon})$$

$$N = 7.54 \times 10^{23} \text{ photons}$$

where

I = the intensity in watts/m^2
P = the source power in watts
A = the light beam's cross-sectional area in m^2

EM radiation travels in straight lines, so the intensity of light decreases even though the source power remains constant. If the light source is a very small point and the EM radiation propagates in all directions, we have a condition of maximum divergence. These point sources have an intensity that decreases as the inverse square of the distance from the point. You can find the intensity at a distance r by dividing the total power by the surface area of a sphere of radius r from the source. Since the surface area of a sphere of radius r is given by the equation, $A = 4\pi r^2$, the intensity over the entire surface surrounding a point source is defined as follows:

$$I = \frac{P}{A} = \frac{P}{4\pi r^2} \tag{8-19}$$

Equation 8-19 shows that the light intensity of a point source decreases as the inverse square of the distance from the source.

Example 8-7 illustrates how light intensity varies with the distance from a source.

EXAMPLE 8-7

Problem: Find the intensity of a 100-W point source at 1 meter and 2 meters from the source.

Solution: The intensity at 1 meter from the source is as follows:

$$I = \frac{P}{4\pi r^2}$$

$$I = \frac{100W}{4\pi(1m)^2} = 7.96W/m^2$$

The intensity at 2 meters from the source is

$$I = \frac{100W}{4\pi(2m)^2} = 1.98W/m^2$$

Photodetectors

How to measure or detect radiation is an important aspect of the use of electromagnetic sources in analytical measurement. In most analytical instruments, the radiation lies in the range from infrared through ultraviolet, including the visible range. The measurement devices used in these applications are normally called *photodetectors* to distinguish them from other spectral ranges of EM radiation such as RF (radio frequency) detectors. In this section, we will discuss three common photodetectors: photoconductive sensors, photovoltaic sensors, and photomultiplier tubes.

Photoconductive Sensor

A common type of photodetector is based on the change in conductivity that occurs in a semiconductor material with radiation intensity. Since resistance is the inverse of conductivity, these devices are also called *photoresistive* cells.

A semiconductor is a material in which there is an energy gap between conduction electrons and valence electrons. In the semiconductor-based photoconductive sensor, a photon is absorbed and in turn excites an electron from the valence to the conduction band. When a large number of electrons are excited into the conduction band, the semiconductor resistance decreases, which makes the resistance an inverse function of radiation intensity. For the photon to make such an excitation it must carry at least as much energy as the gap. From Equation 8-17, this implies a maximum wavelength as follows:

$$E_p = \frac{hc}{\lambda} = \Delta E_g \tag{8-20}$$

$$\lambda_{max} = \frac{hc}{\Delta E_g} \tag{8-21}$$

where

λ_{max}	=	the maximum detectable radiation wavelength in meters
ΔE_g	=	the semiconductor energy gap in joules
h	=	Planck's constant in joules
c	=	the speed of light in m/s

It is important to note that any radiation with a wavelength greater than that predicated by Equation 8-21 cannot cause any resistance change in the semiconductor.

Example 8-8 shows how to find the maximum wavelength for resistance change by photon absorption for a semiconductor.

EXAMPLE 8-8

Problem: The semiconductor material germanium has a band gap of 1.072×10^{-19} joules. Find the maximum wavelength for resistance change by photon absorption.

Solution: Using Equation 8-21, we find the maximum wavelength as follows:

$$\lambda_{max} = \frac{hc}{\Delta E_g}$$

$$\lambda_{max} = \frac{(6.63x10^{-34} J - s)(3x10^{8} m/s)}{(1.072x10^{-19} - J)}$$

$$\lambda_{max} = 1.86 \mu m$$

The two most common photoconductive semiconductor materials are cadmium sulfide (CdS), which has an energy band gap of 2.42 eV, and cadmium selenide (CdSe), which has a 1.74 eV band gap. Note that one electron volt (eV) is equal to 1.6 x 10^{-19} joules of energy. Because of the large gap energy for both semiconductors, they have a very high resistivity at room temperatures. This gives bulk samples of these semiconductors a resistance much too high for practical applications. To compensate for this, a special construction is used, as shown in Figure 8-14, that minimizes resistance while providing maximum surface area for the detector. This photodetector construction is based on the equation for resistance discussed earlier.

$$R = \frac{\rho l}{A} \tag{8-22}$$

where

R	=	the resistance in ohms
ρ	=	resistivity in ohm-m
l	=	length in meters
A	=	the cross-sectional area in meters squared

A side view of the photoconductive cell is shown in Figure 8-14a. This arrangement produces a minimum length, l, and a maximum area, A, to produce a low resistance value according to Equation 8-22. The area of photoconductive material is the length of the material times its thickness. By using a thin narrow strip of material and by winding the material back and forth as in the front view shown in Figure 8-14b, we obtain the maximum surface area to detect photons.

A photoconductive cell resistance decreases nonlinearly with light intensity, much like a thermistor. Generally, this change in resistance is several hundred orders of magnitude from dark to normal daylight. A readout of light intensity can be obtained by using an electronic circuit to convert the nonlinear resistance of the photoconductive cell to a light intensity value.

Photovoltaic Sensors

An important type of photodetector is the *photovoltaic* cell, which generates a voltage that is proportional to the incident EM radiation intensity. These sensors are called photovoltaic cells because of their voltage-generating capacity, but the cells actually convert EM energy into electrical energy. Photovoltaic cells are very important in instrumentation and control applications because they are used both as light detectors and in power sources that convert solar radiation into electrical power for remote-measuring systems. Our emphasis here is on their use in analytical instruments.

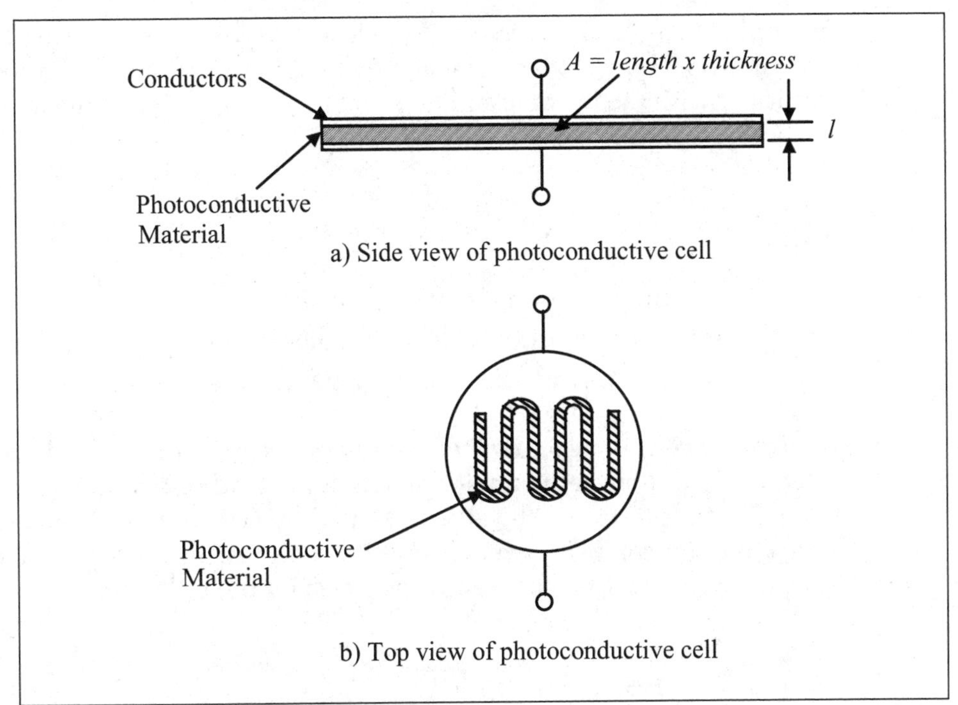

Figure 8-14. Photoconductive cell

The operating principle of the photovoltaic cell is illustrated in Figure 8-15. The cell is a large exposed diode that is constructed using a *pn* junction between appropriately doped semiconductors. Photons hitting the cell pass through the thin p-doped upper layer and are absorbed by electrons in the n-doped layer. This causes conduction electrons and holes to be created.

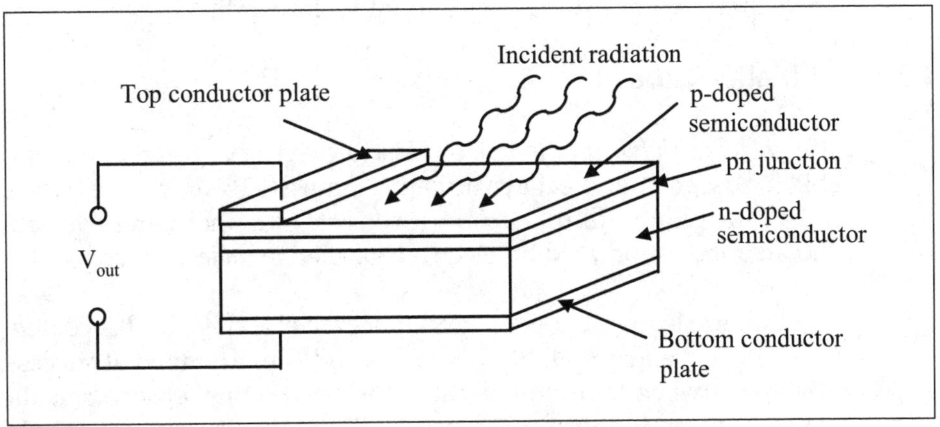

Figure 8-15. Photovoltaic cell

The upper terminal is positive and the lower negative. In general, the open-circuit voltage V that is developed on a photovoltaic cell varies logarithmically with the incident radiation intensity according to the following equation:

$$V = V_o \ln(I_R) \tag{8-23}$$

where

I_R = the radiation intensity in W/m^2
V_o = the calibration voltage in volts
V = the unloaded output voltage in volts

Photovoltaic cells have a low internal resistance, so when they are connected to a measurement circuit that has some load resistance, the cell voltage is reduced from the value indicated by Equation 8-23. Since the internal resistance, R_i, is in series with the load resistance, R_L, the actual current, I, that is delivered to a load is given by the following:

$$I = \frac{V}{R_i + R_L} \tag{8-24}$$

The cell calibration voltage V_o is a function of the cell material only, and this indicates that the voltage produced is independent of the cell geometry. The current that is produced by a photovoltaic depends on the radiation intensity and also the cell surface area. Cells are generally arranged in series and parallel combinations to obtain the desired voltage and current output.

Example 8-9 illustrates how to calculate the internal resistance and the open-circuit voltage for a typical photovoltaic cell.

Photomultiplier Tube

The *photomultiplier tube* is one of the most sensitive photodetectors available for use in analytical instruments. Figure 8-16 shows the basic structure of the photomultiplier tube. It consists of a photoemissive cathode and an anode separated by electrodes called *dynodes*.

The cathode is maintained at a high negative voltage and is coated with a photoemissive material. Numerous dynodes maintained at successively more positive voltages follow the cathode. The final electrode is the *anode*, which is grounded through a resistor R. When a light photon strikes the photoemissive cathode with sufficient energy, several electrons are ejected from the surface, and the voltage potential difference accelerates them to the first dynode. Each electron from the cathode that strikes the first

EXAMPLE 8-9

Problem: A photovoltaic cell generates 0.4 volts open-circuit when exposed to 20 W/m^2 of radiation intensity. A current of 2 mA is delivered into a 100 Ω at that intensity. Calculate (a) the internal resistance of the cell and (b) the open-circuit voltage at 50 W/m^2.

Solution:

a. Using Equation 8-24, we can calculate the internal cell resistance as follows:

$$I = \frac{V}{R_i + R_L}$$

Solving for internal resistance, we obtain

$$R_i = \frac{V - IR_L}{I}$$

$$R_i = \frac{0.4V - (2mA)(100\Omega)}{2mA} = 100\Omega$$

b. To find the open-circuit voltage at 50 W/m^2, we use Equation 8-23 to first calculate V_o, as follows:

$$V_o = \frac{V}{\ln(I_R)}$$

$$V_o = \frac{0.4volts}{\ln 20} = 0.134volts$$

So that at 50 W/m^2, we obtain

$$V = (0.134 \text{ volts})\ln(50)$$

$$V = 0.524 \text{ volts}$$

dynode ejects several electrons. All of these electrons are accelerated to the second dynode where each one strikes the surface with sufficient energy to again eject several electrons. This process is repeated for each dynode until the electrons that reach the anode have greatly multiplied and they produce a large current flow through the output resistor. The voltage pro-

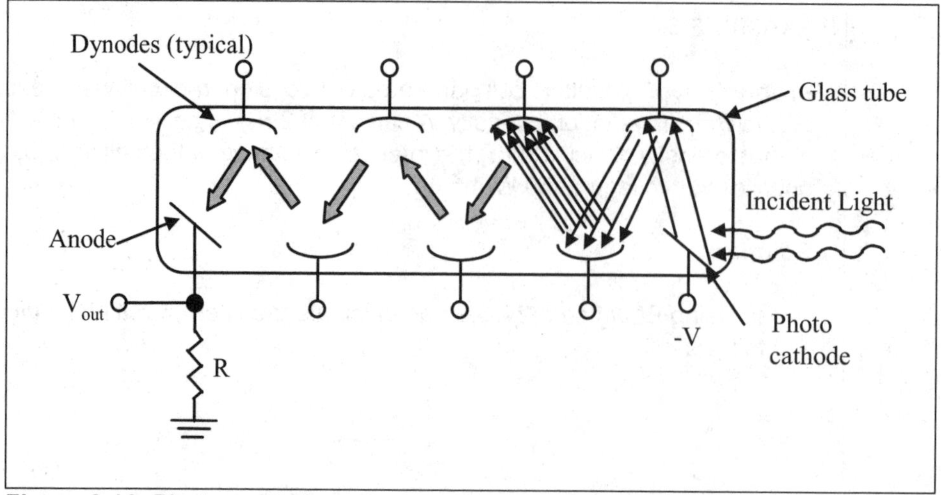

Figure 8-16. Photomultiplier tube

duced across the output resistor R is directly proportional to the light striking the cathode.

The number of dynodes and material from which they are constructed determines the gain of the photomultiplier tube detector. Typical gains are in the range of 10^5 to 10^7 for photoelectrons from the cathode to electrons at the anode. The spectral response of a phototube is determined by two factors: the spectral response of the cathode photoemissive coating material and the transparency of the window through which the light must pass. By using various materials it is possible to construct different types of photomultipliers, which span wavelengths from 0.12 μm to 0.95 μm.

Turbidity Analyzer

A typical application for photodetectors in analytical measurement is as a *turbidity analyzer*. The cloudiness of a liquid, called *turbidity*, is caused by the presence of finely divided suspended material. Turbidimetric methods involve measuring the light transmitted through a medium.

Turbidity can be caused by a single substance or by a combination of several chemical components. For example, the amount of silica in a liquid may be determined in approximate concentrations of 0.1 to 150 ppm of SiO_2. Sometimes composite material turbidities are expressed as being equivalent to silica.

In the typical application shown in Figure 8-17, a turbidity value is developed from a test sample under controlled conditions. In this application, a laser beam is split and passed through two mediums to matched photode-

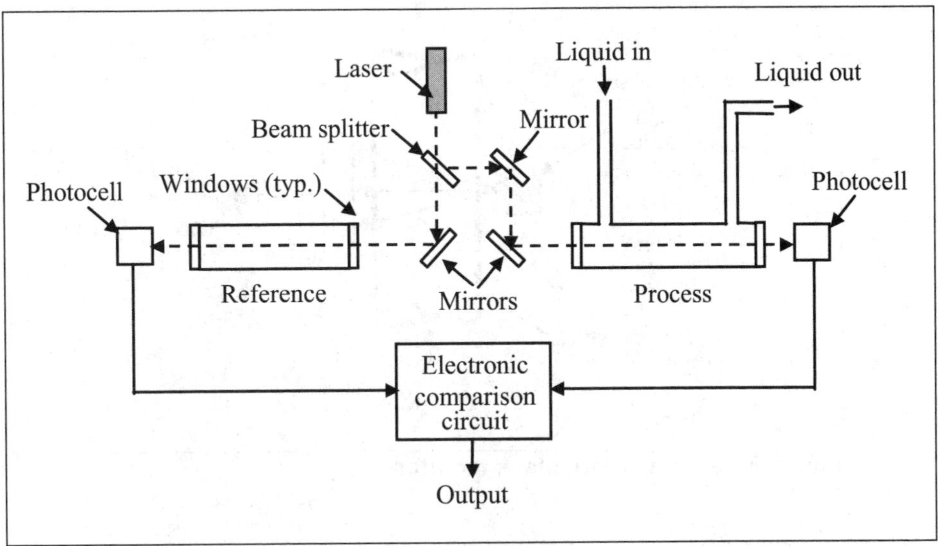

Figure 8-17. Turbidity measurement application

tectors. One medium is a carefully selected standard sample of fixed turbidity. The other medium is an in-line process liquid. If the in-line process liquid attenuates the laser beam more than the standard or reference sample, the electronic circuit triggers an alarm or takes some appropriate control action to reduce turbidity.

Gas Analysis

Continuous measurements of a gas stream to determine the concentration of one or more components are widely made in the process industries. The following sections describe several common analyzers that are used for this purpose.

Particulates

Particle emissions are usually monitored by using light transmission and detection techniques. The normal objective is to indicate particle density, but some applications may also measure particle size.

Opacimeters are most commonly used to measure particulates. These instruments transmit measurement beams across process plant stacks, as shown in Figure 8-18. Opacimeter light sources and detectors are sometimes housed together with passive reflectors across the stack. A distance of 50 ft can be accommodated in these double-pass instruments, but aligning them can be difficult. You may also mount the sources and detectors at opposite ends of assembled pipes to simplify installation.

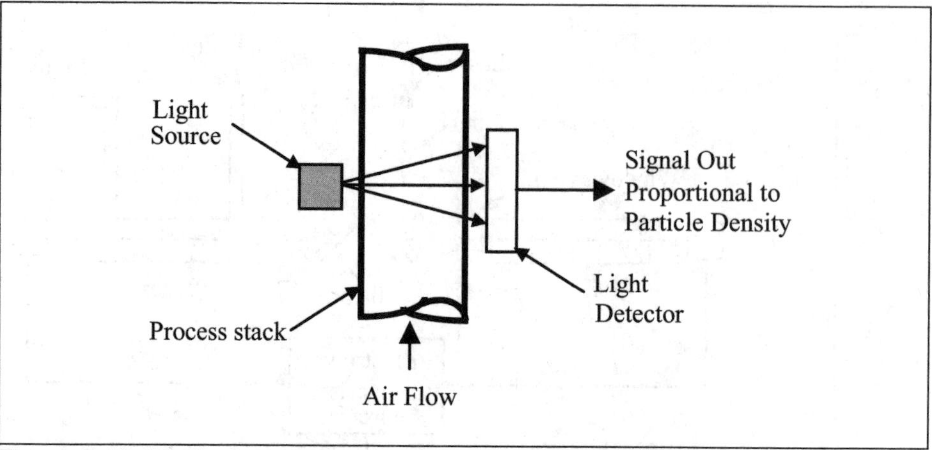

Figure 8-18. Particulate stack monitor

Oxygen Analyzers

Electrochemical zirconium oxide cells are popular for measuring oxygen. Cell output voltages respond to changes in partial pressures of oxygen. The cell output voltages are linearized to produce standardized concentration signals. Ranges are typically from 0 to 10 percent and 0 to 30 percent. Air is used as a reference, and oxygen standards are used for calibration.

Some oxygen analyzers exhibit thermal sensitivity and are usually operated at elevated temperatures for stability. It is advantageous to have sample extraction versions when stack temperatures are low because the cells can be installed in ovens outside the stacks.

Paramagnetic gas sensors are among the earliest oxygen detectors. Analyzers can be specified for ranges from 0 to 1 percent and 0 to 100 percent. The instruments respond to the magnetic susceptibility of the sample. The susceptibility of oxygen is two orders of magnitude greater than that of most other stack constituents, so the accuracy of these sensors is high.

Combustibles

Combustible gases and vapors are detected by measuring the thermal properties of samples. Catalytic bead devices, the most common form of measuring equipment, have both active and inactive filaments arranged in bridge circuits. As gas contacts the catalytically treated filaments, combustible materials diffuse into the surfaces and are oxidized. The heat of reaction raises the temperature, inducing resistance differences. You can also detect flammable gases or vapors by mixing samples with hydrogen and burning the product. The concentration of combustibles in the sample is inferred from flame temperature. Combustibles are usually monitored for explosion hazards, so outputs are calibrated to read in terms of percent of lower explosive limit (LEL).

Hydrocarbon Analyzers

Analyzers that determine the total number of hydrocarbons are used for pollution monitoring and control programs, particularly for automobile exhausts and fossil-fueled stationary combustion facilities. They are also used to detect leaks in refrigerant systems and aerosol packaging as well as to detect breakthroughs from carbon adsorption beds.

Flame ionization detectors (FIDs) are often employed to measure total hydrocarbon concentrations in ranges from 0 to 1 ppm and 0 to 12,000 ppm. The samples are mixed with hydrogen and burned. Electrodes around the flames sense the carbon ions released during the combustion. FIDs typically have response times of less than 10 s. FIDs also have good sensitivity, with concentrations as low as 0.01 ppm. Infrared analyzers can be used for selected hydrocarbons, but they cannot measure total hydrocarbon concentrations.

Carbon Monoxide Analyzers

Carbon monoxide (CO) is measured in chemical, gas processing, metallurgical, and many other types of manufacturing facilities. CO analyzers, the measuring instruments for this application, can also be found in sewers, mines, bus terminals, and garages. Likewise, CO monitors are used to control ventilating equipment in vehicular tunnels. The standard threshold limit for CO is 50 ppm. CO monitors will normally trigger audible and visible alarms at 50 ppm or higher.

Many instruments are available for sensing carbon dioxide. One CO detector cell operates on the principle of catalytic oxidation. In the cell, the gas sample passes through an inactive chemical bed and then through an active catalytic bed. This catalytic bed, which consists of a mixture of oxides of copper, cobalt, manganese, and silver, is used to convert carbon monoxide to carbon dioxide. Each bed has a thermistor that is connected to a Wheatstone bridge circuit. Any CO gas present is immediately oxidized in the catalytic half of the sensor. This raises the temperature of the catalytic thermistor and changes its electrical resistance. This in turn causes a resistance imbalance in the bridge circuit and thus a signal that is proportional to the CO gas concentration in the sample.

In another type of CO measurement system, the sensor has an electronic interface circuit and an electrochemical polarographic cell that contains a sulfuric acid electrolyte. Air samples diffuse through a gas-porous membrane and a sintered metal disk and enter a sample area within the cell. The cell electro-oxidizes CO gas to carbon dioxide (CO_2) gas in proportion to the partial pressure of CO in the sample. The oxidation generates an electrochemical signal that is proportional to the concentration of CO gas in the ambient air. The resulting electrical signal is temperature compensated. The signal is also amplified by an electronic circuit to drive a front-panel meter on the instrument for the purpose of indicating the percentage of CO.

Sulfur Dioxide Analyzers

Several types of general-purpose analyzers are available for measuring sulfur dioxide. Most utilize some form of spectrophotometry. Ultraviolet spectrophotometers provide high accuracy and sensitivity. Ranges as narrow as 0 to 100 ppm are encountered, but instruments can also detect concentrations up to 100 percent by volume. Ultraviolet analyzers are capable of fast response, that is, 1 s or less.

Infrared analyzers can also be used to measure sulfur dioxide. The instruments lack the sensitivity and response of ultraviolet devices, but are more versatile and less costly.

Fluorescence analyzers are used for sulfur dioxide monitoring in ranges from 0 to 0.25 ppm and 0 to 5,000 ppm. Such analyzers emit light when exposed to ultraviolet radiation, with an intensity that varies with the concentration of sulfur dioxide.

Nitrogen Oxide Analyzers

Nitrogen oxides are measured with spectral or electrochemical analyzers. Which instrument you select often depends on whether you desire data that show nitric oxide, nitrogen dioxide, or total oxides of nitrogen.

Chemiluminescence instruments are accurate and sensitive. These analyzers respond directly to nitric oxide; nitrogen dioxide must be reduced for detection to be possible. Chemiluminescence occurs when the samples react with ozone. Intensities, which are measured with photomultipliers, are correlated with nitric oxide concentration. Detection ranges vary from 0 to 0.1 ppm and 0 to 10,000 ppm. Instruments can be specified for concentrations as low as 0.5 ppm.

Ultraviolet analyzers are capable of monitoring oxide as well as dioxide. The lower detection limits are only 10 ppm for nitric oxide; therefore, converting the nitric oxide to the dioxide usually raises sensitivity.

Infrared analyzers are sensitive to nitric oxide but not to dioxide. Units are available for measurements from 0 to 1,000 ppm and 0 to 10,000 ppm. Ranges can also be specified for 0 to 1 percent and 0 to 10 percent.

Hydrogen Sulfide Analyzers

Hydrogen sulfide is difficult to monitor accurately and often must be conditioned chemically. Some analyzers expose sample gases to chemically treated paper tape. Hydrogen sulfide reacts with the tape, and the resulting color change is used to infer hydrogen sulfide concentration.

Conventional fluorescence analyzers are also used to monitor hydrogen sulfide, but the hydrogen sulfide is first converted to sulfur dioxide. Automatic titrators are employed to determine hydrogen sulfide concentrations.

Ultraviolet analyzers respond to hydrogen sulfide, but their sensitivities are low. Polarographic instruments can be used, but filters must remove unsaturated hydrocarbons.

Analyzer Measurement Applications

We will close the discussion of analyzers with a measurement system application that uses a gas analyzer. A typical SO_2 stack analyzer instrument loop is shown in Figure 8-19. The system consists of an SO_2 analyzer, a temperature element, and a flow measurement system. The temperature and flow signals are used in this system to obtain the amount of SO_2 in pounds per hour. The computation is made in the control unit (AIT). Note that electrical heat tracing has been used on the analyzer sample line to keep the SO_2 in the gaseous state.

The analyzer's operation is based on the absorption of light by the sample gas. Rigidly defined, light is only that narrow band of electromagnetic radiation visible to the naked eye, as discussed earlier. However, in this discussion the term *light* is used to refer to electromagnetic radiation over the specific wavelengths covered by the analyzer. Wavelengths used for SO_2 analysis are in the 280- to 313-nm range for the measuring channel and 578 nm for the reference channel.

Figure 8-20 shows a block diagram of a typical SO_2 analyzer. The optical system operates as follows: radiation from the light source (A) passes through the sample (B) by flowing through a sample cell. Some light of the measuring wavelength is adsorbed by SO_2 in the sample. Light transmitted through the sample is divided by a semitransparent mirror (C) into two beams (D and H). Each beam then passes through its own optical filter

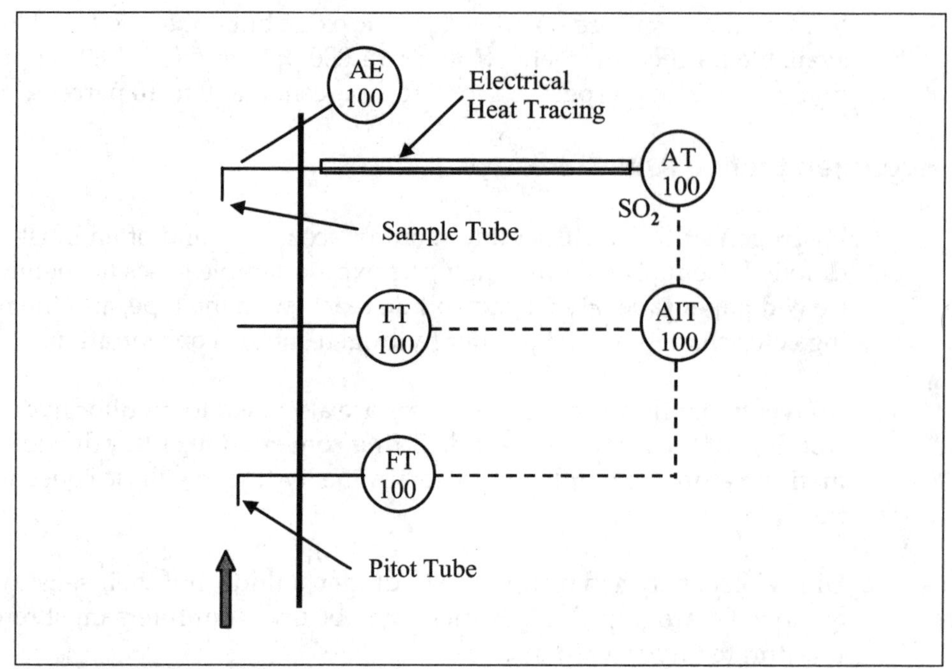

Figure 8-19. SO_2 stack gas analyzer

(E or I). Each filter permits only a particular wavelength to reach its associated phototube (G or K).

Optical filters in one beam permit only radiation at the measuring wavelength (J) to pass through. The measuring wavelength is chosen so that light intensity reaching the photomultiplier tube or phototube (K) varies greatly when SO_2 concentration changes.

The optical filter in the second beam permits only light at the reference wavelength (F) to pass through. The reference wavelength is chosen so that light intensity reaching the reference phototube (G) varies little or not at all when SO_2 concentration changes. Each phototube sends a current to its logarithmic amplifier (log amp) that is proportional to the intensity of the light striking the phototube. The signal output of the analyzer circuit is the voltage difference produced by the log amps.

If SO_2 concentration increases, light arriving at the measuring phototube decreases, as does the measuring phototube current. The reference circuit is unaffected. Since voltage generated in the measuring circuit increases with the drops in phototube current, the output voltage (measuring voltage minus reference voltage) rises with a concentration increase.

This analyzer's design also provides inherent compensation for changes in overall light intensity. Factors such as light source variations or dirt on the

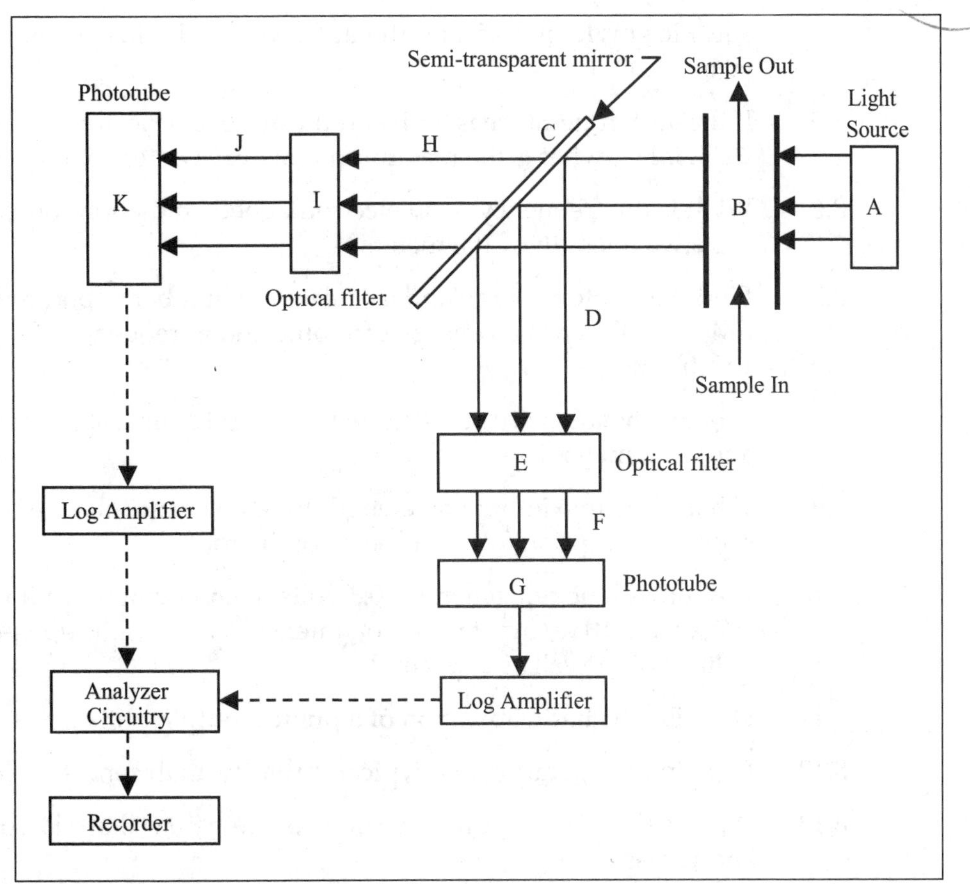

Figure 8-20. SO_2 analyzer block diagram

cell windows, which affect equally the intensities of both the measuring and reference wavelengths, will change the output voltages to an equal extent. Therefore, these variations have minimal net effect on the difference or the final output voltage.

EXERCISES

8.1 Given the following resistance values: (a) 50,000 Ω, (b) 200,000 Ω, (c) 250,000 Ω for a given solution, calculate the conductance of each solution.

8.2 The [OH^-] ion concentration of an aqueous solution is 10^{-11}. What is the value of the H^+ ion concentration and pH? Is the solution basic or acidic?

8.3 The specific gravity of a lead-acid cell in a 12-V (six-cell) battery is 1.24. Calculate the no-load voltage of the battery.

8.4 Calculate the span in inches of a differential-pressure density instrument if the minimum specific gravity is 1.0, the maximum specific gravity is 1.25, and the difference in liquid elevation is 50 in.

8.5 If the air temperature is 80°F and the atmospheric pressure is 29.92 inHg, what is the maximum moisture content of the air?

8.6 What is the frequency of an electromagnetic radiation source that has a wavelength of 100 meters?

8.7 Find the photon energy and calculate the number of photons in an EM pulse that has an energy of 1 joule and a frequency of 1×10^9Hz.

8.8 What is the intensity of a 1000-w point light source at 10 meters and at 20 meters?

8.9 What is the maximum wavelength for a resistance change by photon absorption for a CdS semiconductor?

8.10 A photovoltaic cell generates 0.3 volts open-circuit when it is exposed to 10 w/m^2 of radiation intensity. What is the open-circuit voltage of the cell at 20 w/m^2?

8.11 Describe the basic operation of a photomultiplier tube.

8.12 Explain the operation of a typical turbidity analyzer.

8.13 What is the most popular type of measurement cell used in oxygen analyzers?

9 Flow Measurement

Introduction

We begin this chapter by discussing the basic principles of flow, then move on to derive the basic equations for flow velocity and volumetric flow. The chapter concludes with a discussion of the common types of flow measuring devices and instruments, such as orifice plates, venturi tubes, flow nozzles, wedge flow elements, pitot tubes, annubars, turbine flowmeters, vortex shedding devices, magnetic flowmeters, ultrasonic flowmeters, positive-displacement flowmeters, mass flowmeters, and rotameters.

The study of fluids in motion, or flow, is one of the most complex branches of engineering. This complexity is reflected in such familiar examples as the flow of a river during a flood or a swirling cloud of smoke from a process plant smokestack. Each drop of water or each smoke particle is governed by Newton's laws, but the equations for the entire flow are very complicated. Fortunately, idealized models that are simple enough to permit detailed analysis can represent most situations in process control.

Flow Principles

Our discussion in this chapter will consider only a so-called *ideal fluid*, that is, a liquid that is incompressible and has no internal friction or viscosity. The assumption of incompressibility is usually a good approximation for liquids. A gas can also be treated as incompressible if the differential pressure driving it is low. Internal friction in a fluid gives rise to shearing stresses when two adjacent layers of fluid move relative to each other, or when the fluid flows inside a tube or around an obstacle. In most cases in process control, these shearing forces can be ignored in contrast to gravitational forces or forces from differential pressures.

All flow involves some form of energy. Energy can be expressed in many different forms, including thermal, chemical, and electrical. However, flow measurement focuses on two main types of energy: potential and kinetic. Potential energy (U) is defined as force (F) applied over a distance (d), or

$$U = Fd \tag{9-1}$$

Kinetic energy (K) is defined as one-half the mass (m) times the square of velocity of the body in motion, or

$$K = \frac{1}{2}mv^2 \tag{9-2}$$

Potential Energy

The term *potential energy* probably came from the idea that we give an object the "potential" to do work when we raise it against gravity. The potential energy of water at the top of a waterfall is converted into a kinetic energy of motion at the bottom of the fall.

Potential energy is usually applied to the work that is required to raise a mass against gravity. Force is defined as mass (m) times acceleration (a)

$$F = ma \tag{9-3}$$

Therefore, the work (W) required to raise a mass through a height (h) is expressed as follows:

$$W = Fh = mgh \tag{9-4}$$

where g is the acceleration of the object due to gravity. The term mgh is called gravitational potential energy, or

$$U = mgh \tag{9-5}$$

This energy can be recovered by allowing the object to drop through the height h, at which point the potential energy of position is converted into the kinetic energy of motion.

Work and Kinetic Energy

The work that a force does on a body is related to the resultant change in the body's motion. To develop this relationship further, consider a body of mass m being driven along a straight line by a constant force of magnitude F that is directed along the line. Newton's second law gives the acceleration of a body as follows:

$$F = ma \tag{9-6}$$

Suppose the speed increases from v_1 to v_2 while the body undergoes a displacement d. From standard analysis of motion, we know that

$$v_1^2 = v_2^2 + 2ad \tag{9-7}$$

or

$$a = \frac{v_2^2 - v_1^2}{2d} \tag{9-8}$$

Since $F = ma$,

$$F = m\frac{v_2^2 - v_1^2}{2d} \tag{9-9}$$

therefore,

$$Fd = \frac{1}{2}mv_2^2 - \frac{1}{2}mv_1^2 \tag{9-10}$$

The product Fd is the work (W) done by the force (F) over the distance d. The quantity $1/2mv^2$—that is, one-half the product of the mass of the body and the square of its velocity—is called its *kinetic energy* (KE).

The first term on the right-hand side of Equation 9-10, which contains the final velocity v_2, is the final kinetic energy of the body, KE_2, and the second term is the initial kinetic energy, KE_1. The difference between these terms is the change in kinetic energy. This leads to the important result that the work of the external force on a body is equal to the change in the kinetic energy of the body, or

$$W = KE_2 - KE_1 = \Delta KE \tag{9-11}$$

Kinetic energy, like work, is a *scalar* quantity. The kinetic energy of a moving body, such as fluid flowing, depends only on its speed, not on the *direction* in which it is moving. The *change* in kinetic energy depends only on the work ($W = Fd$) and not on the individual values of F and d. This fact has important consequences in the flow of fluid.

For example, consider the flow of water over a dam with height, h. Any object that falls through a height h under the influence of gravity is said to gain kinetic energy at the expense of its potential energy. Let's assume that water with mass m falls through the distance h, converting all its potential energy (mgh) into kinetic energy. Since energy must be conserved, the kinetic energy must equal the potential energy. Therefore,

$$mgh = \frac{mv^2}{2} \tag{9-12}$$

This equation can be solved for velocity v to obtain the following:

$$v = \sqrt{2gh} \tag{9-13}$$

Equation 9-13 shows that the velocity of water at the base of the dam depends on the height (h) of the dam and on gravity (g). Since gravity is constant at about 32 ft/sec ^{2} or 9.8 m/sec^2 on the earth's surface, the velocity depends only on the height h and not on the mass of the flowing fluid. This is an important property in the study of fluid flow. The following example will illustrate this property.

EXAMPLE 9-1

Problem: A valve is opened on the bottom of a storage tank filled to a height of 4 feet with water. Find the discharge velocity of the water just after the outlet valve is opened.

Solution: The velocity can be found from Equation 9-13 as follows:

$$v = \sqrt{2gh}$$

$$v = \sqrt{2(32\,ft/\sec^2)(4\,ft)} = 16\,ft/\sec$$

Flow in a Process Pipe

Another example of the relationship between energy and fluid velocity is the flow of fluid in a process pipe of uniform and fixed cross section (A), as shown in Figure 9-1. The differential pressure (ΔP) between the inlet and the outlet causes the fluid to flow in the pipe.

The flow of fluid is maintained by the energy difference between the inlet and the outlet. Let's find the fluid velocity (v) in terms of the inlet pressure P_1 and the outlet pressure P_2, assuming no energy loss in the pipe. Since

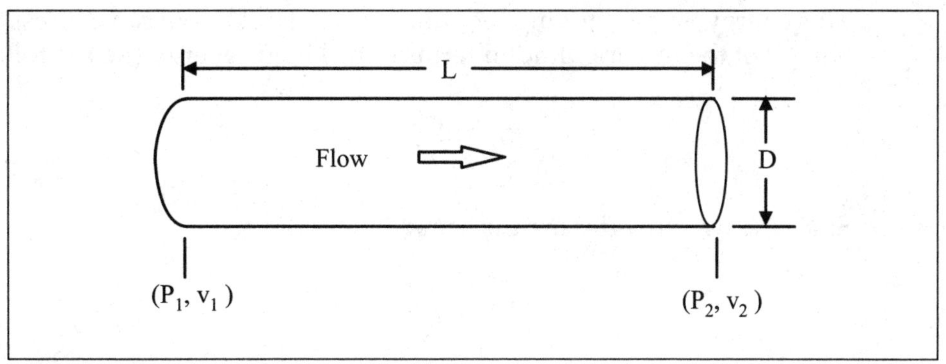

Figure 9-1. Flow in a pipe

the pipe has a uniform area A, the pressure at the inlet is P_1 and the pressure at the outlet is P_2. The total force at the input is $F_1 = P_1A$, and the total force at the output is $F_2 = P_2A$.

The energy (work) required to move the fluid through the distance L is force times distance.

$$(F_1 - F_2)L = P_1AL - P_2AL$$

$$= (P_1 - P_2)AL$$

Since AL is the volume of the pipe, the work is given by the following:

$$\text{Work} = \text{Energy} = (P_1 - P_2)\ (\text{Volume})$$

$$\text{Work} = \Delta P \times V \qquad (9\text{-}14)$$

The complete energy equation for a flow system must include all possible energy terms, including "internal energy" changes (the energy stored in each molecule of the fluid). This energy includes molecular kinetic energy, molecular rotational energy, potential energy binding forces between molecules, and so on. This internal energy is significant only in laminar flow, where high frictional forces can raise the temperature of the fluid. However, in process control we generally encounter turbulent flow, so we can ignore internal energy in most cases.

Assuming that the flow in Figure 9-1 is steady, let's find the energy relationship for flow in a uniform pipe. We have just shown that the work (energy) done in moving a fluid through a section of pipe is as follows:

$$\text{Energy} = \Delta PV \qquad (9\text{-}15)$$

This energy is spent giving the fluid a velocity of v. We can express this energy of the moving fluid in terms of its kinetic energy (*KE*) as follows:

$$KE = \frac{mv^2}{2} \tag{9-16}$$

Since the two energies are the same,

$$\Delta PV = \frac{mv^2}{2} \tag{9-17}$$

However, by definition, mass m is equal to volume V time's density ρ, so we can replace mass in the equation with $V x \rho$ to obtain the following:

$$\Delta PV = \frac{V \rho v^2}{2} \tag{9-18}$$

If we cancel the volume term from both sides of the equation, we obtain

$$\Delta P = \frac{\rho v^2}{2} \tag{9-19}$$

Then, solving for velocity and taking the square root of both sides of the equation, we obtain the general equation for the velocity of any fluid in a pipe.

$$v = \sqrt{\frac{2\Delta P}{\rho}} \tag{9-20}$$

This velocity is expressed in terms of the pressure differential and density of the fluid.

Volumetric flow is defined as the volume of fluid that passes a given point in a pipe per unit of time. This is expressed as follows:

$$Q = Av \tag{9-21}$$

where

Q	=	the volumetric flow
A	=	the cross-sectional area of the flow carrier (e.g., pipe)
v	=	the fluid's velocity

We can also define mass flow rate (W) as the mass or weight flowing per unit time. Typical units are pounds per hour. This is related to the volumetric flow by the following:

$$W = \rho Q \tag{9-22}$$

where

W	=	the mass flow rate
ρ	=	the density
Q	=	the volumetric flow rate

Reynolds Number

The basic equations of flow assume that the velocity of flow is uniform across a given cross section. In practice, flow velocities at any cross section approach zero in the boundary layer adjacent to the pipe wall and vary across the diameter. This flow velocity profile has a significant effect on the relationship between flow velocity and the development of pressure difference in a flowmeter. In 1883, the English scientist Sir Osborne Reynolds presented a paper before the Royal Society that proposed a single dimensionless ratio, now known as the Reynolds number, as a criterion for describing this phenomenon. This number R_e, is expressed as follows:

$$R_e = \frac{vD\rho}{\mu} \tag{9-23}$$

where

v	=	the flow velocity
D	=	the inside diameter of the pipe
ρ	=	the fluid density
μ	=	fluid viscosity

The Reynolds number expresses the ratio of internal forces to viscous forces. At a very low Reynolds number, viscous forces predominate and inertial forces have little effect. Pressure difference approaches direct proportionality to average flow velocity as well as to viscosity. A Reynolds number is a pure, dimensionless number, so its value will be the same in any consistent set of units. The following equations are used in the United States to more conveniently calculate the Reynolds number for liquid and gas flow through a process pipe:

$$R_e = \frac{3160 Q_{gpm} SG}{\mu D} \quad \text{(Liquid)} \tag{9-24}$$

or $$R_e = \frac{50.6 Q_{gpm} \rho}{\mu D} \quad \text{(Liquid)} \tag{9-25}$$

EXAMPLE 9-2

Problem: Water at 60°F is pumped through a pipe with a 1-in. inside diameter at a flow velocity of 2.0 ft/s. Find the volumetric flow and the mass flow. The density (ρ) of water is 62.4 lbs/ft^3 at 60°F.

Solution: The flow velocity is given as 2.0 ft/s, so the volumetric flow can be found as follows:

$$Q = Av$$

The area of the pipe is given by the following:

$$A = \frac{\pi d^2}{4}$$

so that

$$A = \frac{\pi(1\,\text{in.} \times 1\,\text{ft}/12\,\text{in.})^2}{4} = 0.0055\,\text{ft}^2$$

The volumetric flow is as follows:

$$Q = Av$$

$$Q = (0.0055\ \text{ft}^2)\ (2\ \text{ft/s})\ (60\ \text{s/min})$$

$$Q = 0.654\ \text{ft}^3/\text{min}$$

The mass flow rate is found using Equation 9-22

$$W = \rho Q$$

$$W = (62.4\,\text{lb/ft}^3)(0.654\,\text{ft}^3/\text{min})$$

$$W = 40.8\,\text{lb/min}$$

$$R_e = \frac{379 Q_{acfm} \rho}{\mu D} \quad \text{(Gas)} \tag{9-26}$$

where

ρ = density in pounds per cubic foot

D = the pipe inside diameter, is in inches

EXAMPLE 9-3

Problem: An incompressible fluid is flowing through a process pipe with an inside diameter of 12 inches under a pressure head of 16 in. Calculate the fluid velocity and volumetric flow rate.

Solution: The fluid velocity is found as follows:

$$v = \sqrt{2gh}$$

where g = 32.2 ft/s^2 and h = (16 in.) (1 ft/12 in.) = 1.33 ft.

Thus,

$$v = \sqrt{2(32 \text{ ft/sec}^2)(1.33 \text{ ft})} = 9.23 \text{ ft/s}$$

The volumetric flow rate is obtained as follows:

$$Q = Av$$

$$Q = r[(1\ ft)^2/4]\ (9.23\ ft/s)$$

$$Q = 7.25 \text{ ft}^3/\text{s}$$

As shown in Figure 9-2, three flow profile types are encountered in process pipes: laminar, transitional, and turbulent flow. At high Reynolds numbers, inertial forces predominate, and viscous drag effects become negligible. At low Reynolds numbers, flow is laminar and may be regarded as a group of concentric shells. Moreover, each shell reacts in the manner as viscous shear on adjacent shells; the velocity profile across a diameter is substantially parabolic. At high Reynolds numbers flow is turbulent, and eddies form between the boundary layer and the body of the flowing fluid and then propagate through the stream pattern. A very complex, random pattern of velocities develops in all directions. This turbulent, mixing action tends to produce a uniform average axial velocity across the stream.

Flow is in the laminar region when the Reynolds number is less than 2,000, while flow is generally turbulent if the Reynolds numbers are greater than 4,000. Transitional flow occurs in the range of 2,000 to 4,000. Since the Reynolds number only reflects fluid effects and disregards factors such as pipe bends, pipe fittings, and pipe roughness, the boundaries of laminar,

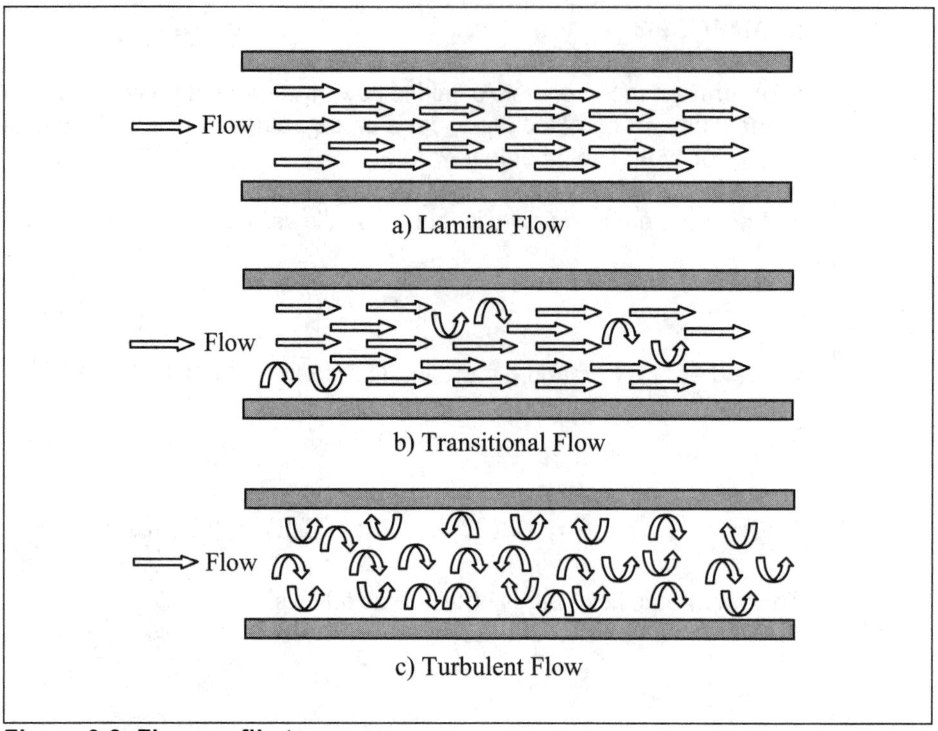

Figure 9-2. Flow profile types

transitional, and turbulent flow are estimates suggested for practical control applications.

In the equation for the Reynolds number for liquid flow, the velocity, v, generally varies in a ten-to-one range; the specific gravity generally ranges from 0.8 to 1.2; and the pipe diameter is constant. However, for some liquids, the viscosity can vary from less than one, to thousands of centipoises. So, in most liquid flow applications viscosity has the most effect on the Reynolds number. While the value of viscosity can be well defined for a given liquid under fixed operating conditions, relatively small changes in temperature can cause order-of-magnitude changes in viscosity. These changes can determine whether the flow is laminar, transitional, or turbulent.

Example 9-4 illustrates a typical Reynolds number calculation to determine flow type.

In the equation for the Reynolds number for gas flow, the velocity, v, generally varies in a ten-to-one range, the density generally varies over a range of less than two to one, and the pipe diameter is constant. However, the viscosity is small and virtually constant in most process applications. So, flow and density have the most effect on the Reynolds number in most

EXAMPLE 9-4

Problem: Water at 80°F is flowing at a rate of 50 gpm through a 2-in. schedule 40 steel pipe that has an inside diameter of 2.067 inches. Calculate the Reynolds number, and determine if the flow is laminar, transitional, or turbulent. Note that at 80°F, the density of water is 62.220 pounds per cubic foot, and the viscosity is 0.85 centipose (cP).

Solution: Using Equation 9-25, calculate the Reynolds number

$$R_e = \frac{50.6 Q_{gpm} \rho}{\mu D}$$

$$R_e = \frac{50.6 x 50 x 62.220}{2.067 x 0.85}$$

$$R_e = 89{,}600$$

Since the Reynolds number is much greater than 4,000, the flow is clearly turbulent.

gas flow applications. Since the flow and density are well established in most applications and the viscosity is low, gas flow is turbulent in properly designed process piping systems.

Example 9-5 demonstrates a typical Reynolds number calculation to determine flow type.

Flow Measuring Techniques

This section discusses the most common types of flow detection devices encountered in process control. The techniques used to measure flow fall into four general classes: differential pressure, velocity, volumetric, and mass.

Differential-pressure flowmeters measure flow by inferring the flow rate from the drop in differential pressure (*dP*) across an obstruction in the process pipe. Some of the common *dP* flowmeters are orifice plates, venturi tubes, flow nozzles, wedge flow meters, pitot tubes, and annubars.

With velocity devices, the flow rate is determined by measuring the velocity of the flow and multiplying the result by the area through which the

EXAMPLE 9-5

Problem: Gas with a density of 0.4 lb/ft^3 and a viscosity of 0.02 cP is flowing at 5 acfm through a 2-in. schedule 40 pipe. Calculate the Reynolds number, and determine if the flow is laminar, transitional, or turbulent.

Solution: Since the inside diameter of 2-in. schedule 40 pipe is 2.067 in., using Equation 9-27 the Reynolds number is calculated as follows:

$$R_e = \frac{379 Q_{gpm} \rho}{\mu D}$$

$$R_e = \frac{379x5x0.4}{2.067x0.02} = 18{,}336$$

Since the Reynolds number is much greater than 4,000, the flow is clearly turbulent.

fluid flows. Typical examples of velocity devices include turbine, vortex shedding, magnetic, and Doppler ultrasonic flowmeters.

Volumetric or positive-displacement (PD) flowmeters measure flow by measuring volume directly. Positive-displacement flowmeters use high-tolerance machined parts to physically trap precisely known quantities of fluid as they rotate. Common devices include rotary-vane, oval-gear, and nutating-disk flowmeters.

Mass flowmeters measure the mass of the fluid directly. An example is the Coriolis mass flowmeter.

Differential-Pressure Flowmeters

One of the most common methods for measuring the flow of liquids in process pipes is to introduce a restriction in the pipe and then measure the resulting differential pressure (ΔP) drop across the restriction. This restriction causes an increase in flow velocity at the restriction and a corresponding pressure drop across the restriction. The relationship between the pressure drop and the rate of flow is given by the following equation:

$$Q = K\sqrt{\Delta P} \tag{9-27}$$

where:

Q = the volumetric flow rate

K	=	a constant for the pipe and liquid type
ΔP	=	the differential pressure drop across the restriction

The constant depends on numerous factors, including the type of liquid, the size of the process pipe, and the temperature of the liquid, among others. The configuration of the restriction that is used will also change the constant in Equation 9-27. As this equation shows, the flow is linearly dependent not on the pressure drop but rather on the square root of the pressure drop. For example, if the pressure drop across the restriction increases by a factor of two when the flow is increased, the flow only increases by a factor of 1.41, the square root of two. Example 9-6 will illustrate this concept.

In the following sections, we will first discuss the four common restriction-type differential-pressure flow-measuring devices: orifice plates, venturi tubes, flow nozzles, and wedge flow elements. Then, we will turn to pitot tubes and annubar flow differential-pressure flow measuring devices.

Orifice Plate

The orifice plate is the most common type of restriction used in process control applications to measure flow. The principle behind the orifice plate is simple. The plate is inserted in a process line, and then the differential pressure ($\Delta P = P_{high} - P_{low}$) developed across the orifice plate is measured to determine the flow rate (see Figure 9-3). To maintain a steady flow through the orifice plate, the velocity must increase as it passes through the orifice. This increase in velocity, or kinetic energy, comes about at the expense of pressure, or potential energy.

The pressure profile across the orifice plate in Figure 9-3 shows a decrease in pressure as the flow velocity increases through the restriction. The lowest pressure occurs where the velocity is the highest. Then, farther downstream as the fluid expands back into a larger area, velocity decreases and pressure correspondingly increases. The pressure downstream of the orifice plate never returns completely to the pressure that existed upstream because the restriction created friction and turbulence that caused energy loss.

The differential pressure across the orifice plate is a measure of the flow velocity. The greater the flow, the larger the differential pressure across the orifice plate.

Figure 9-4 shows the three most common types of orifice plates: concentric, eccentric, and segmental. The concentric orifice plate (Figure 9-4a) is the most widely used type. The eccentric orifice plate (Figure 9-4b) is

EXAMPLE 9-6

Problem: A liquid is flowing past a restriction in a process pipe that has a volumetric flow of 2 ft^3 /sec. This causes a pressure drop of 1 in. of water column across the restriction. Calculate the volumetric flow if the pressure drop across the restriction increases to 4-inH_2O, that is, four times greater.

Solution: First, we calculate the constant K in Equation 9-27 using the original flow, Q_1, and the original pressure drop, ΔP_1, as follows:

$$K = \frac{Q_1}{\sqrt{\Delta P_1}} = \frac{2\ \text{ft}^3/\text{s}}{(1\ \text{in.})^{1/2}}$$

$$K = (2 ft^3 /(s)(in.^{1/2})$$

Then, we can calculate the second volumetric flow using Equation 9-27 and the value calculated for the constant K:

$$Q_2 = K\sqrt{\Delta P_2}$$

$$Q_2 = \frac{2\ \text{ft}^3}{(\text{s})(\text{in.}^{1/2})}\sqrt{\Delta P_2}$$

$$Q_2 = \frac{2\ \text{ft}^3}{(\text{s})(\text{in.}^{1/2})}\sqrt{4\ \text{in.}}$$

$$Q_2 = (4\ \text{ft}^3/\text{s})$$

exactly like the concentric plate except that the hole is bored off center. The segmental orifice plate (Figure 9-4c) has a hole that is a segment of a circle. You must install it so the circular section is concentric with the pipe and no portion of the flange or gasket covers the hole.

Eccentric and segmental orifices are preferable to concentric orifices for measuring slurries or dirty liquids as well as gas or vapor where liquids may be present, especially large slugs of liquid. Where the stream contains particulate matter, the segmental orifice may be preferable because it provides an open path at the bottom of the pipe. However, when conditions permit, the eccentric orifice is preferable because it is easier to manufac-

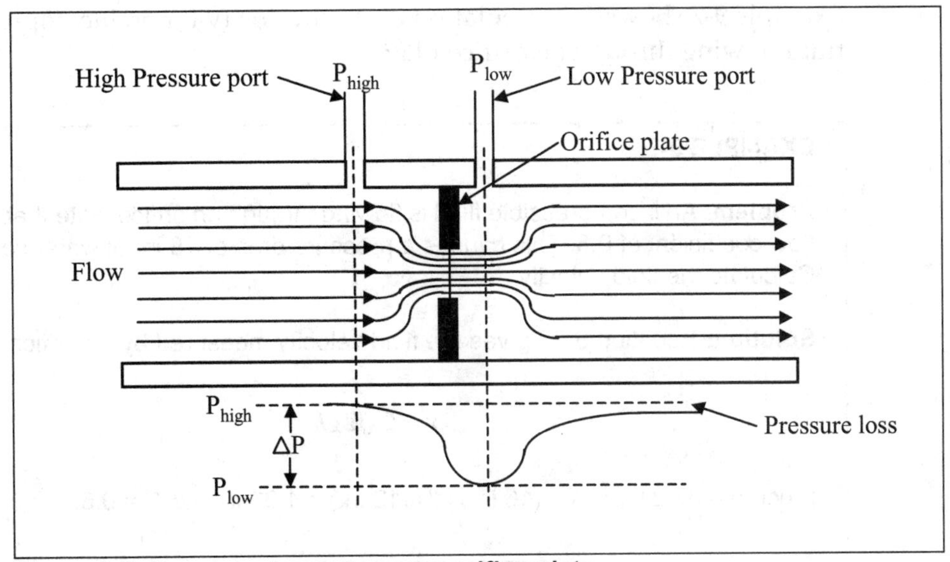

Figure 9-3. Pressure drop across an orifice plate

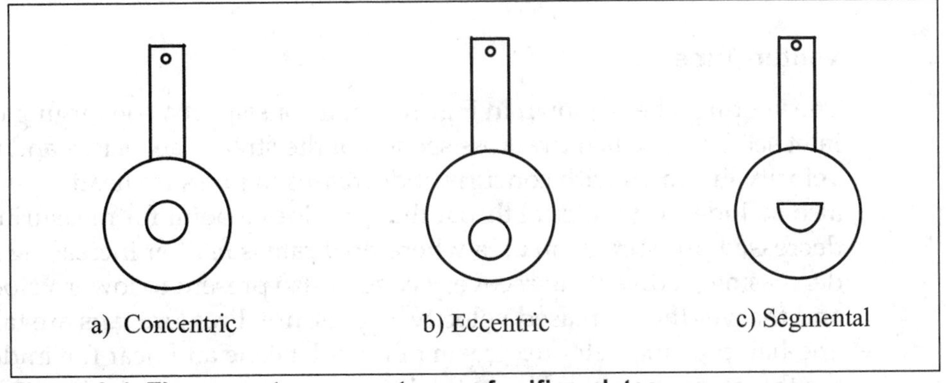

Figure 9-4. Three most common types of orifice plates

ture to precise tolerances and offers generally more accurate and repeatable performance.

The basic flow equation for liquids in a process pipe that we discussed earlier was developed on the assumption that internal energy (pipe losses, molecular-energy losses, etc.) was zero. Since these losses are not zero and there is no direct way to measure them, empirical correction factors, based on flow experiments with various pipe sizes, are tabulated in the literature. These factors are called *flow coefficients* or *discharge coefficients*. Combining all the correction factors into a single factor called the orifice flow constant (C) provides a practical method for computing flow through an orifice plate:

$$v = C\sqrt{2gh} \tag{9-28}$$

Example 9-7 shows the calculation of fluid velocity for an incompressible fluid flowing through an orifice plate.

EXAMPLE 9-7

Problem: An incompressible fluid is flowing through an orifice plate that has a flow coefficient of 0.6. This causes a pressure drop of 16 in. of water column. Calculate the fluid velocity.

Solution: Equation 9-28 gives the fluid velocity measured by an orifice:

$$v = C\sqrt{2gh}$$

Since g = 32.2 ft/s^2, h = (16 in.) (1 ft/12 in.) = 1.33 ft, and C = 0.6.

$$v = 0.6\sqrt{2(32.2\,ft/\sec^2)(1.33\,ft)} = 5.56\,ft/s$$

Venturi Tube

The venturi tube is shown in Figure 9-5. It consists of a converging conical inlet section in which the cross section of the stream decreases and the velocity increases with concurrent decreases in pressure head. The tube also includes a cylindrical throat that provides a point for measuring this decreased pressure in an area where flow rate is neither increasing nor decreasing. A diverging recovery cone is also present to lower velocity and recover the decreased velocity as pressure. Pressure taps are taken at one-half pipe diameter upstream of the inlet cone and near the middle of the throat, as shown in Figure 9-5.

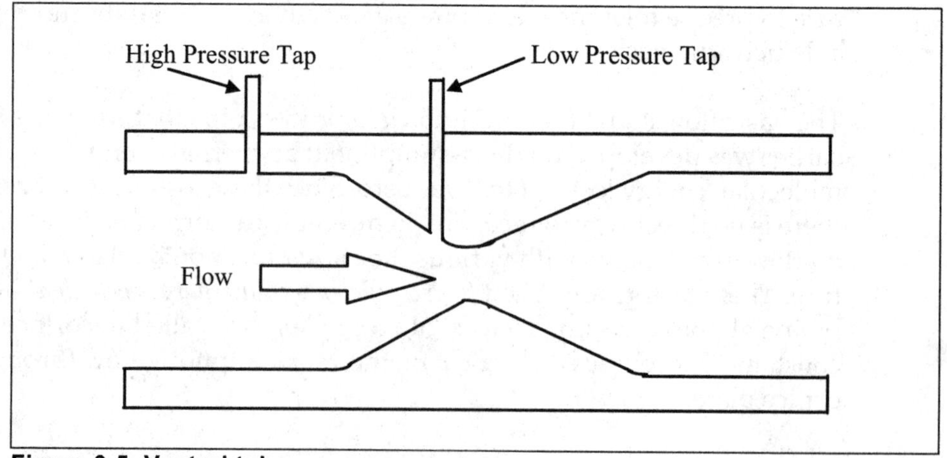

Figure 9-5. Venturi tube

The venturi tube has no sudden changes in shape, no sharp corners, and no projections into the fluid stream. It can therefore be used to measure slurries and dirty fluids, which tend to build up on or clog orifice plates. The major disadvantages of the venturi tube are its cost, both for the tube itself and in terms of the frequency of installation changes required in the larger sizes. A venturi tube is much more difficult to inspect in place than is an orifice. For several reasons, a venturi tube may provide less accuracy of measurement than an orifice plate unless it is flow-calibrated in place.

Flow Nozzles

The flow nozzle shown in Figure 9-6 consists of a restriction with an elliptical, or near elliptical, contour approach section that terminates in tangency with a cylindrical throat section. Flow nozzles are commonly used to measure steam flow and other high-velocity fluid flows where erosion may be a problem. Since the exact contour of the nozzle is not particularly critical, the flow nozzle can be expected to retain precise calibration for a long time under hostile conditions.

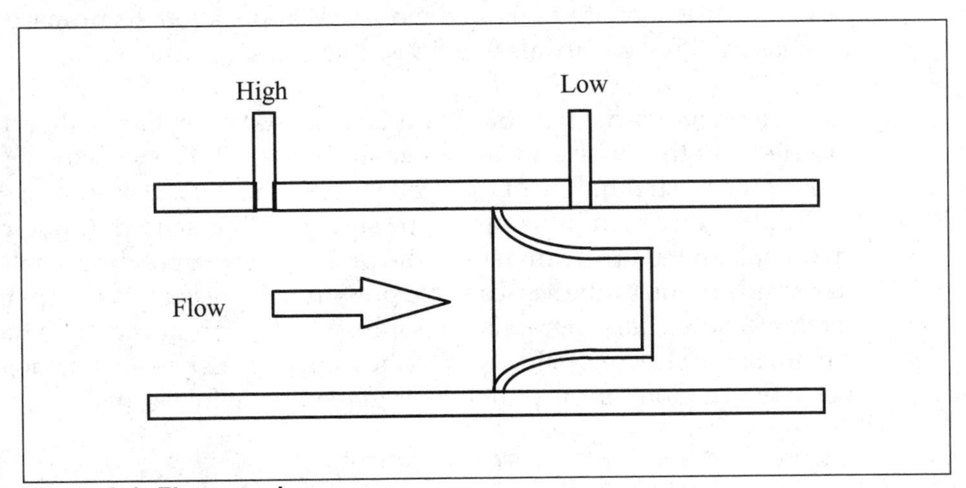

Figure 9-6. Flow nozzles

Because of their streamlined shape, flow nozzles tend to sweep solids through the throat. They are not recommended for measuring fluids that have a large percentage of solids. When your process has appreciable solids, you should mount the flow nozzle in a vertical pipe with the flow going downward.

Wedge Flow Elements

The wedge flow element shown in Figure 9-7 also produces a differential pressure by placing a wedged obstruction in the flow stream. The wedge has no sudden changes in shape and no sharp corners. Therefore, it can be used to measure flow for slurries and dirty fluids, which tend to build up on, or clog, orifice plates.

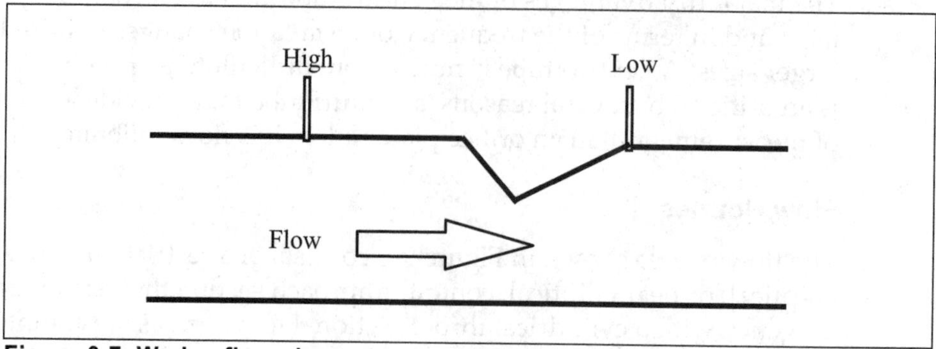

Figure 9-7. Wedge flow element

Pitot Tubes

Although the pitot tube represents one of the earliest developments in flow measurement, it has limited industrial application. Its primary use is to measure low-velocity air flow in ventilation systems.

A common industrial type of pitot tube consists of a cylindrical probe inserted into the air stream, as shown in Figure 9-8. The velocity of fluid flow at the upstream face of the probe is reduced very near to zero. Velocity head is converted into impact pressure, which is sensed through a small hole in the upstream face of the probe. A corresponding small hole in the side of the probe senses static pressure. A pressure-sensitive instrument measures the differential pressure, which is proportional to the square of the stream's velocity near the impact pressure-sensing hole. The velocity equation for the pitot tube is given by the following:

$$v = C_p\sqrt{2gh} \tag{9-29}$$

where C_P is the pitot tube coefficient.

Example 9-8 shows how to calculate the air velocity for a typical pitot tube.

The pitot tube causes practically no pressure loss in the flow stream. It is normally installed through a nipple in the side of a process pipe or ventilation duct. It is often installed through an isolation valve, so it can be moved back and forth across the stream to establish the profile of flow velocity.

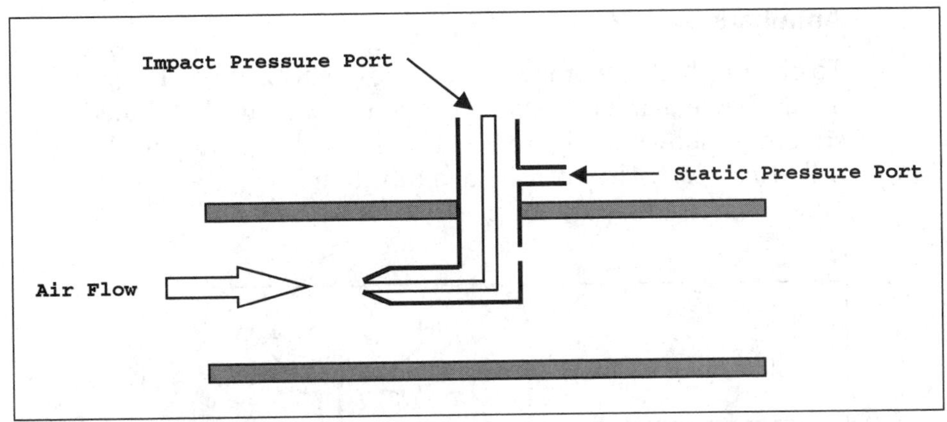

Figure 9-8. Pitot tube

EXAMPLE 9-8

Problem: A pitot tube with a coefficient of 0.98 is used to measure the velocity of air in a ventilation duct. The differential pressure head is 3 ft. What is the air velocity?

Solution: Applying Equation 9-29 for pitot tubes yields the following:

$$v = C_p\sqrt{2gh}$$

$$v = 0.98\sqrt{(2)(32.2 \text{ ft/sec}^2)(3 \text{ ft})}$$

$$v = 13.6 \text{ ft/sec}$$

Certain characteristics of pitot-tube flow measurement have limited its industrial application. To achieve true measurement of flow, it is essential to establish an average value of flow velocity. To obtain this with a pitot tube, you must move the tube back and forth across the stream to establish velocity at all points and then take an average.

For high-velocity flow streams, the measurement system must be designed to provide the necessary stiffness and strength. A pitot tube inserted in a high-velocity stream has a tendency to vibrate, which causes mechanical damage. As a result, pitot tubes are generally used only in low- to medium-flow gas applications where high accuracy is not required.

Annubars

To obtain a better average value of flow, special two-chamber flow tubes are available that have several pressure openings distributed across the stream, as shown in Figure 9-9. These annular averaging elements are called *annubars*. They consist of a tube that has high- and low-pressure holes with fixed separations.

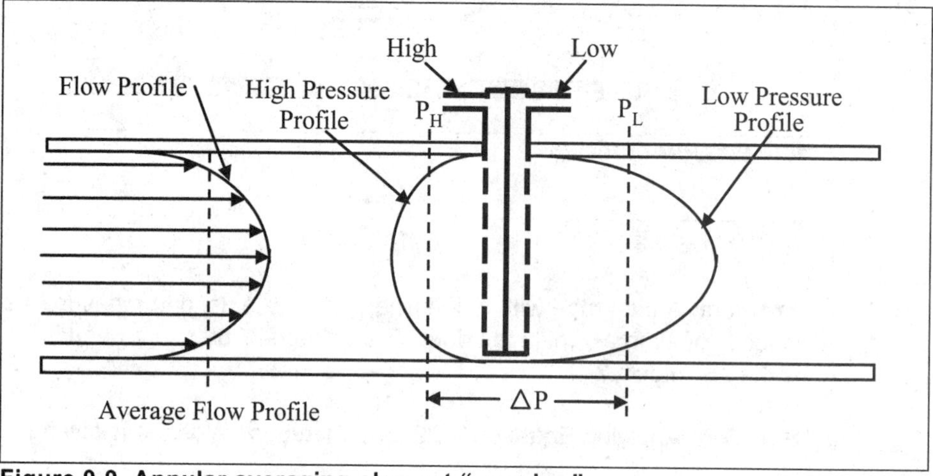

Figure 9-9. Annular averaging element "annubar"

An annubar flow sensor produces a differential-pressure (ΔP) signal that is the algebraic difference between the average value of the high-pressure signal (P_H) and the average value of the low-pressure signal (P_L), as shown in Figure 9-9.

A high-pressure profile is produced when the flow-velocity profile impacts on the upstream side of the sensing tube. An average high-pressure signal is obtained inside the high-pressure chamber by correctly placing the sensing ports in the flow tube. The flow that passes the sensor creates a low-pressure profile. Downstream ports directly behind the high-pressure ports sense this pressure profile. Operating by the same principle as the high-pressure side, an average low-pressure signal is produced in the low-pressure chamber.

Velocity-Type Flowmeters

In this section we will discuss the four common velocity-type flowmeters: turbine flowmeter, vortex shedding flowmeter, magnetic flowmeter, and ultrasonic flowmeter.

Turbine Flowmeters

The turbine flowmeter, shown in Figure 9-10, provides a frequency or pulse output signal that varies linearly with volumetric flow rate over specified flow ranges. The fluid to be measured enters the flowmeter, then passes through a rotor. The fluid passing the rotor causes it to turn with an angular velocity that is proportional to the fluid linear velocity. Therefore, the volumetric flow rate is linear within given limits of flow rate.

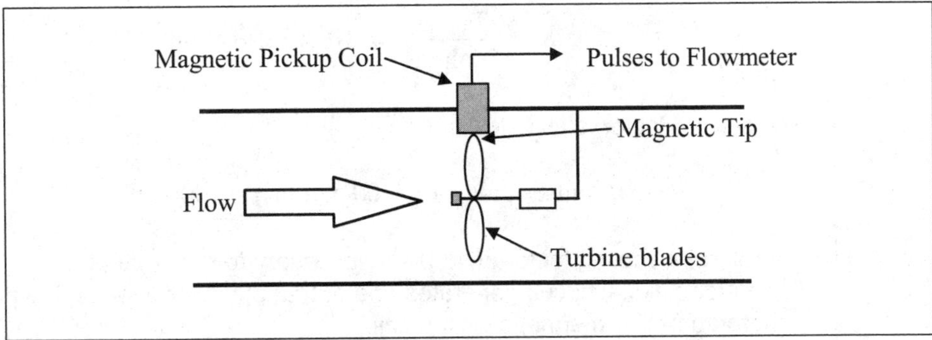

Figure 9-10. Turbine flowmeter

The pickup probe converts the rotor velocity into an equivalent frequency signal. Variable reluctance pickup assemblies are the type most commonly used. In this system, the meter housing must be nonmagnetic and, so, is usually stainless steel. The rotor must also be stainless steel.

The pickup probe consists of a small, powerful permanent magnet and a coil winding. The field of the magnet is influenced by the moving turbine blades of the rotor, which are made of a permeable material. As a rotor blade passes through the field of the magnet, it provides an easier path for the field. The field distorts and thus moves across the coil winding. The relative motion between the magnetic field and the coil winding generates an AC voltage, the frequency of which is proportional to flow rate. This can be stated in equation form as follows:

$$K = \frac{\text{Cycles/time}}{\text{Volume/time}} = \frac{\text{Pulses}}{\text{Volume}} \tag{9-30}$$

This characteristic of turbine flowmeters, called the meter coefficient (K), is used to develop a precisely known number of pulses for a given volume being measured.

Example 9-9 shows the calculation to determine meter coefficient and scaling for a digital turbine flowmeter.

EXAMPLE 9-9

Problem: A digital turbine flowmeter generates 10 pulses per gallon of liquid passing through it. Determine the meter coefficient and calculate the scaling factor needed to develop an output in which each pulse would represent 100 gallons.

Solution: The meter coefficient is as follows:

$$K = \frac{\text{Pulses}}{\text{Volume}} = 10 \text{ pulses/gallon}$$

The scaling factor is as follows:

$$(10 \text{ pulses/gallon}) \times (100 \text{ gallons}) = 1000 \text{ pulses}$$

Therefore, a scaling factor of 1,000 is necessary to ensure that the flowmeter's digital circuit generates one output pulse for every 1,000 pulses generated by the magnetic pickup coil.

The output signal from a turbine flowmeter is a frequency that is proportional to volumetric flow rate. Each pulse generated by the turbine flowmeter is, therefore, equivalent to a measured volume of liquid. Generally, flow rates are converted into flow totals by totalizer-type instruments. For the totalization to be valid the value of each pulse must be essentially constant. Therefore, the turbine flowmeter must be linear. The turbine flowmeter is generally used over its linear range. Totalization is also used for turbine flowmeters that are linear over only a part of the operating range.

Totalizers are available in two general configurations. One form simply either totalizes pulses or does the necessary scaling in direct-reading units. (Scaling means factoring the frequency information so each pulse is equal to a unit volume or decimal part of a volume.) The second configuration not only totalizes but also predetermines the number of counts or unit volumes that are proportional to a given batch size. It then provides a signal, generally a contact closure, to control the process. Batching totalizers that have a ramping function provide an analog output to (1) open a control valve to a given position, (2) control volumetric flow rate, and (3) program the shutdown of the valve to some reduced flow rate at a predetermined point in the batch. This rate is then maintained until the process is terminated.

The flow rate can be indicated digitally or in analog form. Digital counters that have an adjustable time base indicate flow rate either in terms of frequency or in direct-reading units (such as gallons per minute), depending on the time base that has been established. Analog indicators require an analog signal that is proportional to frequency.

Vortex Shedding Flowmeter

The vortex shedding flowmeter is shown in Figure 9-11. Its operating principle is fairly simple. As fluid flows past a bluff body, or shedder, at low velocity, the flow pattern remains streamlined. However, as velocity increases, the fluid separates from each side of the shedder and swirls to form vortices downstream of the shedder. A vortex is an area of swirling motion with high local velocity and thus lower pressure than the surrounding fluid. The amount of vortex generation is directly proportional to the velocity of the fluid. You can, therefore, use the relationship $Q = Av$ to obtain the flow rate.

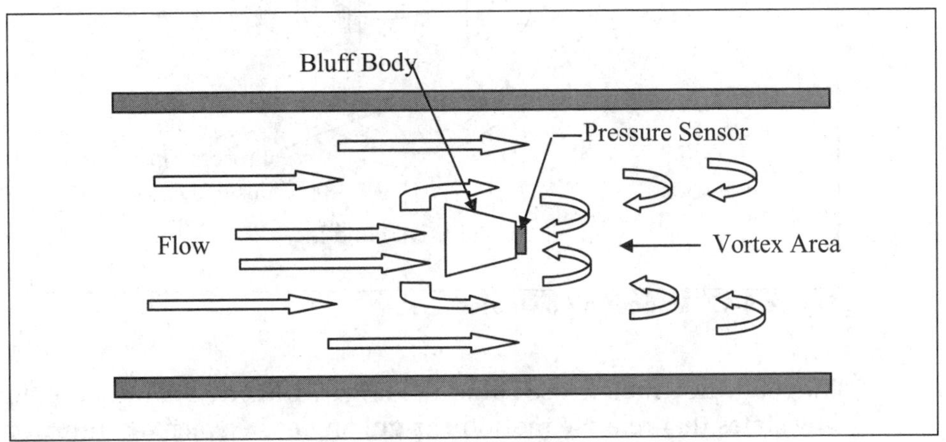

Figure 9-11. Vortex shedding flowmeter

A pressure sensor that is mounted on the downstream side of the flow shedder detects the pressure that is exerted on the shedder by the formation of vortices. The signal from the pressure sensor is converted into a calibrated flow signal by an electronic circuit in a flowmeter.

Magnetic Flowmeters

The magnetic flowmeter is constructed of a nonmagnetic tube that carries the flowing liquid, which must have a minimum level of conductivity. Surrounding the metering tube are magnetic coils and cores that provide a magnetic field across the full width of the metering tube when electric current is applied (see Figure 9-12). The fluid flowing through the tube is the conductor, and as the conductor moves through the magnetic field a voltage is generated that is proportional to the fluid velocity, which in turn is proportional to the volumetric flow rate. This voltage is perpendicular to both the magnetic field and to the direction of the flowing liquid.

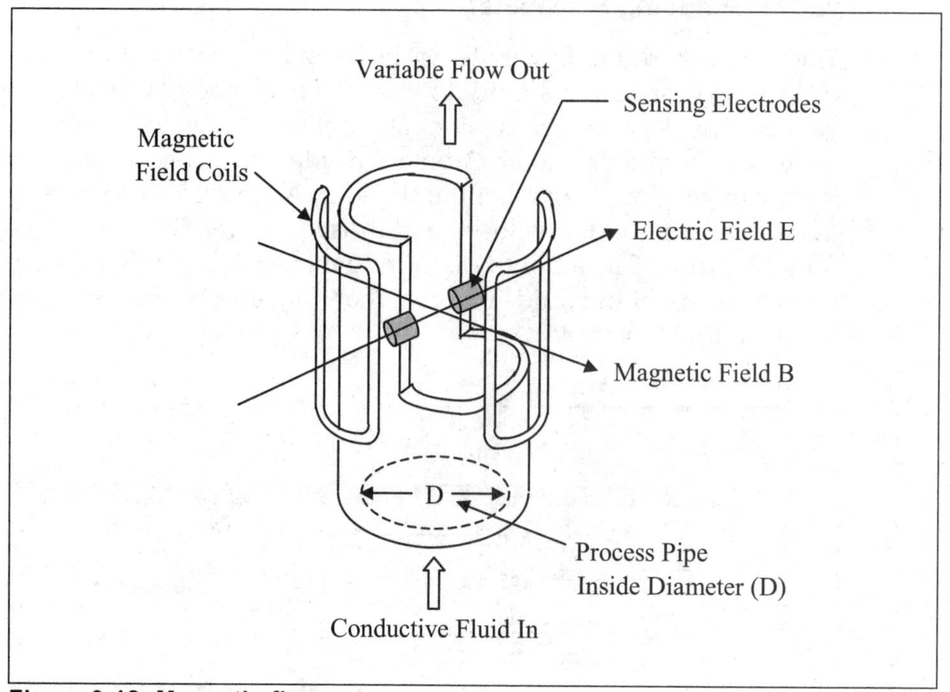

Figure 9-12. Magnetic flowmeter

Magnetic flowmeters use Faraday's law of induction to measure flow. This law states that relative motion at right angles between a conductor and a magnetic field will develop a voltage in the conductor. The induced voltage is proportional to the relative velocity of the conductor and the magnetic field. This is the principle used in DC and AC generators. The most common magnetic flowmeters are a modified form of AC generators. In the magnetic flowmeter, the fluid itself must have some minimum conductivity and acts as the conductor.

Fluid is the conductor in the magnetic flowmeter shown in Figure 9-12. The fluid's length is equivalent to the inside diameter of the flowmeter (D). The fluid conductor moves with an average velocity (v) through a magnetic field (B). The volumetric flow (Q) is proportional to the electric field (E) that the constant magnetic field induces in the conductive fluid. The mathematical relationship for the magnetic flowmeter is as follows:

$$Q = \frac{CE}{BD} \tag{9-31}$$

where

C	=	the meter constant
B	=	the magnetic field strength
D	=	the diameter of the flowmeter

The magnetic field generated lies in a plane that is mutually perpendicular to the axis of the instrument and the plane of the electrodes. The velocity of the fluid is along the longitudinal axis of the detector body. The voltage induced within the fluid is perpendicular to both the velocity of the fluid and the magnetic field, and is generated along the axis of the meter electrodes. The fluid can be considered as a series of fluid conductors that are moving through the magnetic field. An increase in flow rate will result in a greater relative velocity between the conductor and the magnetic field, and as a result a greater instantaneous value of voltage will be generated.

The instantaneous voltage generated at the electrodes represents the average fluid velocity of the flow profile. The output signal of the meter is equal to the continuous average volumetric flow rate regardless of flow profile. Therefore, magnetic flowmeters' measurements are independent of viscosity changes. It is always absolutely essential that the meter be full because the meter senses velocity as being analogous to volumetric flow rate.

Ultrasonic Flowmeters

The operating principle of ultrasonic flowmeters is to measure the velocity of sound as it passes through the fluid flowing in a pipe. The most common approach is shown in Figure 9-13. In this configuration piezoelectric crystals (barium titanate or lead zirconate-titanate) are used as sound transmitters to send acoustic signals through the fluid flowing in the pipe to receivers that are also piezoelectric crystals. The fluid flows through the pipe at a velocity v. The distance between each transmitter-receiver pair is d. The velocity of the sound through the fluid is v, and the path of the sound lies at an angle α from the pipe wall.

The velocity of sound from transmitter A to receiver B (increased by the fluid velocity) is $v_S + v \cos \alpha$, and its frequency is as follows:

$$f_a = \frac{v_s + v\cos\alpha}{d} \qquad (9\text{-}32)$$

The velocity of sound from transmitter B to receiver A (reduced by fluid velocity) is given by $v_S - v \cos \alpha$, and its frequency is as follows:

$$f_b = \frac{v_s - v\cos\alpha}{d} \qquad (9\text{-}33)$$

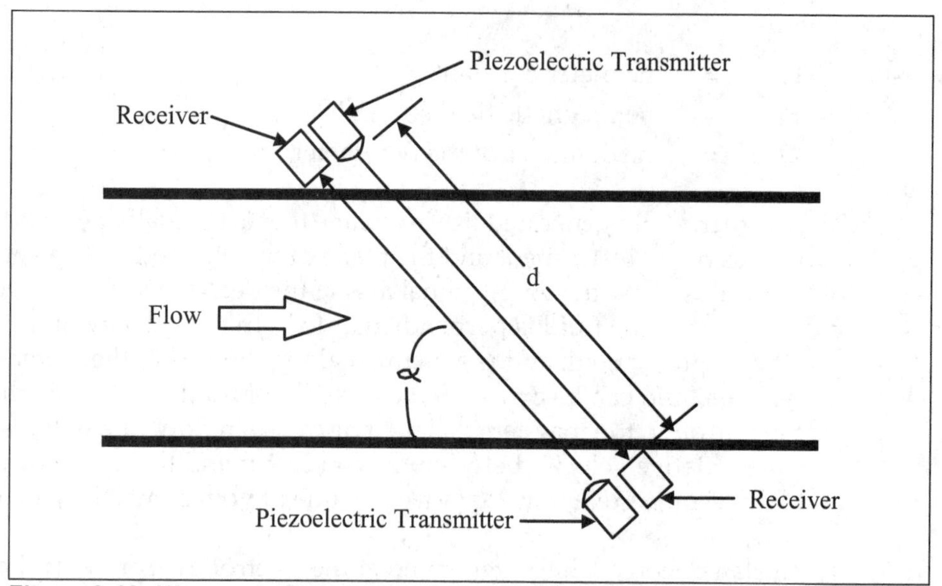

Figure 9-13. Ultrasonic flowmeter

The difference between the two frequencies or beat frequency ($\Delta f = f_a - f_b$) is given by

$$\Delta f = \frac{2v\cos\alpha}{d} \qquad (9\text{-}34)$$

Since α and d are constant, you can obtain the flow velocity by measuring this beat frequency.

Solving Equation 9-34 for flow velocity we obtain the following:

$$v = \frac{(\Delta f)(d)}{2\cos\alpha} \qquad (9\text{-}35)$$

You obtain the volumetric flow rate by multiplying the flow velocity by the cross-sectional area of the pipe.

The beat frequency is measured by using an electronic mixer. The purpose of the mixer is to translate the higher frequencies to a lower frequency level, where it is possible to amplify and select them more efficiently. In general, the design of a mixer is similar to that of a radio- frequency (*rf*) amplifier except that the latter includes an oscillator frequency. The combination of two oscillators and a mixer is referred to as a beat-frequency oscillator (*bfo*). Example 9-10 shows how to calculate fluid velocity for an ultrasonic flowmeter.

EXAMPLE 9-10

Problem: Given a beat frequency (Δf) of 100 cps for an ultrasonic flowmeter, the angle (α) between the transmitters and receivers is 45° and the sound path (d) is 12 in. Calculate the fluid velocity in feet per second.

Solution: Using Equation 9-34 for the velocity based on the beat frequency gives us the following:

$$v = \frac{(\Delta f)(d)}{2\cos\alpha}$$

$$v = \frac{(100\ \text{cycles/sec})(1\ \text{ft})}{2\cos 45^\circ}$$

$$v = 70.7\ \text{ft/s}$$

Ultrasonic flowmeters are normally installed on the outside of liquid-filled pipes. This is so the measuring element is nonintrusive and will not induce a pressure drop or disturbance into the process stream. Ultrasonic flowmeters generally cost more than standard flow measuring devices, such as orifice plates or venturi tubes. However, they can be easily attached to the outside of existing pipes without having to shut down the process or use special pipe sections or isolation valves. For that reason, their overall cost compared to conventional flowmeters is generally less than alternative meters in the larger pipe sizes.

Positive Displacement Flowmeters

Positive displacement (PD) flowmeters continuously entrap a known quantity of fluid as it passes through the meter. Since both the number of times the fluid is entrapped and the volume of the entrapped fluid are known, you can easily determine the amount of fluid that has passed through the meter. This section discusses the three common types of PD flowmeters encountered in process control: rotary vane, oval gear, and nutating disk.

Rotary Vane PD Flowmeters

Rotary vane PD flowmeters are widely used in liquid processes where accuracy is important. This type of flowmeter converts the entrapment of liquid into a rotational velocity that is proportional to the flow through the device. The forces exerted by the flowing fluid rotate blades in the flowmeter on a center shaft. The blades and inside body of the flowmeter are machined to close tolerances during manufacture since they must form a tight seal with each other over the life of the flowmeter.

Oval Gear PD Flowmeters

Oval gear PD flowmeters are generally used on very viscous liquid, which is difficult to measure using other flowmeters. The liquid flow through the flowmeter applies a force on a pair of oval gears, which causes them to rotate.

As Figure 9-14 shows, in position 1 uniform forces are applied equally on the top and bottom of oval gear B, so this gear does not rotate. Rotor A has entrapped a known quantity of liquid between the rotor and the meter body, and there is a balanced force on the top of the gear. However, there is force on the bottom of gear A, which causes it to rotate clockwise (CW). This causes gear B to rotate in a counterclockwise (CCW) direction to position 2.

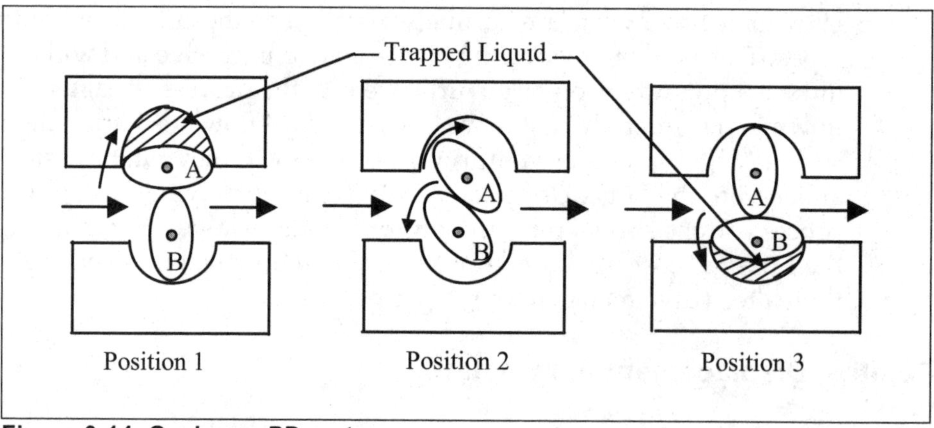

Figure 9-14. Oval gear PD meter

In position 2, fluid enters the space between gear B and the flowmeter body, as the fluid that was entrapped between gear A and the body simultaneously leaves the area of entrapment. The higher upstream pressures oppose the lower downstream pressures at the ends of gear A and gear B. This causes both gears to continue to rotate in CW and CCW directions, respectively, to position C.

In position 3, a known amount of fluid has been entrapped between gear B and the flowmeter body. This operation is then repeated, with each revolution of the gears representing the passage of four times the amount of fluid that fills the space between the gear and the flowmeter body. Therefore, the fluid flow is directly proportional to the rotational velocity of the gears.

Nutating Disk PD Flowmeters

Nutating disk PD flowmeters are generally used in water service to obtain low-cost flow measurement where high accuracy is not required. The nutating disk flowmeter uses a cylindrical measurement chamber, in which a disk is allowed to wobble, or nutate, as fluid flows through the flowmeter, causing the spindle to rotate. This rotation can be used to drive an indicator or transmitter. Since this PD flowmeter entraps a fixed amount of fluid each time the spindle is rotated, the rate of flow is directly proportional to the rotational velocity of the spindle.

Coriolis Mass Flowmeters

The operating principle of Coriolis mass flowmeters is based on the force exerted by the Coriolis acceleration of a fluid. The flowmeter consists of a vibrating tube in which the Coriolis acceleration is created and measured. A typical flow tube is shown in Figure 9-15.

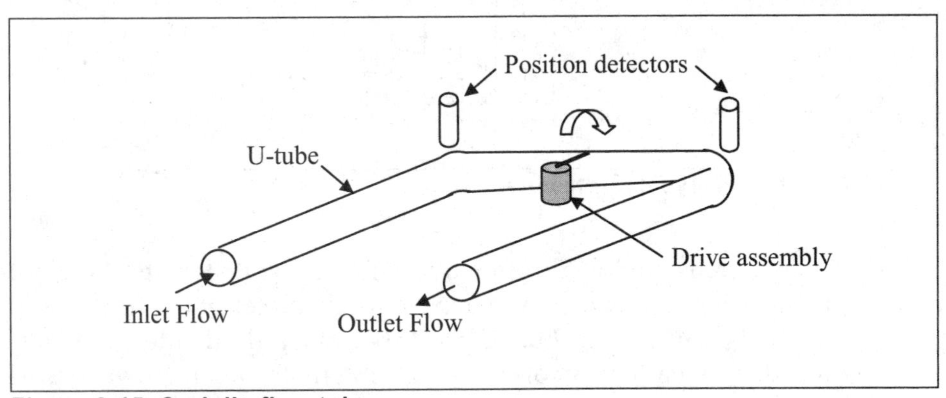

Figure 9-15. Coriolis flow tube

The flow tube is designed and built to have predictable vibration characteristics. A drive assembly connected to the center of the tube causes the tube to twist as shown in Figure 9-15. This vibrates the tube. Position-sensing coils on each side of the flow tube sense this twisting. Since the frequency of the vibration of the tube varies with the density of the fluid inside the tube, the computer inside the electronics unit of the Coriolis flowmeter can calculate a density value. Coriolis flowmeters can be used on virtually any liquid or gas that flows at a mass great enough to operate the meter.

Rotameter

The rotameter is a type of variable-area flowmeter that consists of a tapered metering tube and a float, which is free to move up and down within the tube. The metering tube is mounted vertically, with the small end at the bottom. The fluid to be measured enters at the bottom of the tube, passes upward around the float, and out at the top. Figure 9-16 shows a typical rotameter.

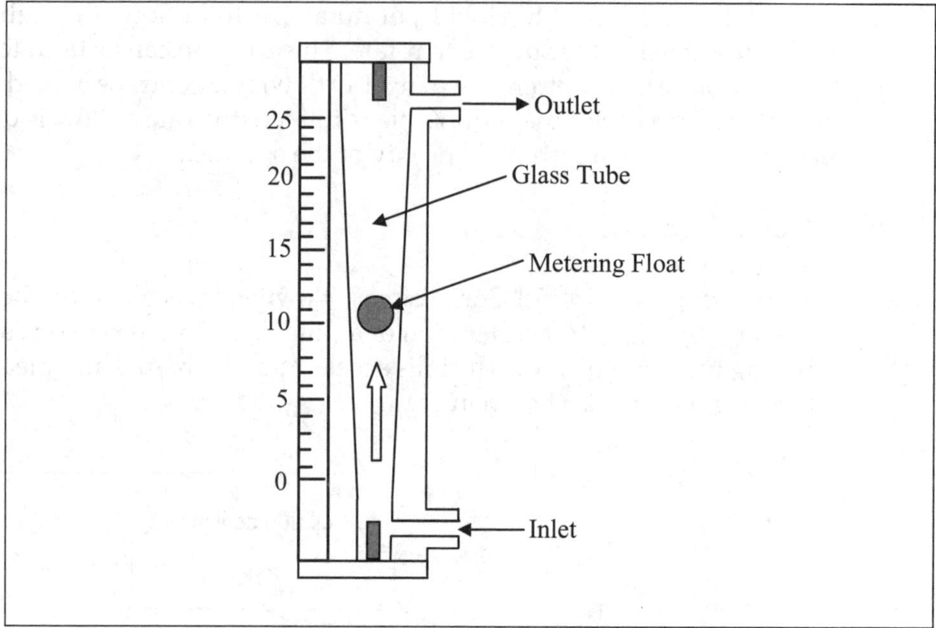

Figure 9-16. Typical rotameter

When there is no flow through a rotameter, the float rests at the bottom of the metering tube, where the maximum diameter of the float is approximately the same as the bore of the tube. When fluid enters the metering tube, the buoyant effect of the fluid lightens the float. However, the float has a greater density than the fluid, and the buoyant effect is not sufficient to raise it.

There is a small annular opening between the float and the tube. The pressure drop across the float increases and raises the float. This increases the area between the float and tube until the upward hydraulic forces acting on it are balanced by its weight, less the buoyant force.

The float moves up and down in the tube in proportion to the fluid flow rate and the annular area between the float and the tube. It reaches a stable position in the tube when the forces are in equilibrium. When the float moves upward toward the larger end of the tapered tube, the annular opening between the tube and the float increases. As the area increases, the pressure differential across the float decreases. The float will assume a position of dynamic equilibrium when the pressure differential across the float plus the buoyancy effect balances the weight of the float. Every float position corresponds to one particular flow rate and no other for a fluid of a given density and viscosity. You take the flow reading from a calibrated scale on the tube.

Glass rotameters are used in many applications, but metal rotameters are used where glass is unsatisfactory. In these cases, you must determine the position of the float indirectly by using either magnetic or electrical techniques. Indirect float position sensors can also provide functions other than direct visual indication. Some rotameters output pneumatic, electronic, or pulse signals.

You can use fluid mechanics theory to derive the following basic equation for liquid flow through a rotameter:

$$Q = CA_a\sqrt{\frac{\rho_F - \rho_f}{\rho_f}} \tag{9-36}$$

where

Q	=	the volumetric flow rate
C	=	a meter flow constant
A_a	=	the annular area between the tube and the float
ρ_F	=	the density of the float
ρ_f	=	the density of the fluid

The rotameter is an inexpensive instrument for measuring gas flow. The pressure drop across the meter is essentially constant over the full ten-to-one operating range. The pressure drop is low, generally less than 1 psi.

The position of the float in the metering tube varies in a linear relationship with flow rate. This is true over ranges up to 10:1. Rotameters can directly measure flows as high as 4,000 gal/min. Higher flow rates can be economically handled by using the bypass-type rotameter. Replacing the float with a different-sized float within limits can change the capacity of the rotameter. By using the same housing, but changing both the metering tube and the float, you can achieve a gross change in capacity. These changes can account for both a change in flow rate and a change in fluid density.

The rotameter tends to be self-cleaning. The velocity of the flow past the float and the freedom of the float to move vertically enable the meter to clean itself by eliminating some buildup of foreign material. Liquids that have fibrous materials are an exception, and you should not meter them with rotameters. Generally, particle size, particle type (whether fibrous or particulate), and particulate abrasiveness determine the suitability of the rotameter for a given application. Other factors include the percentage of solids by weight or by volume and the density of the solids.

EXERCISES

9.1 A valve is opened on the bottom of a process storage tank that is filled with liquid to a height of 10 feet. Find the discharge velocity of the liquid just after the outlet valve is opened.

9.2 Water at 80°F is pumped through a process pipe with a 2-in. inside diameter at a flow velocity of 4.0 ft/s. Find the volumetric flow and the mass flow. The density (ρ) of water is 62.2 lbs/ft^3 at 80°F.

9.3 An incompressible fluid is flowing in a process pipe that has an inside diameter of 4 in. under a pressure head of 20 in. Calculate the fluid velocity and volume flow rate.

9.4 Water at 100°F is flowing at a rate of 40 gpm through a 2-in. schedule 40 steel pipe. Calculate the Reynolds number and determine if the flow is laminar, transitional, or turbulent. Note that at 100°F, the density of water is 61.996 lbs/ft^3, and the viscosity is 0.74 centipose (cP).

9.5 Gas that has a density of 0.4 lb/ft^3 and viscosity of 0.02 cP is flowing at 1 acfm through a 2-in. schedule 40 pipe. Calculate the Reynolds number and determine if the flow is laminar, transitional, or turbulent.

9.6 A liquid is flowing past a restriction in a process pipe with a volumetric flow of 10 ft^3/sec, causing a pressure drop of 2 in.H_2O across the restriction. Calculate the volumetric flow if the pressure drop across the restriction increases to 5 in.H_2O.

9.7 An incompressible fluid is flowing through an orifice plate with a flow coefficient of 0.9, causing a pressure drop of 10 in.H_2O. Calculate the fluid velocity.

9.8 Explain the advantages and disadvantages of the three main types of orifice plates.

9.9 What is the basic principle used by venturi tubes, flow nozzles, and wedge flow elements to measure flow?

9.10 A pitot tube with a coefficient of 0.95 is used to measure the velocity of air in a ventilation duct. The differential pressure head measured by the pitot tube is 1 ft. What is the air velocity?

9.11 A digital turbine flowmeter generates 2 pulses per gallon of liquid passing through the meter. Calculate the meter coefficient, and calculate the scaling factor that is necessary to develop an output in which each pulse would represent 10 gallons.

9.12 Given a beat frequency (Δf) of 50 cps for an ultrasonic flowmeter, the angle (α) between the transmitters and receivers is 45°, and the sound path (d) is 6 in. Calculate the fluid velocity in feet per second.

10 Final Control Elements

Introduction

In Chapter 1, we introduced the concept of process control and defined its three elements: measurement, evaluation, and final control. The final control element is probably the most important because it exerts a direct influence on the process. Final control devices contain the essential pieces of equipment to convert the control signal (generated by a process controller) into the action needed to correctly control the process.

In this chapter, we discuss the fundamentals of final control devices, such as control valves, motors, and pumps. Because control valves are the single most common type of final control element in process control, we will discuss them first and in the greatest detail.

Control Valve Basics

A control valve is simply a variable orifice that is used to regulate the flow of a process fluid according to the requirements of the process. Figure 10-1 illustrates a typical globe-type control valve body in both the fully open and fully closed positions. In a control valve, an actuator that is connected to the valve's plug stem moves the valve between the open and closed positions to regulate flow in the process. The valve body is mounted in the process fluid line and is used to control the flow of fluid in the process.

The *body* of a control valve is generally defined as the part of the valve that comprises the main boundary, including the connecting ends. Valves are classified into two general types based on the movement of the valve's closure part: *linear* and *rotary*.

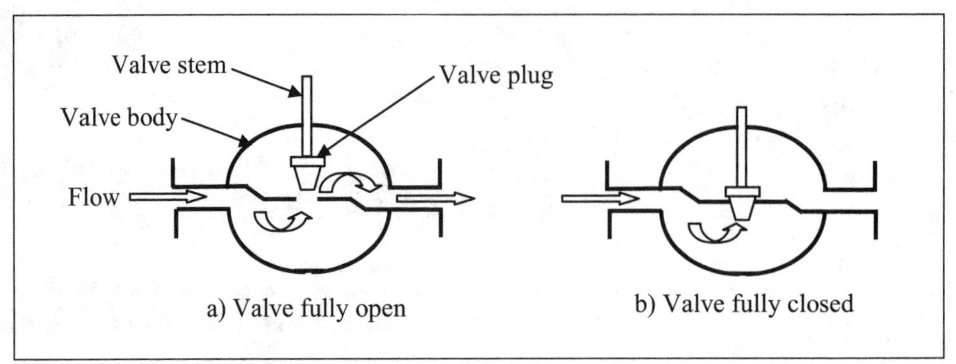

Figure 10-1. Typical globe-type control valve

Types of Control Valves

Though there are many kinds of valves, the most common types are globe, gate, diaphragm, butterfly, and ball valves. The *globe* valve, which is of the linear movement type, is the most common of these five types. In a globe valve, the plug is attached to a stem, which is moved linearly in a cavity with a somewhat globular shape to regulate flow (see Figure 10-1).

A flat or wedge-shaped plate that is moved into or out of the flow path to control flow characterizes the *gate* valve. These valves are widely used for manual on/off service, but a few designs are used in throttling service.

Diaphragm valves are linear-motion valves with flexible diaphragms that serve as flow closure members. Diaphragm valves are mainly used with difficult fluids such as corrosive liquids or slurries. The valve body can be lined with glass, plastic, or Teflon. The diaphragm is normally rubber, but in some cases it is Teflon, which, however, requires a high valve closure force.

The *butterfly* valve is by far the most common rotary-motion control valve. Butterfly valves range in size from one-half inch to over two hundred inches. In the very large pipe sizes, the butterfly valve is the only cost-effective solution for the control valve application.

The *ball* valve is also a rotary-motion valve. The part that closes the flow is a sphere with an internal passageway. The ball valve is the most widely used control valve after the globe valve. Advances in seal design and sealing material enable the ball valve to offer tight shutoff. Because of this feature it is now widely used in on/off service for batch processes.

Control Valve Characteristics

The valve's flow characteristic is the relationship of the change in the valve's opening to the change in flow through the valve. The most frequently used characteristics are quick-opening, linear, and equal percentage, as shown in Figure 10-2.

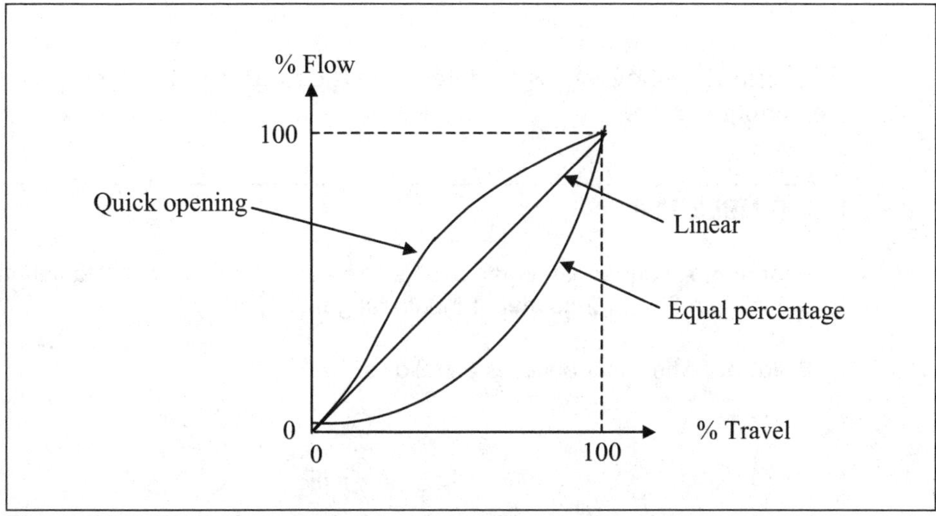

Figure 10-2. Flow characteristic curves of common valves

The *quick-opening valve* is predominantly used for on/off control applications. A relatively small movement of the valve stem causes the maximum possible flow rate through the valve. For example, a quick-opening valve may allow 85 percent of the maximum flow rate with only 25 percent stem travel.

The *linear valve* has a flow rate that varies linearly with the position of the stem. This relationship can be expressed as follows:

$$\frac{Q}{Q_{max}} = \frac{X}{X_{max}} \tag{10-1}$$

where

Q	=	flow rate
Q_{max}	=	maximum flow rate
X	=	stem position
X_{max}	=	maximum stem position

The *equal percentage valve* is manufactured so that a given percentage change in the stem position produces the same percentage change in flow.

Generally, this type of valve does not shut off the flow completely in its limit of travel. Thus, Q_{min} represents the minimum flow when the stem is at one limit of its travel. At the fully open position, the control valve allows a maximum flow rate, Q_{max}. So we define a term called *rangeability* (R) as the ratio of maximum flow (Q_{max}) to minimum flow (Q_{min}):

$$R = \frac{Q_{max}}{Q_{min}} \tag{10-2}$$

Example 10-1 shows how to determine the rangeability of a typical equal percentage valve.

EXAMPLE 10-1

Problem: An equal percentage valve has a maximum flow of 100 gal/min and a minimum flow of 2 gal/min. Find its rangeability (R).

Solution: The rangeability is as follows:

$$R = \frac{Q_{max}}{Q_{min}}$$

$$R = \frac{100 \text{ gal/min}}{2 \text{ gal/min}} = 50$$

Control Valve Actuators

Control valve actuators translate a control signal (normally 3-to-15 psi or 4-to-20 mA) into the large force or torque that is needed to manipulate a valve. There are two common types of actuators: electric and pneumatic. *Electric motor actuators* are used to control the opening and closing of smaller rotary-type valves such as butterfly valves. However, *pneumatic actuators* are used more widely because they can effectively translate a small control signal into a large force or torque. The force generated by the pneumatic valve actuator is based on the definition of pressure as force per unit area:

$$F = \Delta PA \tag{10-3}$$

where

ΔP = the differential pressure (Pa)

A = the area (m^2)

F = the force (N)

If we need to double the force for a given pressure, we only need to double the area over which the pressure is applied. Very large forces can be developed by the standard 3-15 psi (20–100 KPa) control signal. Many types of pneumatic valve actuators are available, but the most common is the diaphragm type shown in Figure 10-3. The pneumatic diaphragm valve actuator can be designed as a direct-acting type (as shown in Figure 10-3) or as reverse-acting (as shown in Figure 10-4).

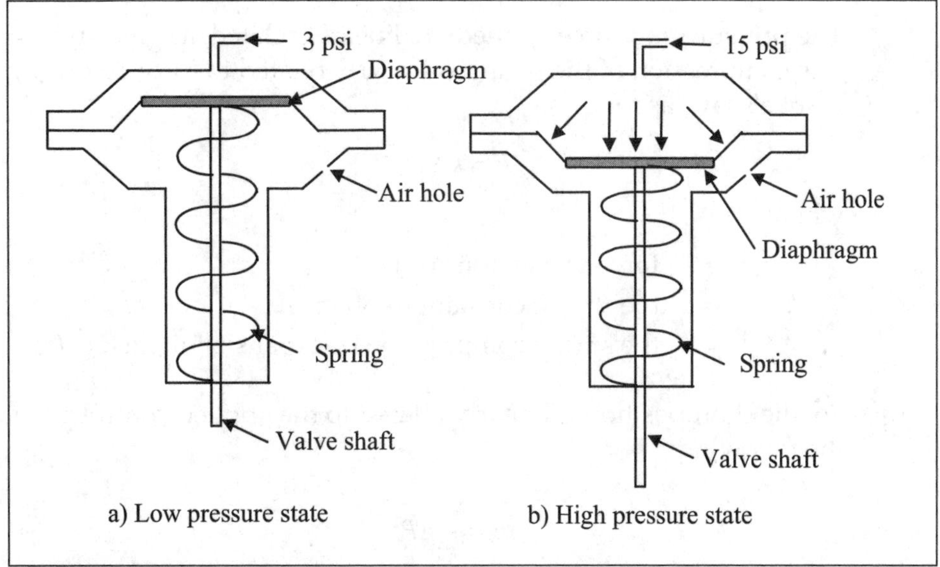

Figure 10-3. Direct-acting pneumatic valve actuator

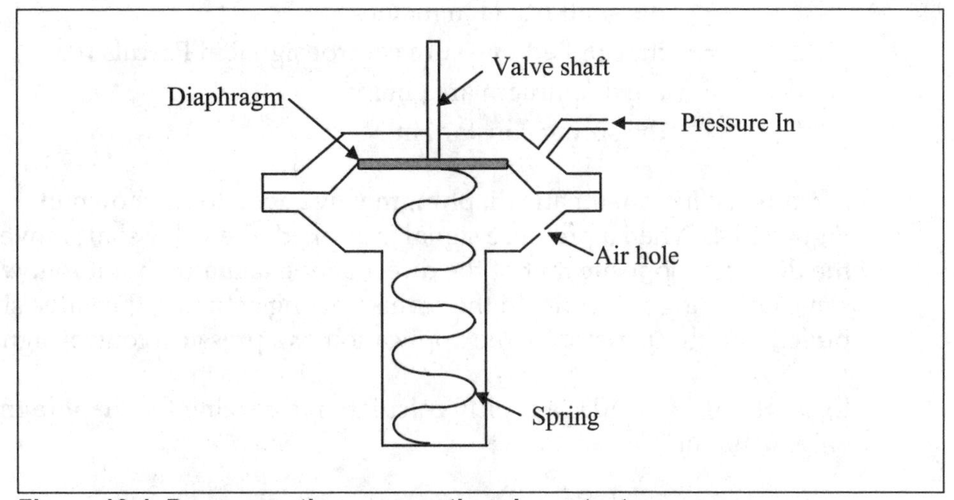

Figure 10-4. Reverse-acting pneumatic valve actuator

Figure 10-3a shows the actuator with a low-pressure (3 psi) signal applied. This signal ensures that the spring maintains the diaphragm and the connected valve shaft in the position shown. The pressure on the opposite or spring side of the diaphragm is maintained at atmospheric pressure because there is a hole opened to the atmosphere on the bottom right side of the actuator. Increasing the control signal pressure applies a force on the diaphragm that moves the diaphragm and the connected valve shaft down against the spring. Figure 10-3b shows this action in a case where the maximum control signal of 15 psi (100 kPa) produces the maximum travel for the valve shaft.

The pressure and force applied are linearly related, as given by $F = \Delta PA$. The compression of the spring is linearly related to force according to Hooke's law, as follows:

$$F = k\Delta d \tag{10-4}$$

where

F	=	force in Newtons (N)
k	=	the spring constant in N/m
Δd	=	the spring compression or expansion in meters (m)

So, the shaft position is linearly related to the applied control pressure by the following:

$$\Delta d = \frac{A}{k}\Delta P \tag{10-5}$$

where

Δd	=	the shaft travel in meters
ΔP	=	the applied pressure control signal in Pascals (Pa)
A	=	the diaphragm area in m^2
k	=	the spring constant in N/m

A *reverse-acting* pneumatic diaphragm valve actuator is shown in Figure 10-4. When a pressure signal is applied, the valve shaft moves in the direction opposite that of the direct-acting actuator, but it follows the same operating principle. In the reverse-acting actuator, the valve shaft is pulled into the actuator by the application of a pressure control signal.

Example 10-2 provides a typical calculation for sizing the diaphragm of a valve actuator.

Valve Sizing for Liquids

Sizing a control valve incorrectly is a mistake both technically and economically. A valve that is too small will not pass the required flow, and

EXAMPLE 10-2

Problem: Assume that a force of 500 N is required to fully open a control valve that is equipped with a pneumatic diaphragm valve actuator. The valve input control signal for the actuator has a range of 3 to 15 psi (20 to 100 kPa). Find the diaphragm area that is required to fully open the control valve.

Solution: The area is calculated using Equation 10-3, as follows:

$$F = \Delta P A$$

Since the pressure signal required to open the valve fully is 100 kPa we have the following:

$$A = \frac{F}{\Delta P} = \frac{500N}{1x10^5 Pa} = 5x10^{-3} m^2$$

this will impact the process. A valve that is oversized will be unnecessarily expensive, can lead to instability, and can make it more difficult to control flow in a process.

To select the correct-sized valve for a given application you must know what process conditions the valve will actually encounter in service. The technique used to size control valves is a combination of fluid flow theories and flow experimentation.

Daniel Bernoulli was one of the first scientists to take an interest in the flow of liquids. Using the principle of conservation of energy, he discovered that as a liquid flows through an orifice the square of the fluid velocity is directly proportional to the pressure differential across the orifice and inversely proportional to the density of the fluid. In other words, the greater the pressure differential, the higher the velocity; the greater the fluid density, the lower the velocity. Equation 10-6 gives this relationship.

$$v = K\sqrt{\frac{\Delta P}{\rho}} \tag{10-6}$$

In addition, as we saw in Chapter 9, you can calculate the volumetric flow of liquid by multiplying the fluid velocity by the flow area, or $Q = Av$. As a result, the volumetric flow through an orifice is given by the following:

$$Q = KA\sqrt{\frac{\Delta P}{\rho}} \tag{10-7}$$

If Equation 10-7 is expressed in U.S. engineering units, then volumetric flow (Q) is in gallons per minute (gpm), pressure differential (ΔP) is in psi, and specific gravity (G) and the flow area (A) are in square inches. Letting the constant (C) account for the proper units of flow, we obtain the following:

$$Q = CA\sqrt{\frac{\Delta P}{G}} \tag{10-8}$$

Although Equation 10-8 has a strong theoretical foundation, it does not take into account the energy losses caused by turbulence and friction as the fluid passes through the orifice. We can compensate for this by adding a discharge coefficient (C_d) that is different for each type of flow orifice:

$$Q = C_d CA\sqrt{\frac{\Delta P}{G}} \tag{10-9}$$

Since the flow area is also a unique function of each type of flow orifice, we can combine all three of these terms into a single coefficient. When applied to valves, this coefficient is called the valve-sizing coefficient (C_v). Replacing the three terms C_d, C, and A with C_v we obtain the equation for liquid flow through a control valve:

$$Q = C_v\sqrt{\frac{\Delta P}{G}} \tag{10-10}$$

Example 10-3 illustrates how to calculate the volumetric flow through a control valve.

EXAMPLE 10-3

Problem: Water flows through a 2-in. control valve with a pressure drop of 4 psi. Find the volumetric flow rate if the control valve has a C_v of 55.

Solution: Using Equation 10-10

$$Q = C_v\sqrt{\frac{\Delta P}{G}}$$

we obtain the following:

$$Q = 55\sqrt{\frac{4}{1}}\ \text{gpm} = 110\ \text{gpm}$$

The valve-sizing coefficient C_v is experimentally determined for each different size and style of valve by using water in a test line under carefully controlled standard conditions. Figure 10-5 shows the standard test piping arrangement established by the Fluid Controls Institute (FCI) to measure C_v data uniformly. Using this test setup, control valve manufacturers determine and publish C_v values for their valves. These published C_v make it relatively easy to compare the capacities of the various valves offered by manufacturers.

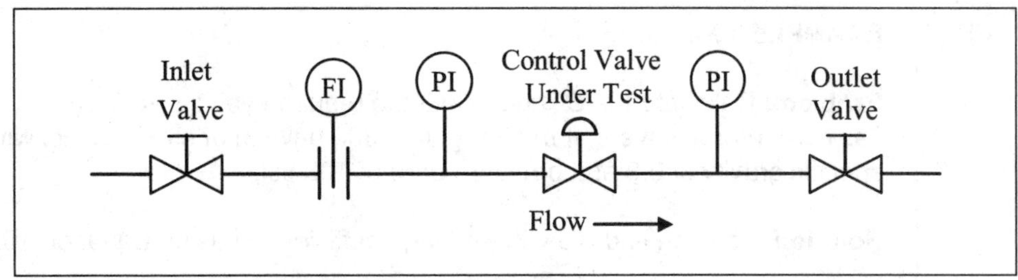

Figure 10-5. Test piping for C_v measurements

To determine the C_v for a valve that controls the flow of water or other liquids that behave like water, we rearrange the liquid valve sizing equation (Equation 10-10) as follows:

$$C_v = Q\sqrt{\frac{G}{\Delta P}} \tag{10-11}$$

This equation is based on ideal liquids, but viscous conditions can result in significant sizing errors since manufacturers' published C_v values are based on test data using water. Although most valve applications involve liquids whose viscosity corrections can be ignored, you should consider liquid viscosity in each application that requires you to select a valve.

A careful review of the basic valve liquid flow equation (Equation 10-10) will help you develop a feel for what C_v really means. Consider a case where water at 60°F flows through a valve. Here the specific gravity (G) is equal to 1. Let's also assume that a 1 psi pressure differential is maintained across the valve. Under these conditions, the entire square root factor becomes 1. This specific example shows that C_v is numerically equal to the number of U.S. gallons of water that will flow through the valve in one minute when the water temperature is 60°F and the pressure differential across the valve is 1 psi. Thus, C_v provides an index for comparing the liquid flow capacities of different types of valves under a standard set of operating conditions.

The valve-sizing coefficient C_v varies with both the size and style of valve. By combining published C_v data, the basic liquid sizing equation, and your actual service conditions, you can select the correct valve size for any given application. Typical values of C_v for different-sized valves are shown in Table 10-1.

Example 10-4 illustrates the sizing and selection process for a typical liquid service application.

EXAMPLE 10-4

Problem: Calculate the C_V and select the required valve size from Table 10-1 for a valve that must regulate 300 gal/min of ethyl alcohol with a specific gravity of 0.8 at a pressure drop of 100 psi.

Solution: You can find the valve-sizing coefficient by using Equation 10-11.

$$C_v = Q\sqrt{\frac{G}{\Delta P}}$$

Then, substitute the process parameters into this equation to obtain C_V.

$$C_v = 300\sqrt{\frac{0.8}{100}}$$

$$C_V = 26.8$$

Based on this value of C_V, you should select a 1½ in. valve for this application.

Table 10-1. Typical Valve-Sizing Coefficients

Valve Size, in.	C_V
¼	0.3
½	3.0
1	15
1 ½	35
2	55
3	110
4	175
6	400
8	750

Flashing and cavitation within a control valve are two other conditions that can significantly affect the sizing and selection of control valves. These two related phenomena limit flow through control valves under certain physical conditions. You must therefore take them into account when selecting the proper valve for a given application.

Flashing and Cavitation

Flashing and cavitation involve a change in the form of the fluid media from the liquid to the vapor state. The change is caused by an increase in fluid velocity at or just downstream of the flow restriction, normally at the valve port or inside the body.

To simplify our discussion of flashing and cavitation, let's imagine a process in which there is a simple restriction in the line. As the flow passes through the physical restriction, there is a necking down, or contraction, of the flow stream. The point at which the cross-sectional area of the flow stream is smallest is called the *vena contracta*, which is just a short distance downstream of the physical restriction (Figure 10-6).

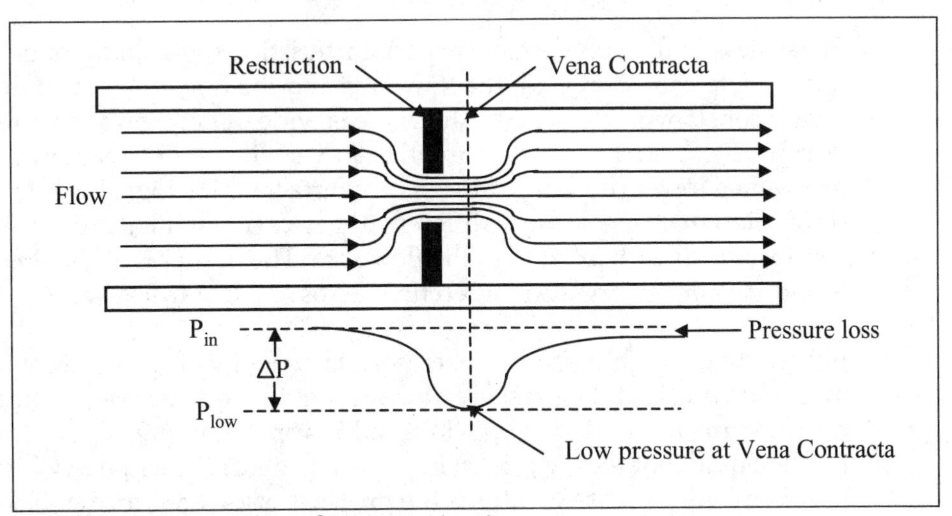

Figure 10-6. Illustration of vena contracta

To understand flashing and cavitation, you first need to understand the interchange between the kinetic and potential energy of a fluid flowing through a valve or other restriction. To maintain a steady flow of liquid through the valve, the velocity must be greatest at the vena contracta. This increase in velocity, or kinetic energy, comes about at the expense of pressure, or potential energy. The pressure profile along the valve shows a sharp decrease in pressure as velocity increases. The lowest pressure, of course, will occur at the vena contracta, where the velocity is the greatest.

Then, farther downstream, as the fluid stream expands into a larger area, velocity will decrease and thus pressure will increase. The pressure downstream of the valve never recovers completely to the level that existed upstream.

The pressure differential that exists across the valve is called the ΔP of the valve. This ΔP is a measure of the amount of energy that was dissipated in the valve. Useful energy is lost in the valve because of turbulence and friction. The more that energy is dissipated in a valve, the greater the ΔP for a given area and flow.

If two valves have the same flow area and upstream pressure and are passing the same flow, then it follows that they must have the same velocities at the vena contracta. This, in turn, means that the pressure drop from the inlet to the vena contracta must also be the same. On the other hand, if one valve dissipates less energy due to turbulence and friction, more energy will be left over for recovery in the form of downstream pressure. Such a valve would be relatively streamlined and classified as a high-recovery valve. In contrast, a low-recovery valve dissipates more energy and, consequently, has a greater ΔP for the same flow.

Regardless of the valve's recovery characteristics, the amount of liquid flow is determined by both the flow area and the flow velocity. If the flow area is constant, such as when the valve is wide open, then any increase in flow must come from an increase in fluid velocity. An increase in velocity results in a lower pressure at the vena contracta. This logic leads to the conclusion that the pressure differential between the inlet and the vena contracta is directly related to the flow rate. The higher that the flow rate is, the greater the pressure differential across the control valve.

If the flow through the valve increases, the velocity at the vena contracta must increase, and the pressure at that point will decrease accordingly. If the pressure at the vena contracta should drop below the vapor pressure for the liquid, bubbles will form in the fluid stream. The rate at which bubbles form will increase greatly as the pressure is lowered further below the vapor pressure. At this stage of development, there is no difference between flashing and cavitation. It is what happens downstream of the vena contracta that makes the difference.

If the pressure at the outlet of the valve is still below the vapor pressure of the liquid, the bubbles will remain in the downstream system creating what is known as *flashing*. If the downstream pressure recovery is sufficient to raise the outlet pressure above the liquid vapor pressure, the bubbles will collapse, or implode, producing *cavitation*. It is easy to visualize why high-recovery valves tend to be more susceptible to cavitation since the downstream pressure is more likely to rise above the vapor pressure.

The implosion of the vapor bubbles during cavitation releases energy that manifests itself as noise and physical damage to the valve. Millions of tiny bubbles imploding near the valve's solid surfaces can gradually wear away the material, causing serious damage to the valve body or its internal parts. It is usually quite apparent when a valve is cavitating; a noise much like gravel flowing through the valve will be audible. Areas damaged by cavitation also appear rough, dull, and cinder-like.

Flashing can also damage valves, but it will produce areas that appear smooth and polished since flashing damage is essentially erosion. The greatest flashing damage tends to occur at the point of highest velocity.

Choked Flow

The noise and physical damage to the valve that cavitation and flashing cause are important considerations when sizing a valve, but even more important is that they limit flow through the valve. During both phenomena, bubbles begin to form in the flow stream when the pressure drops below the vapor pressure of the liquid.

These vapor bubbles cause a crowding condition at the vena contracta that reduces the amount of liquid mass that can be forced through the valve. Eventually, the flow is saturated or choked and can no longer be increased.

The valve-sizing equation (Equation 10-10) implies that there is no limit to the flow you can obtain as long as the ΔP across the valve increases. However, this is not true if flashing or cavitation occur. Assuming that the upstream pressure is constant, the flow that can be achieved by decreasing the downstream pressure is limited.

The flow curve deviates from the relationship predicted by the basic liquid-sizing equation (Equation 10-11) because of the vapor bubbles that appear in the flow stream during flashing or cavitation. If you increase the valve pressure slightly beyond the point where bubbles begin to form, the flow will become choked. Assuming constant upstream pressure, a further increase in ΔP will not increase the flow. This limiting pressure differential is called "ΔP allowable" or ΔP_a. Figure 10-7 shows how the ΔP_a is related to the basic liquid-sizing equation and the choked flow. Equation 10-11 will plot as a straight line if the horizontal axis is the square root of ΔP, with a slope equal to the valve-sizing coefficient (C_v). If you use the actual service ΔP in the basic liquid-sizing equation when

$\Delta P > \Delta P_a$, the equation will predict a much larger flow than will actually exist in service.

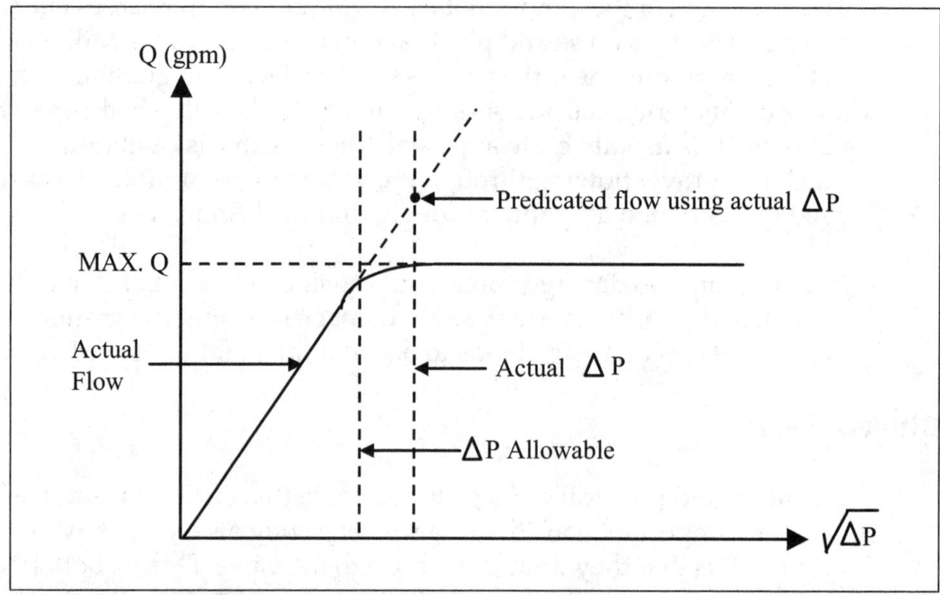

Figure 10-7. Relationship between actual and allowable differential pressure

The choked flow is the flow that will actually exist in service when ΔP is greater than ΔP_a. These two conditions are illustrated in Figure 10-7. The predicted flow, using the actual ΔP, is indicated on the dashed portion of the curve plotted for the liquid-sizing equation. Only the choked flow would exist in service. This example clearly shows why the smaller of the two ΔPs must be used in Equation 10-11 to obtain accurate results.

The allowable differential pressure, ΔP_a, is the pressure drop at which choked flow occurs because of flashing or cavitation and it is a function of the valve's flow geometry. Although the drop can vary widely from one style of valve to the next, it is quite predictable for any given valve and you can easily determine it by test. The experimental coefficient that is used to define the point of choked flow for any valve is called K_m.

For a given flow, the pressure drop across the valve is a measure of the valve-recovery characteristics. At the point of choked flow, the pressure drop is ΔP_a. We showed earlier that the pressure differential from the inlet to the vena contracta is a measure of the amount of flow being forced through the valve. The ratio of these two pressure differentials, when the valve has just reached choked flow, is defined as the valve-recovery coefficient (K_m):

$$K_m = \frac{\Delta P_a}{P_1 - P_{vc}} \tag{10-12}$$

where

P_1	=	inlet pressure (psi)
P_{vc}	=	vena contracta pressure at the point of choked flow (psi)
ΔP_a	=	the allowable differential pressure across the valve

With constant upstream pressure, ΔP_a is the maximum pressure drop across the valve that will effectively increase the flow. From this definition, you can see that K_m relates the valve's pressure-recovery characteristics to the amount of flow passing through the valve.

Consider two valves that are passing the same choked flow but have different recovery characteristics. One is a high-recovery valve and the other is low-recovery. Figure 10-8 shows the pressure profiles for two such valves. Since both valves have just reached the same choked flow, the pressure differential from the inlet to the vena contracta is the same for both of them. On the other hand, the pressure-recovery characteristics, and consequently the ΔP_a values, are significantly different. Relating this to the definition of K_m, it is easy to see that the high-recovery valve will have a much smaller K_m value than the low-recovery valve.

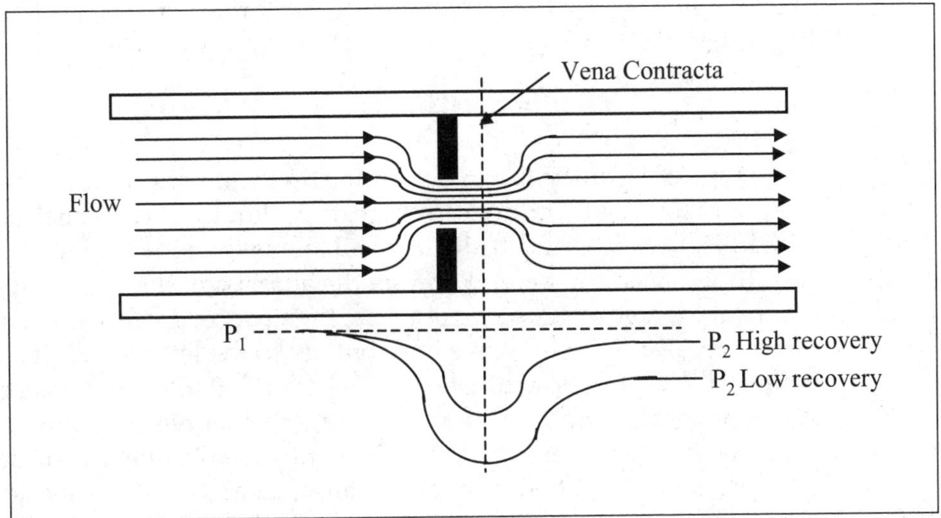

Figure 10-8. Comparison of pressure profiles for high- and low-recovery valves

In effect, K_m describes the relative pressure-recovery characteristics of a valve. High-recovery valves will tend to have low K_m values, and low-recovery valves will tend to have higher K_m values. A typical range of K_m

values is from about 0.3 to 0.9. The values for valves with intermediate-recovery characteristics will be somewhere within this range.

Equation 10-13 shows the K_m expression rearranged into a more useful form.

$$\Delta P_a = K_m(P_1 - P_{vc}) \tag{10-13}$$

where

P_1 = a known process condition
K_m = obtained from the sizing data catalog

If you know the vena contracta pressure, you can calculate the limiting pressure drop for choked flow. A term called the *critical pressure ratio* (r_c) can be defined as follows:

$$r_c = \frac{P_{vc}}{P_v} \tag{10-14}$$

where

P_{vc} = the vena contracta pressure at choked flow
P_v = the vapor pressure of the flowing liquid

This expression is easily rearranged into a form that can be used to replace P_{vc} in Equation 10-13:

$$\Delta P_a = K_m(P_1 - r_c P_v) \tag{10-15}$$

For any type of liquid, you can determine the value of r_c as a function of the vapor pressure and critical pressure of the liquid. Curves such as those shown in Figures 10-9 and 10-10 are available in manufacturers' literature. Figure 10-9 is used for water. Enter on the abscissa at the water vapor pressure at the inlet to the control valve, then proceed vertically to intersect the curve, and finally move horizontally to the left to read the critical pressure ratio, r_c, on the ordinate. Use Figure 10-10 for liquids other than water. First, determine the vapor pressure/critical pressure ratio by dividing the liquid vapor pressure (P_v) at the control valve inlet by the liquid's critical pressure (P_c). Then enter on the abscissa at the ratio just calculated and proceed vertically to intersect the curve. Finally, move horizontally to the left and read the critical pressure ratio, r_c, on the ordinate.

Using Equation 10-15, you can easily calculate the limiting pressure drop for choked flow. Inlet pressure and vapor pressure are part of the known service conditions for the application. The equation determines the flow-limiting pressure drop for either flashing or cavitation. The chief differ-

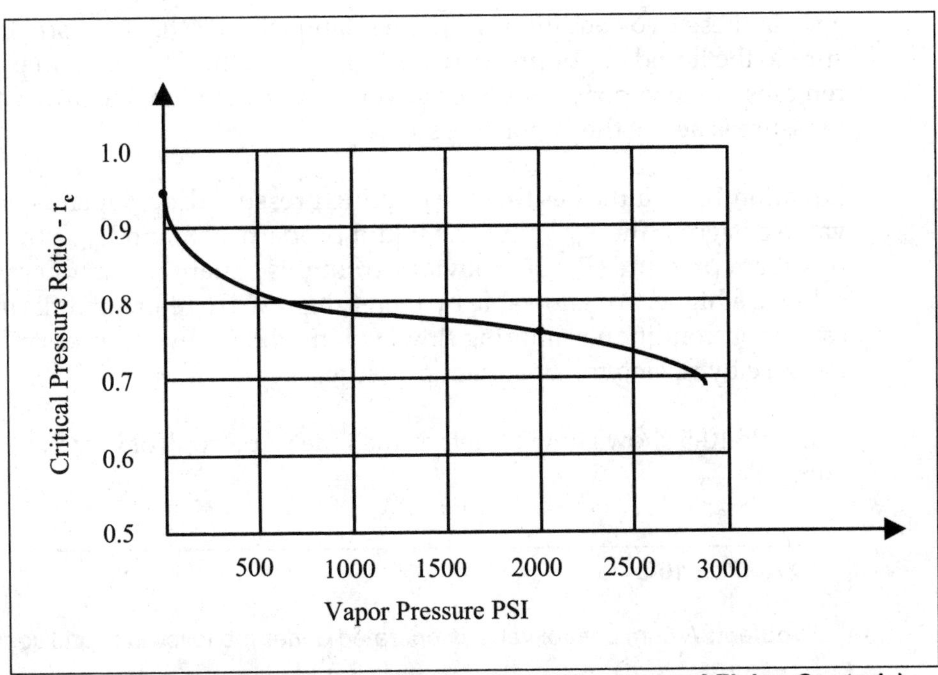

Figure 10-9. Critical pressure ratios for water *(Courtesy of Fisher Controls)*

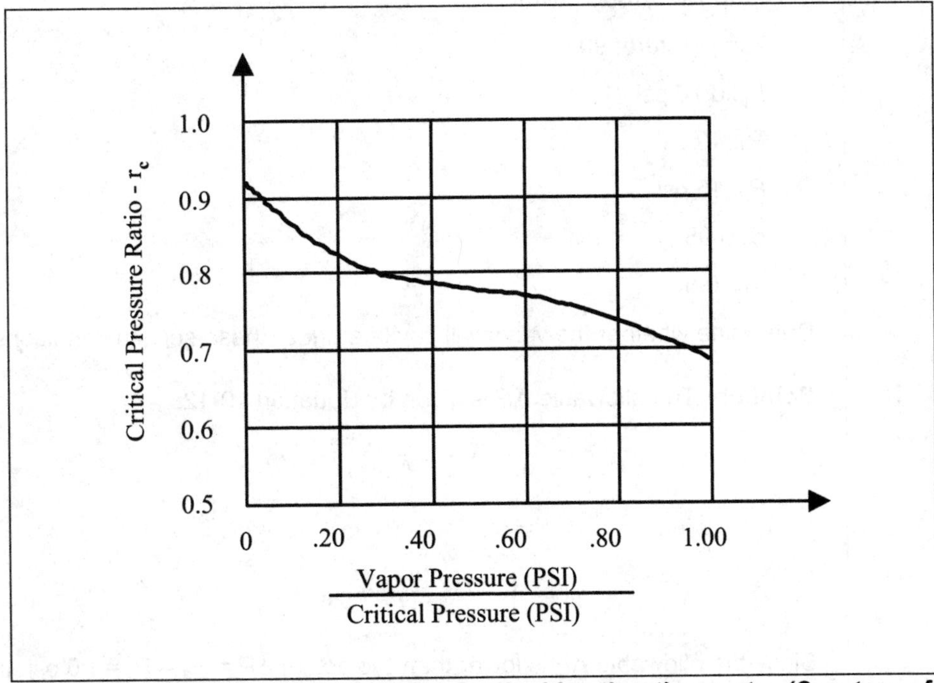

Figure 10-10. Critical pressure ratios for liquids other than water *(Courtesy of Fisher Controls)*

ence in these two conditions is in the relationship of the downstream pressure to the liquid vapor pressure. With flashing, the downstream pressure remains at the vapor pressure or lower. With cavitation, the downstream pressure is above the vapor pressure.

Equation 10-15 indicates that the limiting pressure drop depends on the valve design or the K_m value, the liquid properties (r_c and P_v), and the upstream pressure (P_1). It is obvious from this equation that you can achieve a higher ΔP allowable by increasing P_1. Therefore, if a flashing or cavitation condition is limiting flow, you might achieve a necessary flow increase by raising the upstream pressure.

Example 10-5 shows how to determine whether a control valve will cavitate.

EXAMPLE 10-5

Problem: A 6-in. control valve is operated under the following conditions:

Fluid: water

Flow rate: 1000 gal/min

Temperature: 90°F

P_V: 0.70 psi

P_1: 40 psi

P_2: 15 psi

r_C: 0.95

K_m: 0.5

Determine whether the valve will cavitate under these service conditions.

Solution: The allowable ΔP is given by Equation 10-12:

$$\Delta P_a = K_m(P_1 - r_c P_v)$$

$$= 0.5\ [40 - (0.95)\ (0.7)]\ \text{psi}$$

$$= 19.7\ \text{psi}$$

Since the allowable ΔP is lower than the actual $\Delta P = P_1 - P_2 = 25$ psi, the valve will cavitate.

AC and DC Motors

An electric motor is a device that converts electrical energy into mechanical energy. In many chemical processes, an electric motor imparts rotary or linear motion to a mechanical load. For example, motor-driven stirrers are used to mix chemicals. Electric motors find wide use in automatic process control systems since they can be accurately and continuously controlled over a wide range of speeds and mechanical loads.

Electric motors are available that have power ratings from fractional horsepower to thousands of horsepower. A wide variety of designs are available to meet specific process applications. Generally, motors are classified in terms of whether they operate from either DC or AC voltage.

DC Motors

The DC (direct current) motor is used extensively in low- and high-power control system applications. The motor consists of a field magnet and an armature and operates on the principle that a force is exerted on a current-carrying conductor in a magnetic field. In a DC motor, the conductors take the form of loops (armature), and the force acts as a torque that tends to rotate the armature. The direction of the torque depends on the relative direction of the magnetic flux and the armature current. To provide a unidirectional torque when the armature is rotating, you must ensure that the direction of the current is reversed at appropriate points. You do this by using a commutator on the moving part and brushes connected to the DC power. In the motor, the armature is wound with a large number of loops to provide a more uniform torque and smoother rotation. Therefore, the commutator must have segments cut that correspond to the number of loops in the armature winding.

The torque a DC motor develops is proportional to both the magnetic flux and the armature current. The magnetic flux may be produced by permanent magnets in smaller motors or electromagnetically from windings on the field magnet. The current that produces the magnetic field is called the *field current*.

In the typical control system application, the armature and field are powered from separate sources. You can control either the armature or the field to vary the speed and torque of the motor. The most common method for doing this, shown in Figure 10-11, is to provide a constant field current, I_f, and to apply a variable control voltage, V_a, to the armature. In this method, the control source must supply the motor's full power requirements.

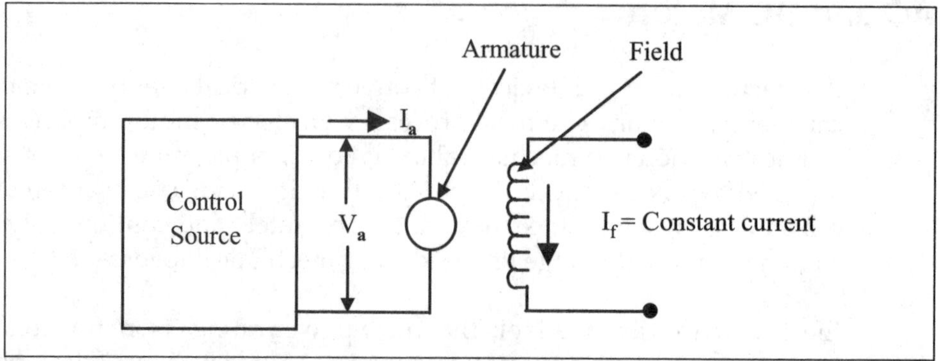

Figure 10-11. Armature control of DC motor

An alternative arrangement is shown in Figure 10-12. In this application, a constant armature current (I_a) is provided and the field current (I_f) is varied. The advantage of this method is that the control source must supply only a relatively small field current, and the torque is closely proportional to the control current. The disadvantage is that the motor's speed of response to a changing input is generally slower than with armature control. It is also more difficult and inefficient to provide a constant current to the armature using this method. Therefore, control of the field is generally limited to low-power DC motors of several horsepower.

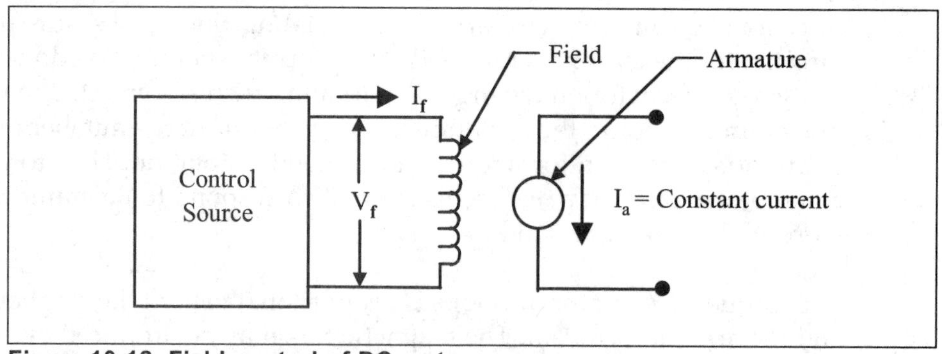

Figure 10-12. Field control of DC motor

Direct current motors used in servo applications are based on the same principles as conventional motors, but with several special design features. They are designed for high torque and low inertia. You can produce a high torque/inertia ratio by reducing the armature diameter and increasing its length compared to that of a standard motor. The main requirement in most servo control system applications is that the motor operate smoothly without jumping or clogging action and that the set point be reached smoothly and quickly.

A disadvantage of using electric motors is that their shaft speeds are normally much higher than the speed desired to move the load. You must use some method to reduce the speed, such as a gear train, to gain efficient control. One type of DC motor that can operate correctly at low speeds and be directly attached to the load is the direct-drive DC torque motor. This motor has a wound armature and a permanent magnet that converts electric current directly into torque. In general, torque motors are used in high-torque positioning systems and in slower-speed control systems.

You can obtain most of the design performance information from the motor's torque and speed specifications. If you keep the field constant, the torque produced by the motor is closely proportional to the armature current, as shown in Figure 10-13. The intercept of the line and the horizontal axis is the value of current that is required to overcome the static-friction torque of the motor. The slope of the line is defined as the torque constant of the DC motor:

$$K_i = \frac{\Delta T}{\Delta I_a} \tag{10-16}$$

where

K_i = the torque constant (oz-in./A)
T = the torque (oz-in.)
I_a = the armature current (A)

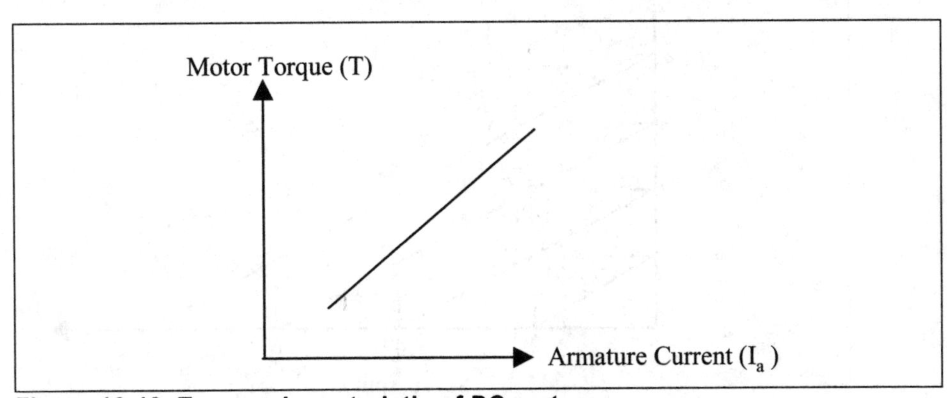

Figure 10-13. Torque characteristic of DC motor

When the armature is rotating, a voltage that is proportional to the speed is generated in the winding:

$$V_g = K\varphi\omega \tag{10-17}$$

where

V_g = the voltage generated
φ = the magnetic flux
ω = the armature speed
K = a constant

This voltage is called a *back electromotive force* (emf) because it opposes the voltage applied to the armature and limits the armature's current. The motor speed will be such that the back emf is just small enough vis-à-vis the applied voltage to allow an armature current to drive the load. If the load on the motor is increased, the armature slows down. As the motor slows, the back emf falls and a higher current flows. This larger current causes the motor to drive the increased load but at a lower speed. If you require the original speed, the control system can automatically increase the applied voltage.

This motor characteristic is shown by the family of curves in Figure 10-14. These curves, known as *speed-torque curves,* give the motor torques as a function of speed for constant values of armature voltage. The zero-torque points on the curve represent the no-load speed of the motor. At this speed, no torque can be obtained from the motor.

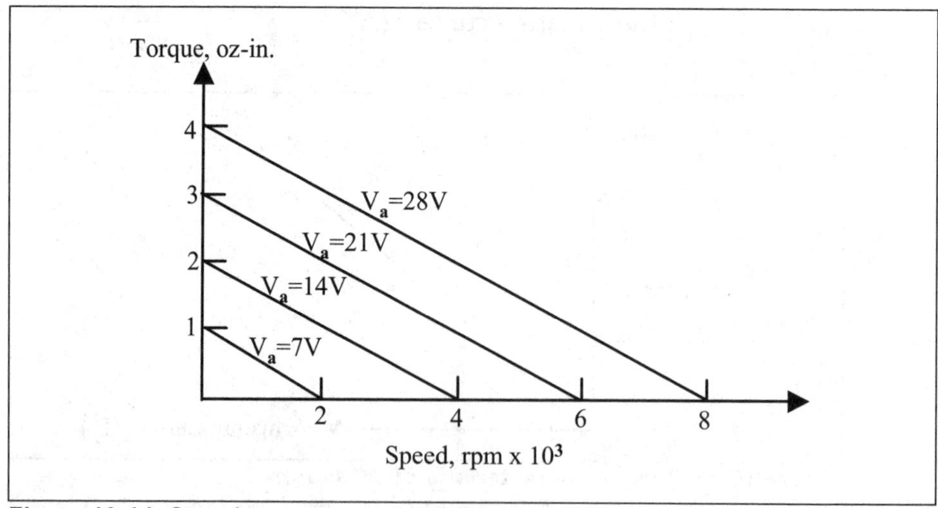

Figure 10-14. Speed-torque curves of DC motor

Although motor torque for armature-controlled motors is proportional to armature current, it is more convenient to define a torque constant in terms of armature voltage. At zero speed, or stall condition, no back emf is generated, and the armature current is determined solely by the armature resistance. Therefore, under stall conditions, the following is true:

$$K_t = \frac{\Delta T_s}{\Delta V_a} \tag{10-18}$$

where

K_t	=	a torque constant (oz-in./V)
T_s	=	stall torque (oz-in.)
V_a	=	armature voltage (V)

The drop in motor speed with increasing torque is characterized by a damping constant that is the slope of the speed-torque curves.

$$D_m = \frac{\Delta T_s}{\Delta S} \tag{10-19}$$

where

D_m	=	a damping constant [oz-in./(rad/s)]
T_s	=	stall torque (oz-in.)
S	=	speed (rad/s)

If the speed-torque curves are linear, as shown in Figure 10-14, the damping constant is given by the following:

$$D_m = \frac{T_s}{S_{nl}} \tag{10-20}$$

where

T_s	=	stall torque at rated voltage (oz-in.)
S_{nl}	=	no-load speed at rated voltage (rad/s)

Example 10-6 illustrates how to determine the torque and damping constants for a typical armature-controlled DC motor.

Using field control of a DC motor, the armature current is supposed to be held constant, and the armature torque and current are unaffected by the back emf. Therefore, the torque should be independent of speed. This is shown by the horizontal lines in Figure 10-15, which illustrates torque versus speed for various values of field current.

In practice, you do not always obtain the armature current from a true constant-current source, and thus the armature current will be affected somewhat by the back emf. Furthermore, mechanical friction increases with speed, further reducing the available torque. Thus, the torque-speed curves for field-controlled DC motors exhibit a slight drop as speed increases. This decrease in torque is generally advantageous in closed-

EXAMPLE 10-6

Problem: Determine the torque and damping constants for the armature-controlled DC motor whose speed-torque curves are shown in Figure 10-14. The armature voltage is 21*V*.

Solution: To find torque constant, use Equation 10-18.

$$K_t = \frac{\Delta T_s}{\Delta V_a}$$

Since the armature voltage is 21*V*, the stall torque (T_S) is 3 oz-in. Thus,

$$K_t = \frac{3 \text{ oz} - \text{in.}}{21 \text{ volts}}$$

$$K_t = 0.143 \text{ oz} - \text{in.}/v$$

You calculate the damping constant using Equation 10-20:

$$D_m = \frac{T_s}{S_{nl}}$$

Armature voltage is 21*V* and S_{nl} is 6,000 rpm according to the information shown in Figure 10-14. Therefore, we have

$$D_m = \frac{3 \text{ oz} - \text{in.}}{\left(6 \times 10^3 \frac{\text{rev}}{\text{min}}\right)\left(\frac{1 \text{ min}}{60 \text{ s}}\right)\left(\frac{2\pi - \text{rad}}{\text{rev}}\right)}$$

$$D_m = 4.8 \times 10^{-3} \text{ oz} - \text{in./rad/s}$$

loop control systems because this damping feature reduces the tendency toward overshoot and instability.

AC Motors

The properties that make an AC (alternating current) motor the choice for most constant-speed industrial drives are not necessarily those needed for control system applications. The majority of the latter applications require a motor that can operate over a wide range of speeds with high starting torque. The main advantage the AC motor has over the DC motor is that it

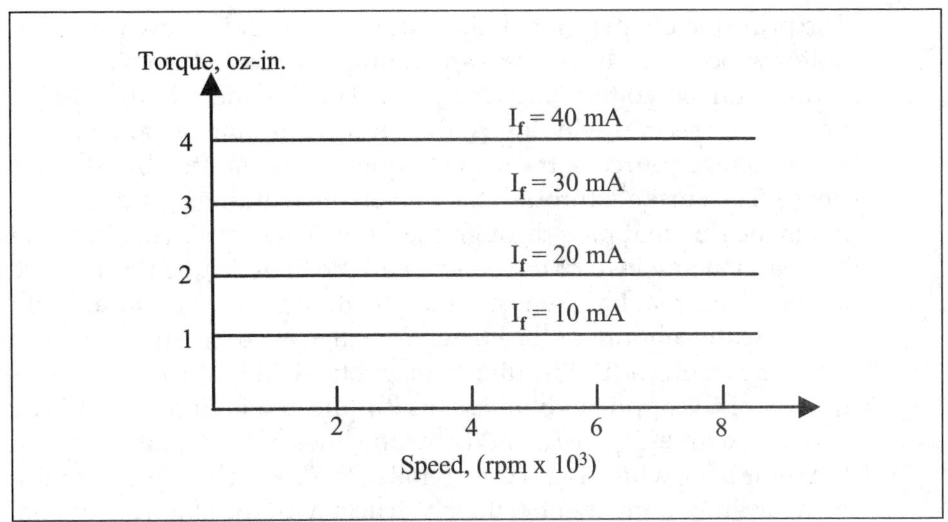

Figure 10-15. Speed-torque curves of DC motor with field control

is compatible with the AC signals from AC drive and control devices. This eliminates the need to convert signals from AC to DC and back.

One type of AC motor that is used extensively in low-power control system applications is the two-phase induction motor shown in Figure 10-16. This motor has two stator windings that are located at 90° with respect to each other. The rotor consists of a slotted cylinder of iron laminate mounted on the motor shaft. Solid copper or aluminum bars are set lengthwise in the slots of the rotor and are connected together at both ends by rings. An elongated, small-diameter rotor is normally used to minimize inertia and produce faster motor response.

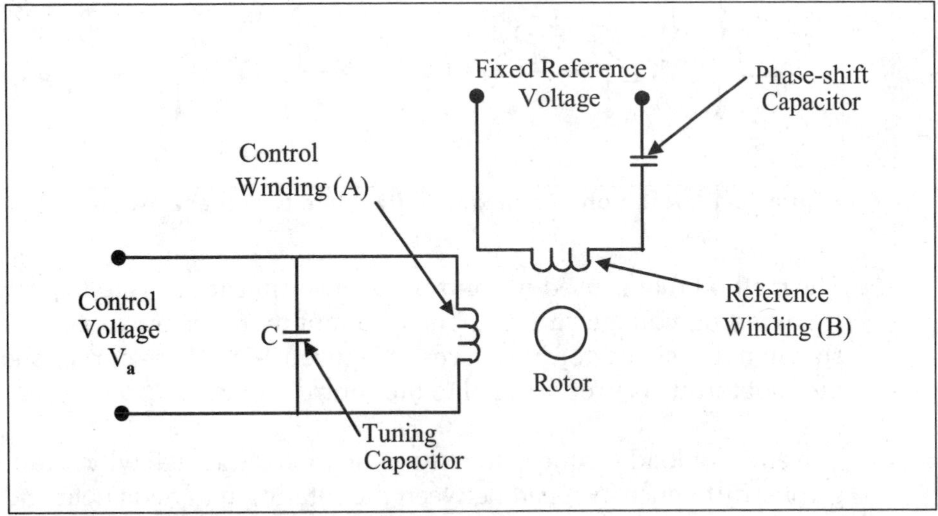

Figure 10-16. Two-phase AC induction motor

The principle of operation of the induction motor is shown in Figure 10-17. Stator winding B, the reference winding, is operated from a sinusoidal source of fixed voltage and frequency. This is generally the 60-Hz AC power. The second winding A, the control winding, is powered from the same voltage source of the same frequency, but shifted by 90° in electrical phase. The currents through the two-stator windings produce lines of magnetic flux that pass through the rotor. These magnetic flux lines combine, and at any instant the effect of all the lines can be represented by a single vector that has some annular position relative to the axis of the rotor. As the amount of the currents change at successive instants, the vector angle changes. The result is a magnetic field that rotates at the frequency of the applied voltage. This is indicated in Figure 10-17 at four successive times: t_1, t_2, t_3, and t_4. As the lines of flux rotate, they cut across the rotor bars, which induce currents in the bars. These electric currents, in turn, produce a magnetic field that reacts with the stator magnetic field and causes the rotor to turn smoothly and continuously.

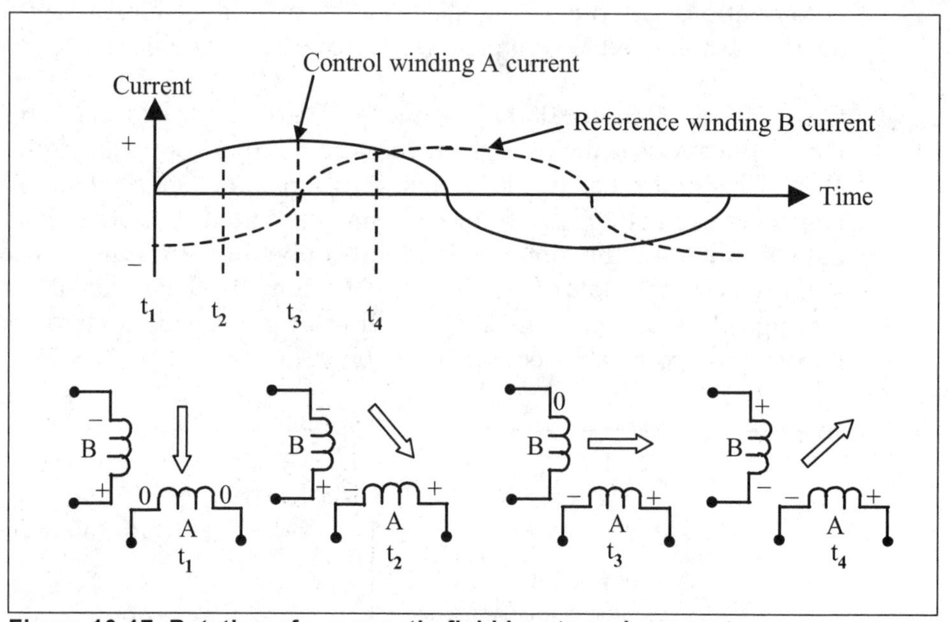

Figure 10-17. Rotation of a magnetic field in a two-phase motor

The method that is used in control system applications is to make the control winding voltage variable. The resultant motor characteristics are shown in the speed-torque curves of Figure 10-18. These curves show that the stall torque is proportional to the control voltage.

The effect of load torque is to reduce the motor's speed, which causes a greater difference in speed between the rotating magnetic field and the motor. The induced magnetic field will increase with this difference, creat-

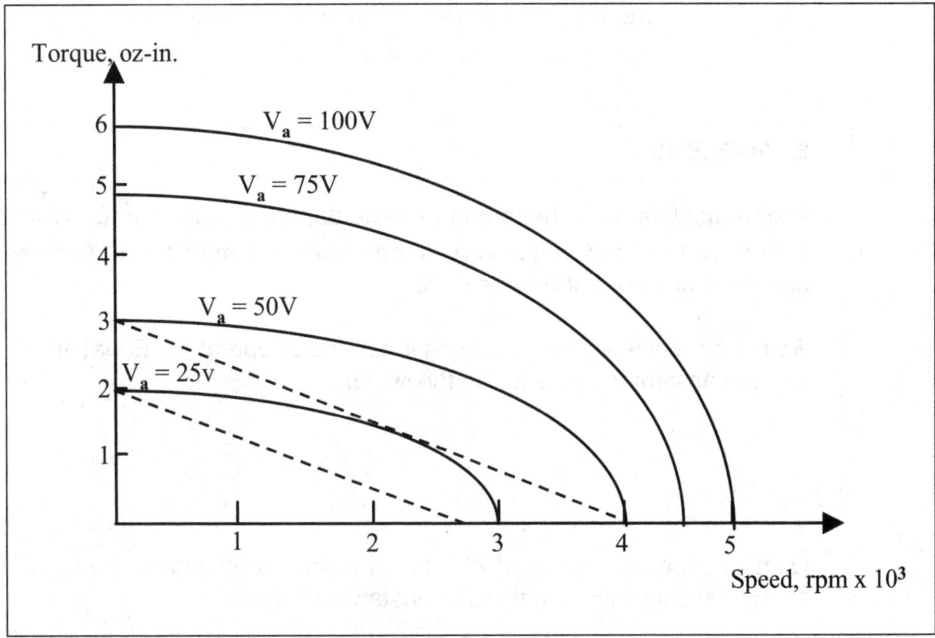

Figure 10-18. Speed-torque curves of AC motor

ing additional torque to pull the load but at a lower speed. Although the internal effects of rotor slowdown are somewhat more complex in the induction motor than in the DC motor, the external effects are about the same.

In the discussion that follows, we will consider the speed-torque characteristics of the AC motor to be linear (see Figure 10-18). Using this assumption, you can derive AC motor constants that are similar to those of the armature-controlled DC motor. This can be demonstrated by Example 10-7.

In two-phase AC induction motors, a tuning capacitor is added in parallel with the control winding (see Figure 10-16) to help match the output impedance of the control circuit to the input of the AC motor. The capacitor is normally chosen to produce a unity power factor, so the parallel resonant circuit will have a very high input impedance (i.e., minimum power drain on the control circuit). The equation for finding the input tuning capacitor that is required to obtain a unity power factor is as follows:

$$C = \frac{X_L}{2\pi f Z^2} \quad (10\text{-}21)$$

where

C = capacitance in farads
X_L = the inductive reactance of the control winding (Ω)

Z = the impedance of the control winding (Ω)
f = the frequency of excitation (Hz)

EXAMPLE 10-7

Problem: Determine the torque (K_t) and damping constants (D_m) for the AC motor whose speed-torque curves are shown in Figure 10-18. Assume linear operation at a control voltage of 25*V*.

Solution: Since we are assuming linear motor operation, Equation 10-18 for torque constant reduces to the following:

$$K_t = \frac{T_s}{V_a}$$

Obtaining the stall torque of 2 oz-in. for a control voltage of 25*V* from Figure 10-16, we determine the torque constant as follows:

$$K_t = \frac{2 \text{ oz} - \text{in.}}{25v}$$

$$K_t = 0.08 \text{ oz} - \text{in.}/v$$

The damping constant is given by Equation 10-20 ($D_m = T_s/S_{nl}$). Obtaining a speed of 2,500 rpm for a control voltage of 25*V* from Figure 10-18, we calculate the damping constant as follows:

$$D_m = \frac{2.0 \text{ oz} - \text{in.}}{\left(2.5 \times 10^3 \frac{\text{rev}}{\text{min}}\right)\left(\frac{1 \text{ min}}{60 \text{ s}}\right)\left(\frac{2\pi - \text{rad}}{\text{rev}}\right)}$$

$$D_m = \frac{2.0oz - in.}{(2.5x10^3 \frac{rev}{\min})(\frac{1\min}{60s})(\frac{2\pi - rad}{rev})}$$

$$D_m = 7.64 \times 10^{-3} \text{ oz} - \text{in./rad/s}$$

Example 10-8 illustrates how to calculate the value of parallel capacitance that is required to make a AC inductive motor winding appear to be purely resistive.

The traditional method for obtaining the necessary 90° phase shift between the two motor currents is to use a phase-shifting capacitor in

EXAMPLE 10-8

Problem: A 60-Hz two-phase induction motor has a control field winding impedance of 100 Ω and a control winding inductive reactance of 150 Ω. Find the value of parallel capacitance that is required to make the winding appear purely resistive.

Solution: Use the equation for the input tuning capacitor that produces a unity power factor as follows:

$$C = \frac{X_L}{2\pi f Z^2}$$

$$C = \frac{150}{2(3.14)(60)(100)^2} = 40\mu \text{ farads}$$

series with the reference winding, as shown in Figure 10-16. The correct value for this capacitor is normally given in the motor's engineering specifications.

Another factor you should consider in AC and DC motor applications is the static friction associated with the armature. This requires that a minimum control voltage be applied to rotate the armature. Static friction represents a form of dead-time error in the control system, which degrades system performance. It causes system insensitivity since the controller must develop enough output voltage to overcome the static friction. The static can be expressed as a torque and is generally between 1 and 10 percent of the rated torque of the motor.

Pumps

Pumps are used extensively in liquid flow control system applications because they are simple to operate, are energy efficient, and can provide both flow and pressure control in a single piece of equipment. The types of pumps most commonly used in process control systems are centrifugal, positive displacement, and reciprocating.

Centrifugal Pumps

The centrifugal pump is the most widely used process fluid-handling device. It is generally operated at a fixed speed to recirculate or transfer process fluids. If your process requires controlled fluid flow, a control valve is generally installed in the discharge line. However, a control valve is an energy-consuming device. In some applications, this energy loss cannot be tolerated, so a variable-speed drive is used to control the pump.

A centrifugal pump imparts velocity to a process fluid. This velocity is then transformed mainly into pressure as the fluid leaves the pump. The pressure head (H) that is developed is approximately equal to the velocity at the discharge of the pump. We expressed this relationship earlier during our discussion of fluid flow in Chapter 9. It is given by the following:

$$H = \frac{V^2}{2g} \tag{10-22}$$

where

H	=	pressure head (ft)
V	=	velocity (ft/s)
g	=	32.2 ft/s

We can obtain the approximate head of any centrifugal pump by calculating the peripheral velocity of the impeller and substituting this velocity into Equation 10-22. The formula for peripheral velocity is as follows:

$$V = \pi d \omega \tag{10-23}$$

where

d	=	diameter (m or in.)
ω	=	angular speed (rad/s or rpm)

In English units, velocity is normally expressed in feet per second and angular velocity in revolutions per minute, so Equation 10-23 can be converted into English units as follows:

$$V[\text{ft/s}] = \frac{3.14\omega d}{\left(\frac{12 \text{ in.}}{1 \text{ ft}}\right)\left(\frac{60 \text{ s}}{1 \text{ min}}\right)}$$

$$V[\text{ft/s}] = \frac{\omega d}{229} \tag{10-24}$$

where

d	=	impeller diameter (in.)
ω	=	the angular velocity (rpm)

The following example illustrates how to use Equation 10-24 to determine the velocity of fluid from a centrifugal pump.

EXAMPLE 10-9

Problem: Determine the total head developed in feet by a centrifugal pump that has a 12-in. impeller and is rotating at 1,000 rpm.

Solution: Use Equation 10-24 to find the fluid velocity in feet per second.

$$V = \frac{\omega d}{229}$$

$$V = \frac{(1000)(12)}{229} = 52.4 \text{ ft/s}$$

Then, use Equation 10-19 to find the total head developed by the pump.

$$H = \frac{V^2}{2g}$$

$$H = \frac{(52.4 \text{ ft/s})^2}{2(32 \text{ ft/s}^2)}$$

$$H = 42.9 \text{ ft}$$

Positive-Displacement Pumps

Positive-displacement pumps can be divided into two main types: rotary and reciprocating.

Rotary Pumps

Rotary pumps function by continuously producing reduced-pressure cavities on the suction side, which fills with fluid. The fluid is moved to the discharge side of the pump, where it is compressed and then discharged from the pump. In most cases, flow is directly proportional to pump speed. Therefore, the control system only needs to control the pump speed to control the process fluid flow. This type of pump provides accurate, uniform flow and a minimized power requirement.

To help maintain a constant ΔP across the pump, you can hold pressure on the suction side constant by using a small storage tank with a level controller, as shown in Figure 10-19. You can hold this discharge pressure constant by pumping fluid to the top of the delivery tank. A low-slip

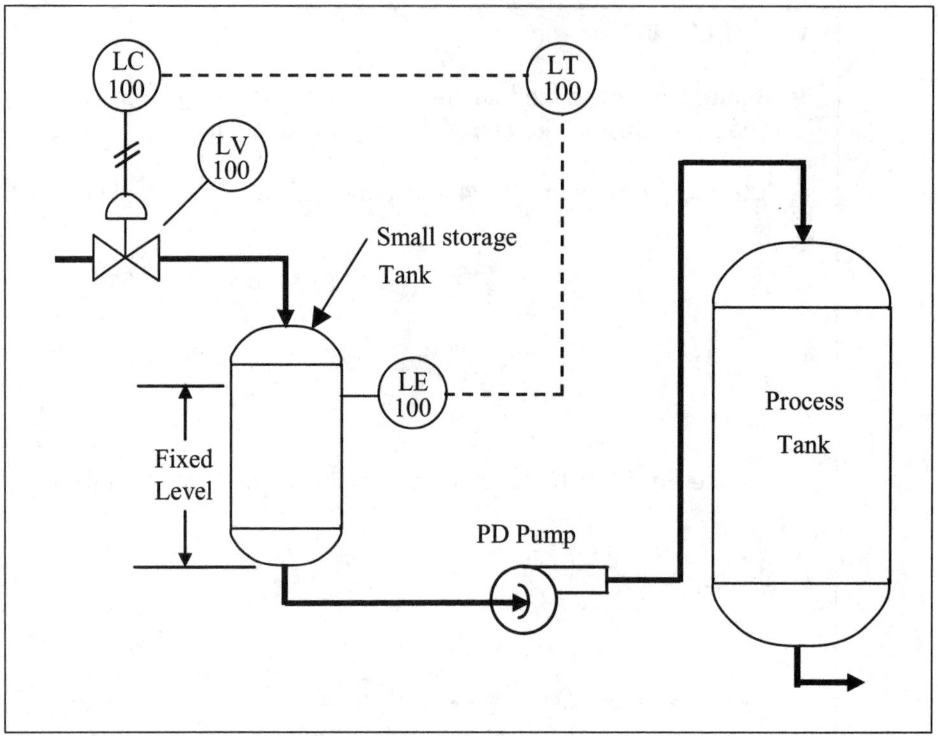

Figure 10-19. Positive displacement (PD) pump application

pump delivers nearly all its internal displacement in each pump revolution and meters fluid accurately.

Three types of rotary pumps are used for flow control: gear pumps, circumferential piston pumps, and progressing cavity pumps. Gear pumps are used to pump high-viscosity fluids or to generate moderately high differential pressures. Because these pumps have close gear-to-gear contact, they are used for clean, nonabrasive lubricating fluids.

Circumferential piston pumps are widely used in sanitary and food applications. This type of pump has two noncontacting fluid rotors. Because the sealing area is relatively long, slip is nearly eliminated at viscosities above 200 centistokes. Flow control applications include pumping dairy and bakery products, plastics and resins, and pharmaceuticals. These pumps cannot be used to pump large solids or extremely abrasive fluids or in high-back-pressure applications.

In the progressing cavity pump, a single helical rotor turns eccentrically in a stationary stator. As the rotor turns, cavities are produced, which are filled at the inlet by the process fluid and move through the stator to the discharge side of the pump.

Reciprocating Pumps

Reciprocating pumps use a linear reciprocal stroke in combination with check valves to pump fluid. Reciprocating pumps are commonly used to control the rate at which a volume of fluid is injected into a process stream or vessel. These pumps are also called metering pumps in some applications because they are highly accurate and consistent in the volume of fluid discharge per cycle. Large reciprocating pumps generally have variable-speed drives. Small, controlled-volume reciprocating pumps used for precise chemical injection normally use a variable-stroke controller.

The two basic types of reciprocating pumps normally used in metering applications are plunger pumps and diaphragm pumps. In the plunger pump, a packed plunger draws in and then expels fluid through a one-way check valve. A diaphragm-reciprocating pump can be mechanically driven, directly coupled to a plunger, or hydraulically actuated. Like the plunger pump, it draws in a precise amount of fluid and discharges it cyclically.

EXERCISES

10.1 An equal percentage valve has a maximum flow of 80 gal/min and a minimum flow of 5 gal/min. Find its rangeability.

10.2 Assume that a force of 200 N is required to fully open a control valve equipped with a diaphragm valve actuator. The valve input control signal connected to the actuator has a range of 3 to 15 psi. Find the diaphragm area that is required to fully open the control valve.

10.3 Describe the procedure valve manufacturers use to determine the valve-sizing coefficients (C_v) for their valves.

10.4 A liquid with a specific gravity of 0.9 is pumped through a pipe that has an inside diameter of 18 in. at a velocity of 10 ft/s. Find the volume flow rate.

10.5 Calculate the C_v, and select the required valve size from Table 10-1 for a valve with the following service conditions:

Fluid: ethyl alcohol ($SG = 0.8$)

Flow rate: 200 gal/min

P_1: 100 psi

P_2: 90 psi

10.6 An 8-in. control valve is operated under the following conditions:

Fluid: water
Flow rate: 800 gal/min
Temperature: 70°F
P_v: 0.36
P_1: 100 psi
P_2: 90 psi
r_c: 0.95
K_m: 0.6

Determine whether the valve will cavitate under these service conditions, and then calculate C_V.

10.7 Describe the basic difference between flashing and cavitation in control valves.

10.8 Determine the torque and damping constants for the armature-controlled DC motor whose speed-torque curves are shown in Figure 10-14. The armature voltage is 28V.

10.9 Determine the torque (K_t) and damping constants (D_m) for the AC motor whose speed-torque curves are shown in Figure 10-18. Assume linear operation at a control voltage of 50V.

10.10 A 60-Hz two-phase induction motor has a control field winding impedance of 200 Ω and a control winding inductive reactance of 250 Ω. Find the value of parallel capacitance that is necessary to make the winding appear purely resistive.

10.11 Calculate the total head in feet that is developed by a centrifugal pump with a 10-in. impeller and that is rotating at 800 rpm.

10.12 Explain the basic operation of a positive-displacement rotary pump.

11

Process Control Computers

Introduction

This chapter discusses the design and application of the process control computers used in industry. The topics covered include the history of process control computers, the basics of the computers used for process control, and the characteristics of programmable logic controllers.

History of Process Control Computers

Before the introduction of computers for industrial process control applications, the standard industrial control system consisted of many single-loop analog controllers, either pneumatic or electronic, as well as their associated field instruments and control devices. This stand-alone method of control provided, and still provides, excellent control of industrial processes. The main disadvantages of stand-alone controllers are that they cannot be easily reconfigured and cannot easily communicate with other plant computers.

Although computer designers long predicted that digital computers would be used for process control applications, the first practical digital computer-based control system was designed for the U.S. Air Force by Hughes Aircraft Company. Using a computer called the DIGITAC, this system controlled an airplane in flight through computed set-point changes to an analog-based automatic pilot. This system was flown successfully in 1954.

The losing company in the competition for the Air Force contract was Ramo-Wooldridge and its computer, the RW-30. Ramo-Wooldridge decided to market its computer as a process control computer, repackaging it as the RW-300. This was a very important entry in the early history of process computers. It is interesting to note that computer control developed not as an initiative of the process and manufacturing industries but as a result of computer and electronics vendors efforts to expand their markets beyond military applications.

The first industrial installation of a computer system was made by the Daystrom Company at the Louisiana Power and Light plant in Sterlington, Louisiana. It was not a closed-loop control system, but rather a supervisory data-monitoring system (see Figure 11-1).

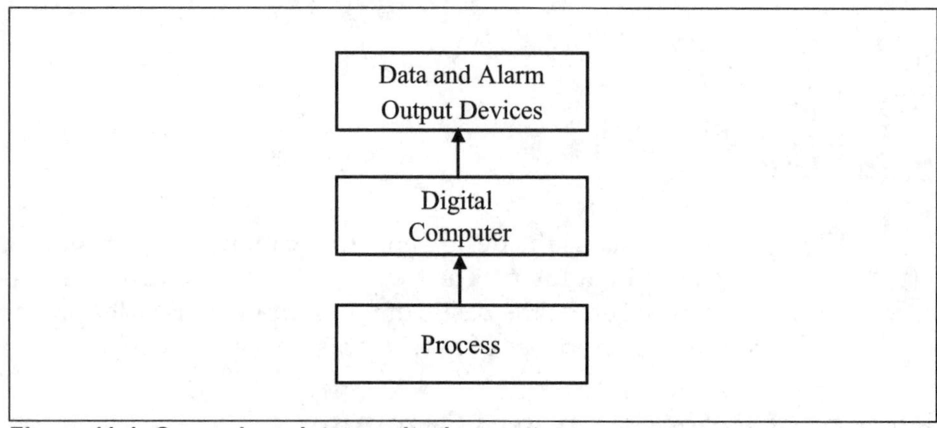

Figure 11-1. Supervisor data-monitoring system

The first industrial closed-loop computer system was introduced in March 1959 at the Texaco Company's Port Arthur, Texas, refinery using an RW-300. The first chemical plant control computer was unveiled at the Monsanto Chemical Company's ammonia plant in Juling, Louisiana, in April 1960. It also used an RW-300 to achieve closed-loop control.

The next major advance in process control computers came with the introduction of programmable logic controllers (PLC). In 1968, a major automobile manufacturer wrote a design specification for the first programmable logic controller. The primary goal was to eliminate the high cost of frequently replacing inflexible relay-based control systems. The automaker's specification also called for a solid-state industrial computer that could be easily programmed by maintenance technicians and plant engineers. It was hoped that the programmable controller would reduce production downtime and provide expandability for future production changes. In response to this design specification, several manufacturers developed logic control devices called *programmable controllers*.

The first programmable controller was installed in 1969, and it proved to be a vast improvement over relay-based panels. Because such controllers were easy to install and program, they used less plant floor space, and they were more reliable than relay-based systems. The first programmable controller did more than meet the automobile manufacturer's production needs. Design improvements in later models also led to the widespread use of programmable controllers in other industries.

Two main factors in the early design of programmable controllers appear to have caused their success. First, designers used highly reliable solid-state components, and the electronic circuits and modules were designed for the harsh industrial environment. The system modules were built to withstand electrical noise, moisture, oil, and the high temperatures encountered in industry. The second important factor was that the programming language designers initially selected was based on standard electrical ladder logic. Earlier computer systems failed because it was not easy to train plant technicians and engineers in computer programming. However, most were trained in relay ladder design, so they could quickly learn to program in a language based on relay circuit design.

When microprocessors were added to PLCs in 1974 and 1975, they greatly expanded and improved the basic capabilities of programmable controllers. They were now able to perform sophisticated math and data-manipulation functions.

In the late 1970s, improved communications components and circuits made it possible to place programmable controllers thousands of feet from the equipment they controlled. As a result, most programmable controllers could now exchange data, meaning they could more effectively control processes and machines. Also, microprocessor-based input and output modules enabled programmable controller systems to evolve into the analog control world.

Programmable controllers can be found in thousands of industrial applications. They are used to control chemical processes and facilities. They are also found in material-transfer systems that transport both raw materials and finished products. PLCs are now used with robots to perform hazardous industrial operations, making it possible for humans to perform more intellectually demanding functions. Programmable controllers are used in conjunction with other computers to collect and report process and machine data, including such uses as statistical process control, quality assurance, and diagnostics. Finally, they are also used in energy-management systems to reduce costs and improve the environmental control of industrial facilities and office buildings.

The introduction of personal computers in the mid-1980s was the next significant advance in the use of computers in process control. The first personal computers were relatively slow and had limited memory and storage. With constant advances in speed, software, and data storage space, however, the personal computer became a major force in process control and data collection.

Direct Digital Control

The goal of the control system engineer in the early days of computer control was to bypass the process controllers and have the computer control the process directly (see Figures 11-2 and 11-3). This was the beginning of direct digital control (DDC).

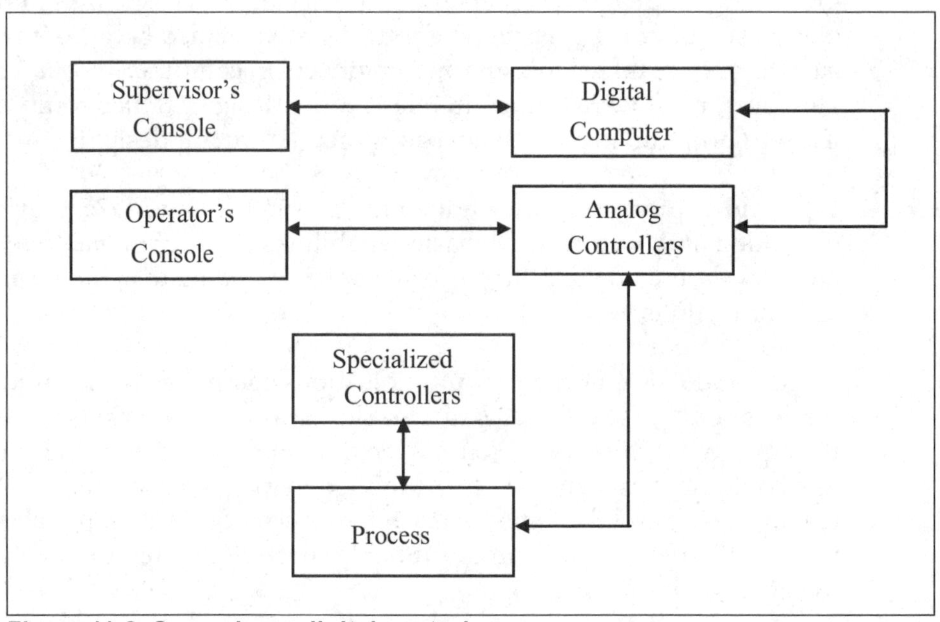

Figure 11-2. Supervisory digital control

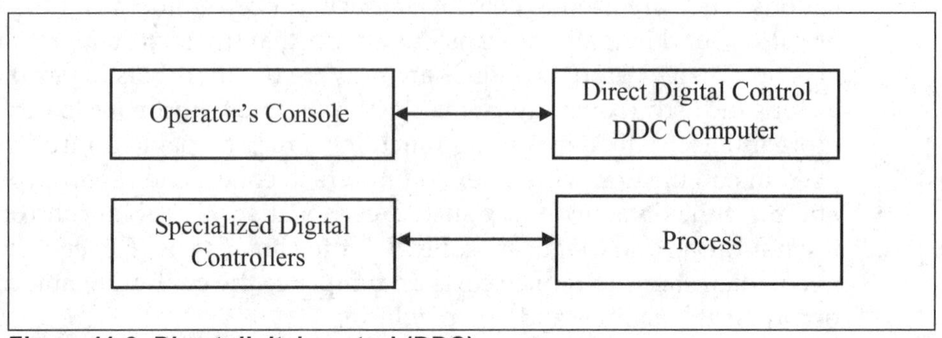

Figure 11-3. Direct digital control (DDC)

Direct digital control was achieved by Imperial Chemicals Industries, Ltd., in England, using an ARGUS 100™ computer built by the Ferranti Company and installed in a soda ash plant in Fleetwood, Lancastershire. At the same time, but independently, the Monsanto Company achieved DDC with an RW-300 computer in its ethylene plant at Texas City, Texas. The Monsanto installation was under closed-loop control in March 1962, where it controlled two distillation columns. The Imperial Chemicals system was installed in the summer of the same year.

The success of these early DDC installations generated a great deal of interest in the user and vendor communities alike. As a result, the Instrument Society of America established the DDC Users Workshop in 1963. ISA's *Guidelines and General Information on User Requirements Concerning Direct Digital Control* is generally credited with having played a major role in the development of the minicomputer for process control.

The DDC system had the potential to be used for an unlimited variety and complexity of the automatic control functions in each and every control loop. However, the vast majority of DDC systems were implemented as digital approximations of the conventional three-mode analog controller.

The Centralized Computer Concept

The early process control computers had many disadvantages. The most significant was their slowness. For example, performing math addition commonly took 4 ms. Another disadvantage was that computer memories were very small; typically, they were able to store only four to eight thousand words of only eight to sixteen bits each. Limited software capability was another problem. All programming had to be done in machine language since assembly language or higher-level languages had not yet been designed or written in the early stages of process computers. A fourth problem was that neither instrument vendors nor user personnel had any experience in computer applications. Thus, it was very difficult to properly size projects within a computer's capabilities; most had to be reduced to fit the available machines. Another very significant disadvantage was that most of the early computer systems were very unreliable, particularly if they used germanium rather than silicon transistors. To operate effectively, computers in this period depended on unreliable mechanical devices (air conditioners).

The response of computer manufacturers to these deficiencies was to design a much larger computer system in which arithmetic functions were designed in the magnetic core memory. These changes made the computers much faster, but, because of the high cost of core memories and the additional electronic circuitry, they also were much more expensive. In addition, to help justify this cost the vendors promoted the incorporation of all types of computer functions, including both supervisory control and DDC, in one computer mainframe at a central control room location (Figure 11-4).

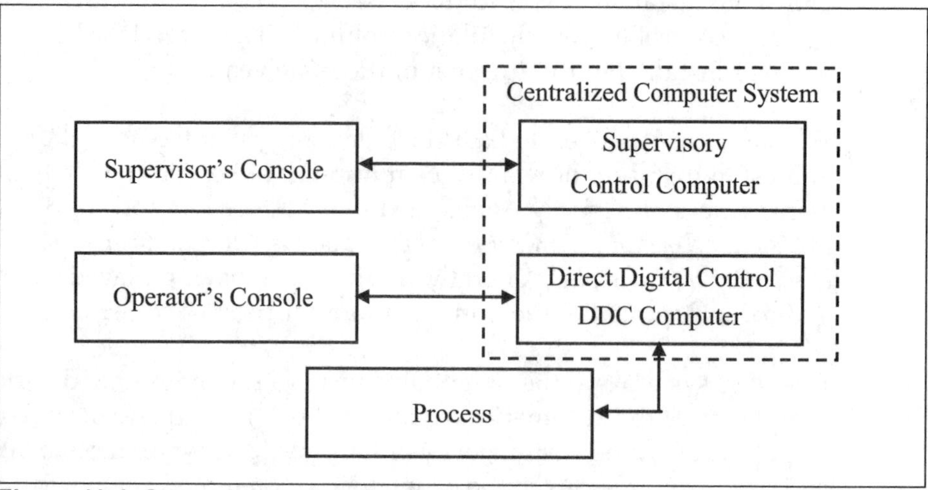

Figure 11-4. Supervisory plus direct digital control

Although these computers greatly alleviated the speed and memory problems of the earlier systems, they led to other problems. Most computer systems were sold and installed before their designs were thoroughly proved or their programming aids (compilers, higher-level languages, etc.) fully developed. Thus, the users who tried to install them experienced many frustrating delays. Because of the centralized location of these computers, it was very expensive to install the vast plant communication system required to bring the plant signals to the computer and return control signals to the field. Moreover, unless this plant communication system was carefully designed and installed, it was prone to electrical noise problems.

Because all the control functions were located in one computer, users feared the ramifications of that computer's failure and demanded a complete analog backup system that paralleled the DDC. The system that resulted is shown in Figure 11-5. To compensate for these high costs, users and vendors alike tried to squeeze the largest possible projects into the computer system, drastically complicating its programming and installation.

Because of these problems and failures many companies' management reacted quite negatively to computer control and the installation of computer systems slowed until about the mid-1970s.

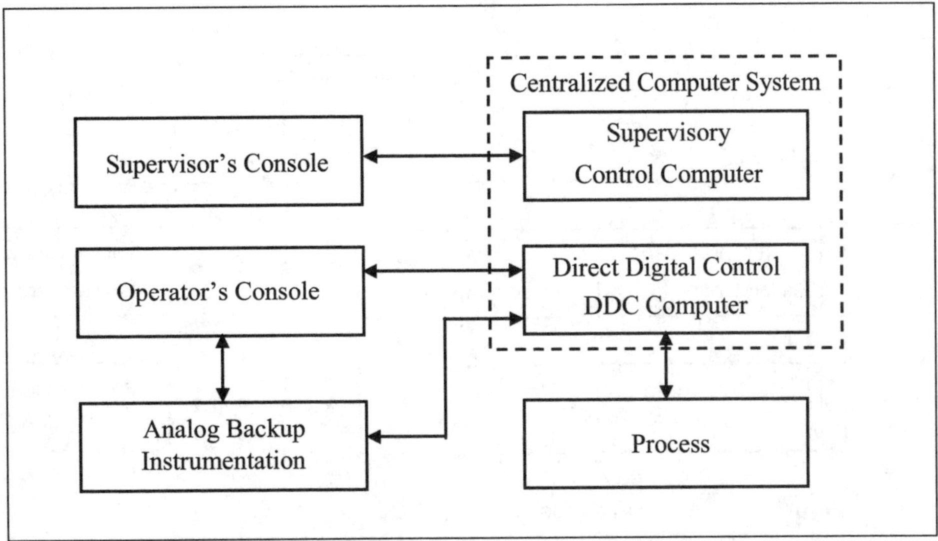

Figure 11-5. Supervisory and direct digital control with analog backup

Distributed Control Systems

Because of the reliability problems and high cost of the process control computer systems of the 1960s, there were few new process computer projects in the early 1970s. The rare projects that were started in this period were based on medium-priced minicomputers that were designed to be small in size, and followed the block diagram shown in Figure 11-4.

At the same time, two developments occurred in electronics that profoundly changed the application of digital computers to process control. The first was the development of integrated circuits and microprocessors. The second was the release of the distributed control system (DCS) by Honeywell in 1969. This new design concept was based on the idea of widely distributing the control to computer modules. Each of these modules controlled several instrument loops, generally one to four. They were connected by a single high-speed data communications link, called a *data highway*, that made possible communications between each of the computer modules and the central operator console. This design allowed the operator to monitor the operation of each local process.

In the mid-1970s, microprocessor-based modules replaced hardwired computer modules. The typical DCS had the configuration shown in Figure 11-6. Today's distributed control systems are much more powerful and faster than the first systems because of improvements in microprocessors, electronic circuits, and software.

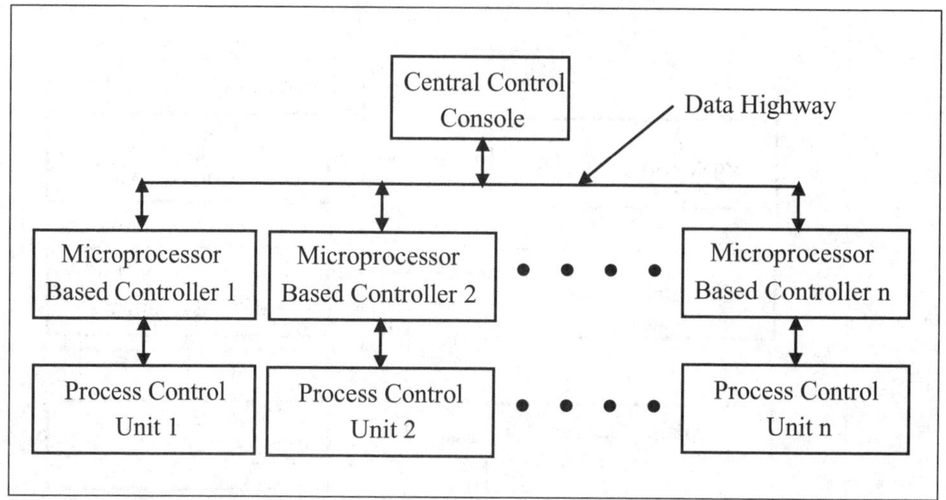

Figure 11-6. Microprocessor-based DCS

Distributed control systems today consist of one or more levels of control and information collection, as shown in Figure 11-7. The lowest level is process control and measurement on the plant floor. At this level, microprocessor-based controllers such as programmable controllers execute loop control, perform logic functions, collect and analyze process data, and communicate with other devices and to other levels in the system.

In Figure 11-7, the process data collected at level 1 is transferred to level 2. At this level, process operators and engineers use operator consoles that have a keyboard, mouse, and video display to view and adjust the various processes being controlled and monitored by the system. Also, at level 3, process and control engineers implement advanced control functions and strategies, and members of the operations management team perform advanced data collection and analysis on process information. The various plant management systems—such as inventory management and control, billing and invoicing, and statistical quality control—exist at level 3. The highest level (level 4) is used in large industrial plants to provide corporate management with extensive process and operations information.

DCS Functions and Features

The typical DCS provides a wide variety of features and functions, such as control and monitoring, an interface to operations, and advanced application software. The most important function, however, is controlling and monitoring the process.

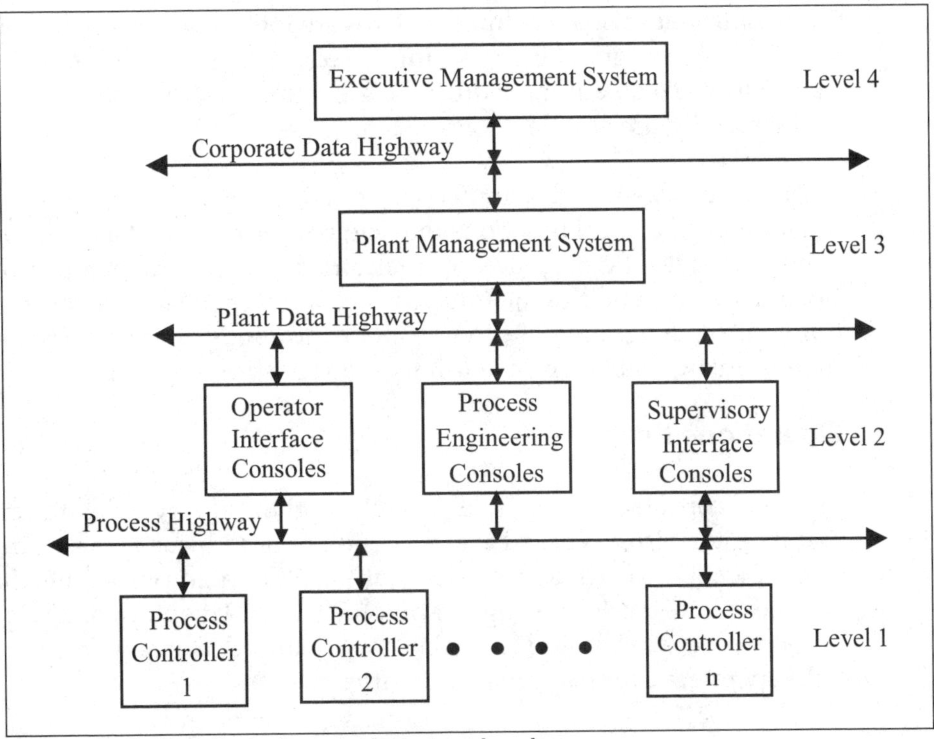

Figure 11-7. Distributed control system levels

Control and Monitoring

Process control and monitoring is performed by DCS controllers, input/output (I/O) systems, and their associated field devices. The main control functions they perform are process interfacing, logic and interlocking, controlling analog loops, and sequencing and batch control. The most important function is handling input and output signals between the process and the DCS. This is performed by the I/O system, which consists of special-purpose electronic circuit cards that are generally mounted in metal enclosures near the process equipment. The purpose of the I/O system is to provide the signal conversions required by a wide variety of measurement devices and control elements.

The typical DCS controller provides control functions from the simplest level, pressure, temperature, and flow loops to complex strategies involving interrelated loops and advanced process calculations. Most of these control functions are part of the DCS software. The operator or engineer only needs to provide information about the I/O point or points to be controlled, and the DCS internal software performs control functions. To add further control and computational power, other specialized functions are available that make it possible to design customized control algorithms and complex calculations for unique applications.

DCS controllers also have built-in software logic functions that make it possible to do interlocking and simple sequencing in addition to continuous control. To obtain the required logic functions the control engineer will generally use simple Boolean statements.

For batch processing, distributed control systems have advanced sequencing and batch control functions that can handle everything from single-product, single-stream processes to multiproduct, multistream applications. These capabilities include advanced logic and sequencing, continuous control, integration of discrete and continuous functions, recipe management, and batch scheduling and reporting.

Operations Interface

The DCS console gives operations personnel the ability either to control and monitor across a broad area of the plant or to focus on a specific area. The typical DCS console provides high-resolution graphics units that can change colors, levels, messages, and other data to display changing process conditions. These systems also use advanced alarming techniques to help operators evaluate process situations.

The console can also send information and instructions to printers to produce periodic operating reports, alarm lists, printouts of graphics screens, and printed copies of process data. They can be configured to print periodically, automatically based on specific events, or at the operator's command.

Application Software

DCS application software can do more than provide advanced control and monitoring of a process plant. It is also able to integrate the control system with plant and business computers. By giving these computers access to the DCS process database, application software facilitates functions such as process analysis, modeling and optimization, production scheduling, batch scheduling and reporting, inventory control, maintenance scheduling, and production reports. Other important features of distributed control systems are long-term data storage, retrieval, and analysis. Such systems can also be used to simulate processes while training operators or developing control strategies.

Programmable Controllers

As we discussed earlier, a widely used type of process control computer is the programmable logic controller (PLC). PLCs were originally designed and manufactured to replace relay-based logic systems and solid-state hardwired logic control panels. However, the modern PLC system is far more complex and powerful.

The most basic function programmable controllers perform is to examine the status of inputs and, in response, control some process or machine through outputs. The logical combination of inputs to produce an output or outputs is called *control logic*. Several logic combinations are usually required to carry out a control plan or program. This control plan is stored in memory using a programming device. Once input into memory, the control plan is periodically scanned by the processor—usually a high-speed microprocessor—in a predetermined sequence. The period required to examine the inputs and outputs, perform the control logic, and execute the outputs is called the *scan time*.

A simplified block diagram of a programmable controller is shown in Figure 11-8. In this diagram, a level switch and panel-mounted push button are wired to input circuits, and the output circuits are connected to an electric solenoid valve and a panel-mounted indicator light. The output devices are controlled by the software program in the logic unit.

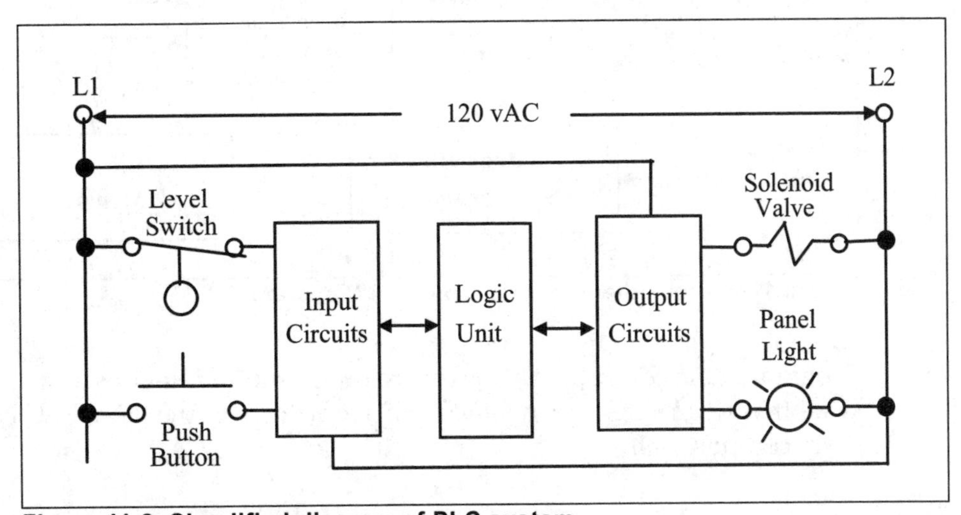

Figure 11-8. Simplified diagram of PLC system

Figure 11-8 shows a typical configuration of the early programmable controller applications, which were intended to replace relay or hardwired logic control systems. The input circuits are used to convert the various field voltages and currents to the low-voltage signals (normally, 0-to-5 volts direct current [DC]) that are used by the logic unit. The output circuits convert the logic signals to a level that will drive the field devices. For example, in Figure 11-8, 120-volt AC power is connected to the field input devices, so the input circuits are used to convert the 120 VAC to the 0-to-5 volt logic signals used by the control unit.

Basic Components of PLC Systems

Regardless of size, cost, or complexity, all programmable controllers share the same basic components and functional characteristics. A programmable controller will always consist of a processor, an input/output system, a power supply, and a communications device or port. The block diagram in Figure 11-9 shows these five basic components, with inputs and outputs to the I/O system.

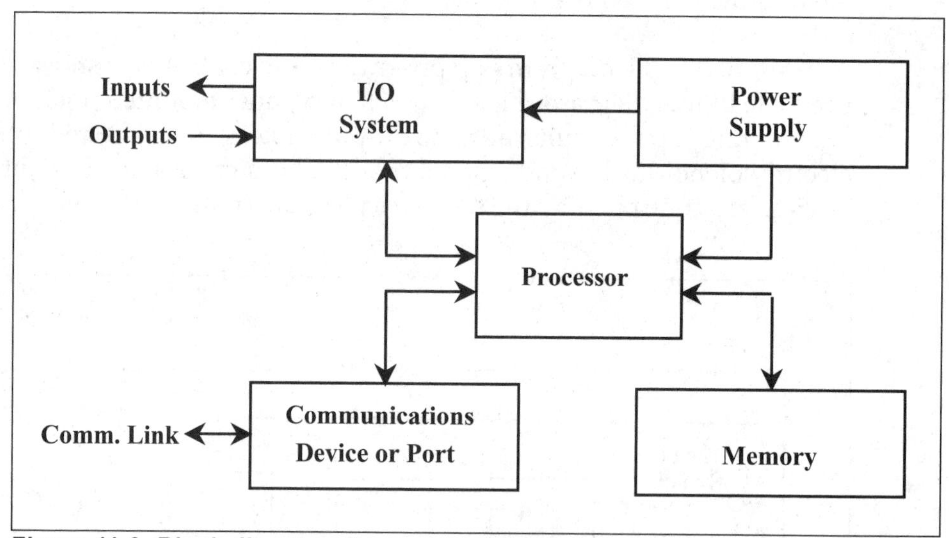

Figure 11-9. Block diagram of a typical PLC system

Programmable controllers also require programming devices and software that will be discussed later. First, we will cover the basic PLC hardware components.

The Processor

The processor consists of one or more standard or custom microprocessors as well as other integrated circuits that perform the logic, control, and memory functions of the PLC system. The processor reads the inputs, executes logic as determined by the application program, performs calculations, and controls the outputs accordingly.

The processor controls the operating cycle or processor scan for the PLC. This operating cycle consists of a series of operations that are performed sequentially and repeatedly. Figure 11-10 shows a typical PLC processor operating cycle.

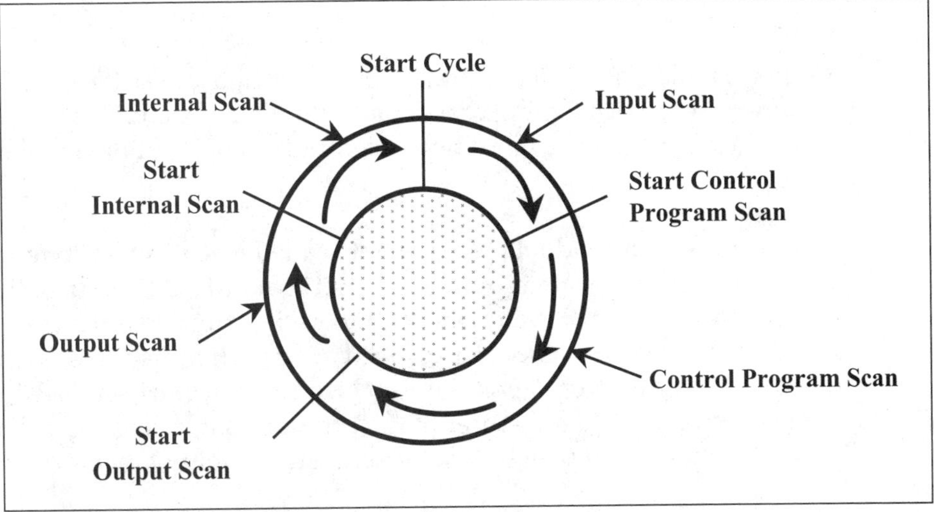

Figure 11-10. PLC processor operating cycle

During the "Input Scan," the PLC examines the external input devices for a signal present or absent (i.e., an on or off state). The status of these inputs is temporarily stored in an input image table or memory file.

During the "Program Scan" cycle, the processor scans the instructions in the control program, uses the input status from the input image file, and determines if an output will or will not be energized. The resulting status of the outputs is written to the output image table or memory file.

Based on the data in the output image table, the PLC energizes or deenergizes its associated output circuits, controlling external devices during the "Output Scan" cycle.

During the "Internal Scan" cycle, the processor performs housekeeping functions such as internal diagnostics and communications. A typical diagnostic function would be to check for math operational errors. The PLC must also check for, and perform, communication operations with its programming device, other PLCs, or other devices connected to its communication port or ports.

This operating cycle typically takes 1 to 25 milliseconds (thousandths of a second). However, the operating cycle time depends on the complexity of the control logic written by the user so that a simple control program might take only 1 ms, but a large and complex program may require a scan time as high as 250 ms. The input, output, and internal scans are normally very short compared to the time taken for the program scan. These scans are continually repeated in a looped process.

Memory

Memory, which is usually located in the same housing as the CPU, is used to store the control program and data for the PLC system. The information stored in memory determines how the input and output data will be processed.

Memory stores individual pieces of data called *bits*. A bit has two states: 1 or 0. Memory units are mounted on circuit boards and are usually specified in thousands or "K" increments, where 1K is 1,024 words (i.e., $2^{10} = 1024$) of storage space. The capacity of programmable controller memory may vary from less than 1,000 words to over 64,000 (64K) words, depending on the manufacturer of the programmable controller. The complexity of the control plan will determine the amount of memory required.

Although there are several different types of computer memory, they can always be classified as *volatile* or *nonvolatile*. Volatile memory will lose its programmed contents if all operating power is lost or removed. Volatile memory is easily altered and is quite suitable for most programming applications when it is supported by battery backup and/or a recorded copy of the program. Nonvolatile memory will retain its data and program even if there is a complete loss of operating power. It does not require a backup system.

The most common form of volatile memory is random access memory, or RAM. RAM is relatively fast and provides an easy way to create and store application programs. If normal power is disrupted, PLCs with RAM use battery or capacitor backups to prevent the loss of programs.

Electrically erasable programmable read-only memory (EEPROM) is a nonvolatile memory. It is programmed through application software running on a personal computer or through a micro-PLC handheld programmer.

The user can access two areas of memory in the PLC system: *program files* and *data files*. Program files store the control application program, subroutine files, and the error file. Data files store data associated with the control program, such as input/output status bits, counter and timer preset and accumulated values, and other stored constants or variables. Together, these two general memory areas are called *user* or *application memory*. The processor also has an *executive* or *system memory* that directs and performs operational activities such as executing the control program and coordinating input scans and output updates. The user cannot access or change this process system memory, which is programmed by the PLC manufacturer.

I/O System

The I/O system provides the physical connection between the process equipment and the microprocessor. This system uses several input circuits or modules to sense and measure the physical quantities of the process, such as motion, level, temperature, pressure, flow, and position. Based on the status it senses or the values it measures, the processor controls output modules. These modules drive field devices such as valves, motors, pumps, and alarms to exercise control over a machine or a process.

Input Types

The inputs from field instruments or sensors supply the data and information that the processor needs to make the logical decisions that control a given process or machine. These input signals come from such devices as push buttons, hand switches, thermocouples, strain gauges, and the like. These signals are connected to input modules where they are filtered and conditioned for use by the processor.

Output Types

The outputs from the programmable logic controller energize or deenergize control devices to regulate processes or machines. These output signals are control voltages from the output circuits, and they are generally not high-power signals. For example, an output module sends a control signal that energizes the coil in a motor starter. The energized coil closes the power contacts of the starter. These contacts then close to start the motor. The output modules are usually not directly connected to the power circuit but rather to devices such as the motor starter and heater contactors that apply high-power (greater than 10 amps) signals to the final control devices.

I/O Structure

PLCs are classified as micro, small, medium, and large mainly based on the I/O count. Micro PLCs generally have an I/O count of 32 or less; small PLCs have less than 256 I/O points; medium-sized PLCs have an I/O count of less than 1,024; and large PLCs have an I/O count greater than 1,024. Micro PLCs are self-contained units in which the processor, power supply, and I/O are all in one package. Because they are self-contained, micro PLCs are also called *packaged controllers*. A modular PLC is one that has separate components or modules.

The advantage of a packaged controller is that the unit is smaller, costs less, and is easy to install. Figure 11-11 shows a typical wiring diagram for a micro PLC, namely, an Allen-Bradley Micro 1000 PLC with nine inputs and five outputs. The unit is powered with 120 volts of alternating current (AC). It has an internal power supply to operate the internal I/O circuits and the built-in microprocessor as well as to generate the 24 volts of direct current (DC) needed for the field input switches and contacts.

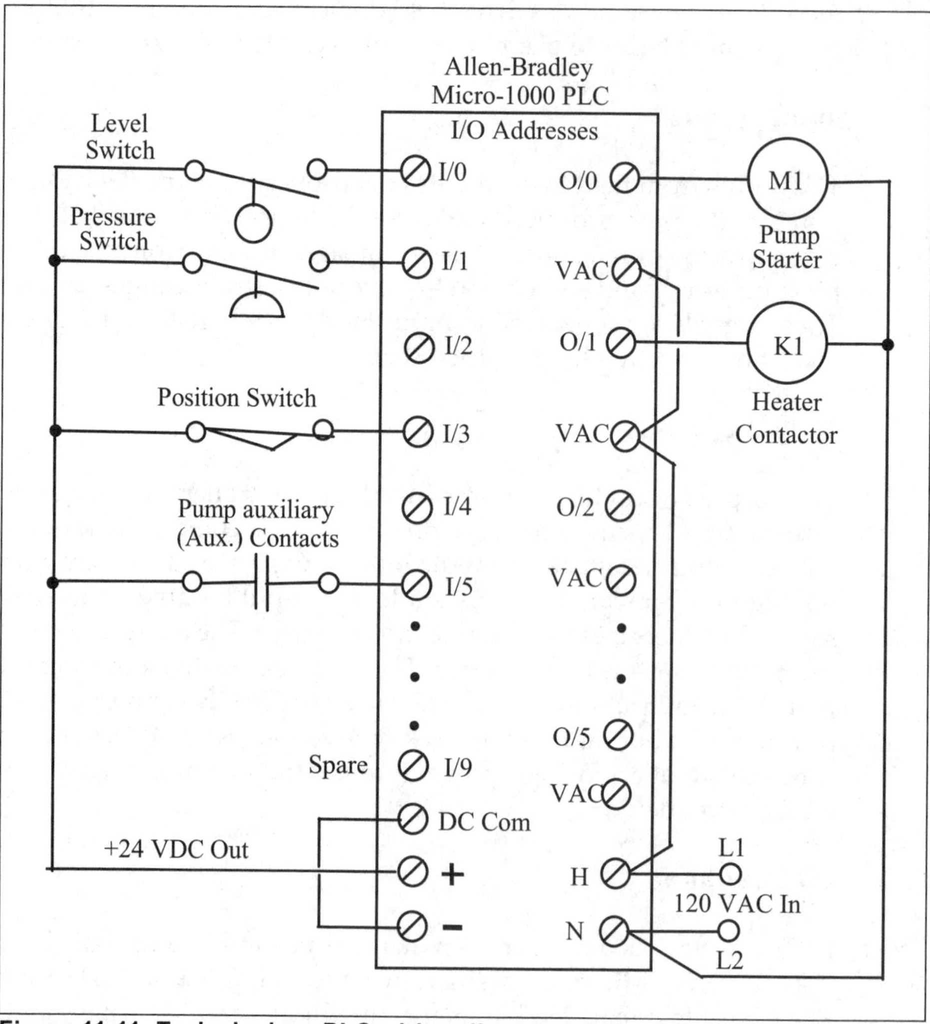

Figure 11-11. Typical micro-PLC wiring diagram

In medium and large PLC systems the I/O modules are normally installed or plugged into a slot in a "universal" modular housing. The term *univer-*

sal in this context means that any module can be inserted into any I/O slot in the housing. Modular I/O housings are also normally designed so that the I/O modules can be removed without turning off the AC power or removing the field wiring.

Figure 11-12 shows some typical configurations for I/O modular housings. The backplane of the housings into which the modules are plugged have a printed circuit card that contains the parallel communications bus to the processor and the dc voltages for operating the digital and analog circuits in the I/O modules. These I/O housings can be mounted in a control panel or on a subpanel in an enclosure. They are designed to protect the I/O module circuits from dirt, dust, electrical noise, and mechanical vibration.

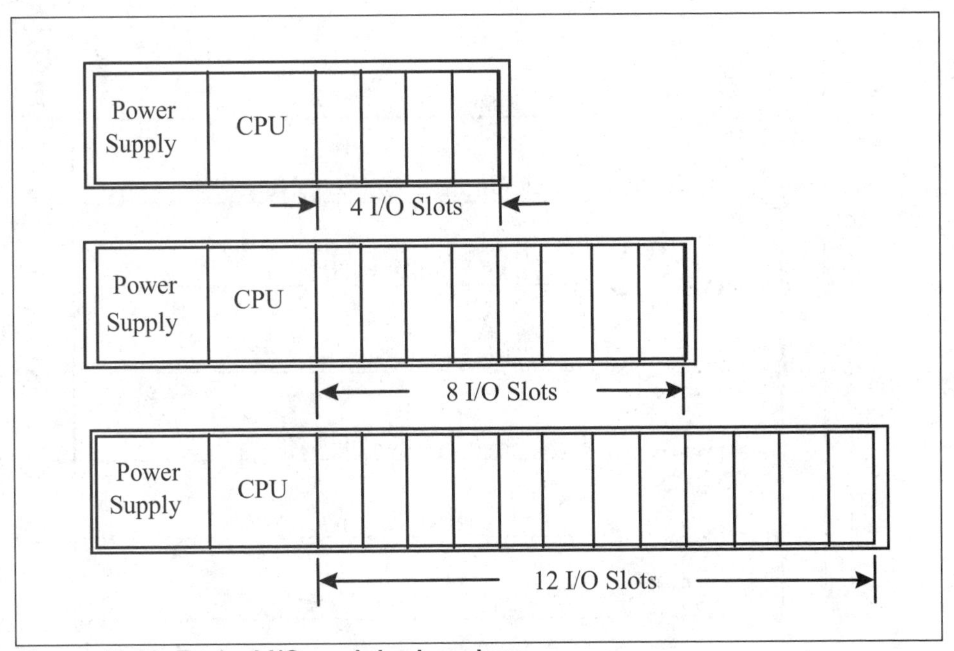

Figure 11-12. Typical I/O modular housings

The backplane of the I/O chassis contains sockets for each module. These sockets provide the power and data communications connection to the processor for each module.

Discrete Inputs/Outputs

Discrete is the most common class of input/output in a programmable controller system. This type of interface module connects field devices that have two discrete states, such as ON/OFF or OPEN/CLOSED, to the processor. Each discrete I/O module is designed to be activated by a field-supplied voltage signal, such as +5 VDC, +24 VDC, 120 VAC, or 220 VAC.

In a discrete input (DI) module, if an input switch is closed, an electronic circuit in the input module senses the supplied voltage. It then converts it into a logic-level signal that is acceptable to the processor in order to indicate the status of that device. For a field input device or switch a logic 1 indicates ON or CLOSED, and a logic 0 indicates OFF or OPENED.

A typical discrete input module is shown in Figure 11-13. Most input modules will have a light-emitting diode (LED) that indicates the status of each input.

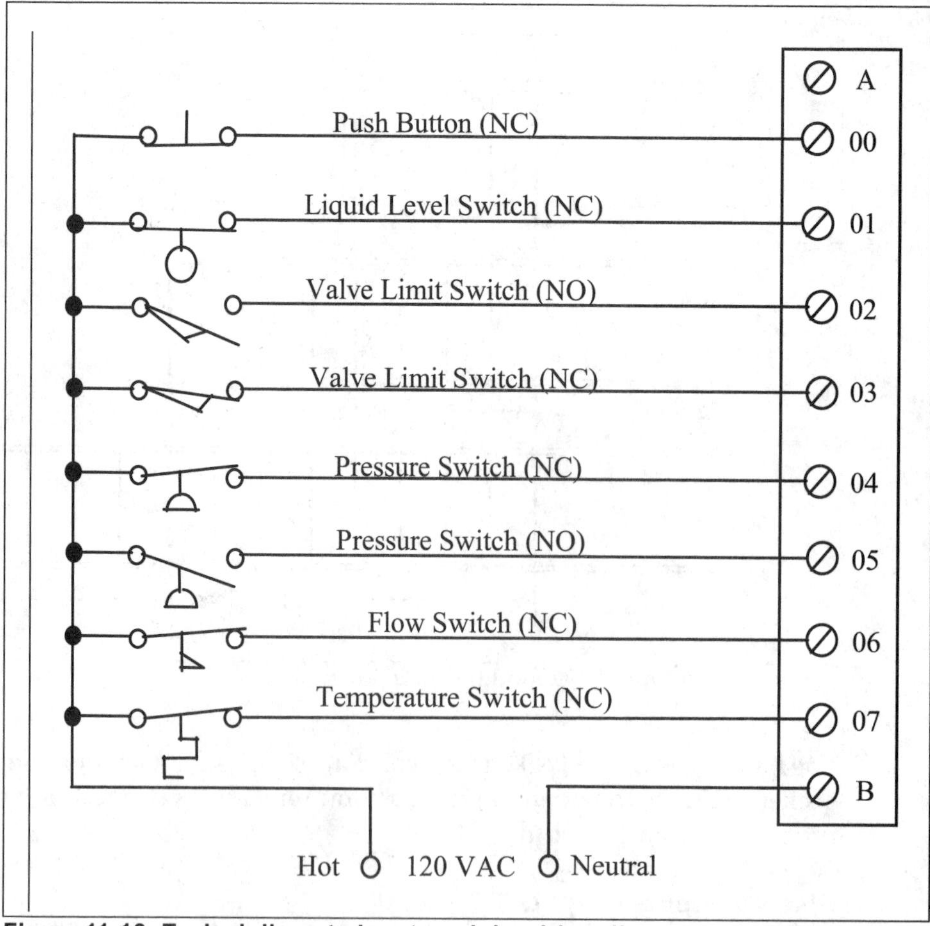

Figure 11-13. Typical discrete input module wiring diagram

In a discrete output (DO) module, the output interface circuit switches the supplied control voltage that will energize or deenergize the field device. If an output is turned ON through the control program, then the interface circuit switches the supplied control voltage to activate the referenced (addressed) output device.

Figure 11-14 shows a wiring diagram of a typical discrete output module. The module can be thought of as a simple switch through which power can be provided to control the output device. During normal operation, the processor sends the output state determined by the logic program to the output module. The module then switches the power to the field device.

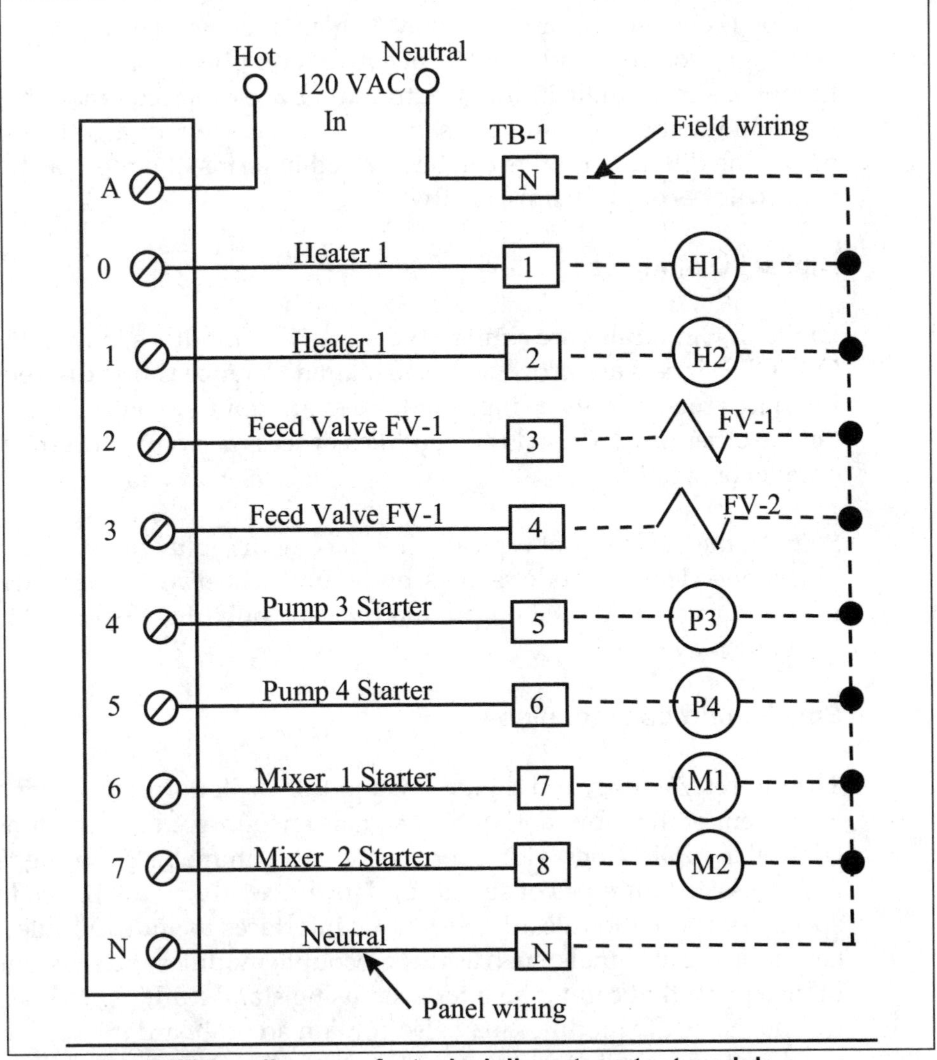

Figure 11-14. Wiring diagram of a typical discrete output module

A fuse is normally provided in the output circuit of the module to prevent excessive current from damaging the wiring to the field device. If it is not, it should be provided during the system design stage.

Analog I/O Modules

The analog I/O modules make it possible to monitor and control analog voltages and currents, which are compatible with many sensors, motor drives, and process instruments. By using analog I/O, you can measure or control most process variables with the appropriate interfacing.

Analog I/O interfaces are generally available for several standard unipolar (single-polarity) and bipolar (negative- and positive-polarity) ratings. In most cases, a single input or output interface can accommodate two or more different ratings and can satisfy either a current or a voltage requirement. The different ratings can be selected in terms of either hardware (i.e., switches or jumpers) or software.

Digital I/O Modules

Digital I/O modules are similar to discrete I/O modules in that discrete ON/OFF signals are processed. The main difference is that discrete I/O interfaces require only a single bit to read an input or control an output. On the other hand, digital I/O modules process a group of discrete bits in parallel or serial form.

Some of the devices that typically interface with digital input modules are binary encoders, bar code readers, and thumbwheel switches. Instruments that are driven by digital output modules include LED displays and intelligent display panels.

Special-purpose Modules

The discrete, analog, and digital I/O modules will normally cover about 90 percent of the input and output signals encountered in programmable controller applications. However, for the programmable controller system to process certain types of signals or data efficiently, you will need to use special-purpose modules. These special interfaces include modules that condition input signals, such as thermocouple modules, pulse counters, or other signals that cannot be interfaced using standard I/O modules. Special-purpose I/O modules may also contain an on-board microprocessor to add intelligence to the interface. Such intelligent modules can perform complete processing functions independently of the CPU and the control program scan.

Another important class of special-purpose I/O modules are communication modules that communicate with distributed control systems (DCSs), other PLC networks, plant computers, or other intelligent devices.

Programming Languages

The programming language allows the user to communicate with the programmable controller by way of a programming device. Programmable controller manufacturers use several different programming languages, but they all convey to the system a basic control plan by means of instructions.

A *control plan* or *program* is a set of instructions that are arranged in a logical sequence to control the actions of a process or machine. For example, the program might direct the programmable controller to turn a motor starter on when a push button is depressed and at the same time turn on a control panel-mounted RUN light when the motor starter auxiliary contacts are closed.

A program is written by combining instructions in a certain order. Rules govern the manner in which instructions are combined and the actual form of the instructions. These rules and instructions combine to form a language.

International Standard for PLC Languages

In the early 1970s, different national and international committees have proposed numerous PLC programming standards to develop a common interface for programmable controllers. Then, in 1979, a working group of international PLC experts was appointed by various national committees to write a first draft of a comprehensive PLC standard. The first committee draft was issued in 1982.

After an initial review of the document by the national committees, it was decided that the standard was too complex to handle as a single document. As a result, the working group was split into five task forces, one for each part of the standard. The subject of each part is as follows: Part 1, General Information; Part 2, Equipment and Testing Requirements; Part 3, Programming Languages; Part 4, User Guidelines; and Part 5, Communications. Each task group consisted of several international experts, each backed by a national advisory group. The International Electrotechnical Commission (IEC) issued their standard for PLC programming languages in March of 1993 and assigned it the number IEC 61131-3.

The IEC standard has three graphical languages: Ladder Diagram (LD), Function Block Diagram (FBD), and Sequential Function Chart (SFC), and two text-based languages: Instruction List (IL) and Structured Text (ST). The PLC language standard allows different parts of an application to be programmed in different languages that can be combined into a single executable program. The three standard languages most commonly used in PLC applications are Ladder Diagram, Instruction List, and Function Block Diagram.

Figure 11-15 shows a simple logic function implemented using the three common types of PLC languages. The logic function shown in the figure is an "And" function—that is, if push button 1 (PB-1) is CLOSED *and* push button 2 (PB-2) is CLOSED then the GO Light is ON.

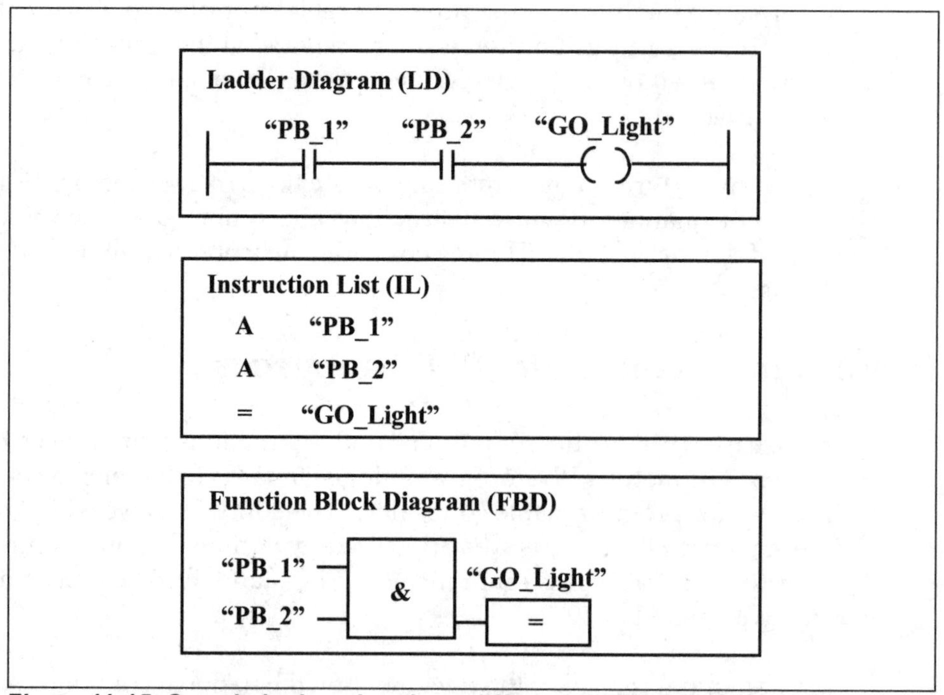

Figure 11-15. Sample logic using three different PLC languages

Ladder Diagram (LD)

Ladder diagram (LD) is the language most commonly used in PLC applications. The reason for this is relatively simple. The early programmable controllers were designed to replace electrical relay-based control systems. These systems were designed by technicians and engineers using a symbolic language called *ladder diagrams*. The ladder diagram consists of a series of symbols interconnected by lines to indicate the flow of current through the various devices. The ladder drawing consists of basically two things. The first is the power source, which forms the sides of the ladder (rails), and the second is the current, which flows through the various logic input devices that form the rungs of the ladder. If there is electrical current flow through the relay contacts in a rung then the output relay coil will be turned on. This is termed "Power Flow" in the ladder rung.

In electrical design, the ladder diagram is intended to show only the circuitry that is necessary for the basic operation of the control system. Another diagram, called the *wiring diagram,* is used to show the physical connection of control devices. The discrete I/O module diagrams shown earlier in this chapter are examples of wiring diagrams. A typical electrical ladder diagram is shown in Figure 11-16. In this diagram, a push button (PBl) is used to energize a pump start control relay (CRl) if the level in a liquid storage tank is not high. Each device has a special symbol assigned to it to make it easier to read the diagram quickly.

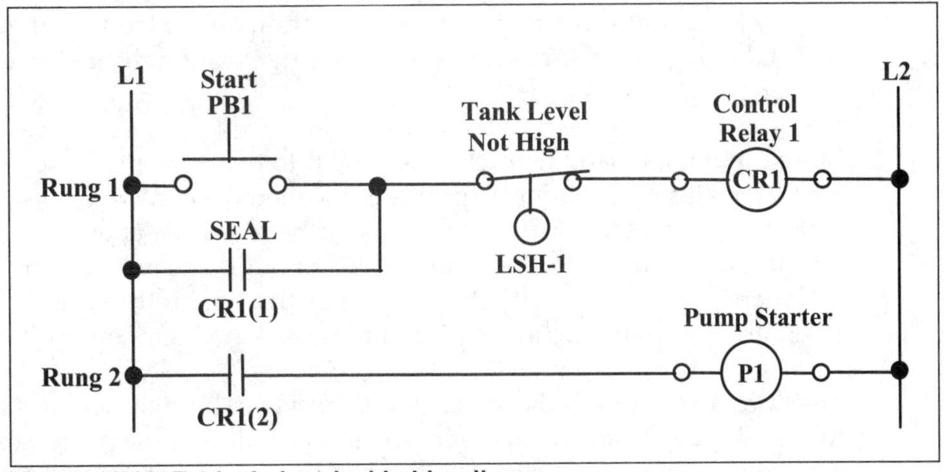

Figure 11-16. Typical electrical ladder diagram

You can implement the same control application using a PLC ladder diagram (LD) program, as shown in Figure 11-17. You read the diagrams in the same way, from left to right, with the logic input conditions on the left and the logical outputs on the right. In the case of electrical diagrams, there must be electrical continuity to energize the output devices; for programmable controller ladder programs, there must be logic continuity to energize the outputs.

In ladder programs, three basic instructions are used to form the program. The first symbol is similar to the normally open (NO) relay contacts used in electrical ladder diagrams; this instruction uses the same NO symbol as in ladder programs. It instructs the processor to examine its assigned bit location in memory. If the bit is ON (logic l), the instruction is true, and

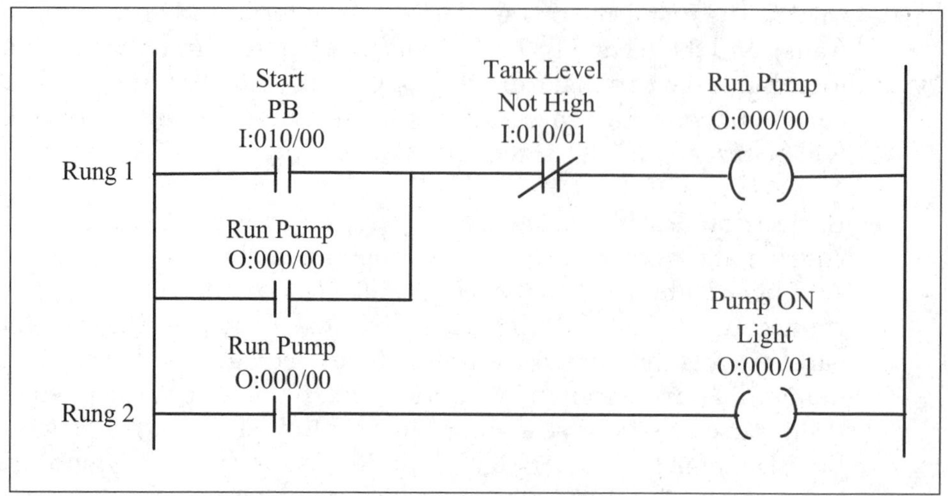

Figure 11-17. Typical ladder diagram (LD) program

there is logic continuity through the instruction on the ladder rung. If the bit is OFF (logic 0), there is no logic continuity through the instruction on the rung.

The second important instruction is similar to the normally closed (NC) contact in electrical ladder diagrams. It is called the *examine off instruction*. Unlike the *examine on instruction*, it directs the processor to examine the bit for logical 0 or the OFF condition. If the bit is OFF, the instruction is true, and there is logic continuity through the instruction. If the bit is ON, the normally closed instruction is false, and there is no logic continuity.

The third instruction is the *output coil instruction*. This instruction is similar to the relay coil in electrical ladder diagrams. It directs the processor to set a certain location in memory to ON or 1 if there is logic continuity in any logic path preceding it. If there are no complete logic continuity paths in the ladder rung, the processor sets the output coil instruction bit to 0 or OFF.

Figure 11-17 shows a typical Allen-Bradley Ladder Program. In this proprietary and non-standard ladder programming language, the letters O or I followed by a 5-digit number above the instructions are the reference addresses for the logic bits. The letter "I" before the 5-digit number indicates an input bit and the letter "O" before the 5-digit number indicates an output bit. The reference address indicates where in the memory the logic operation will take place.

In the ladder diagram program shown in Figure 11-17, the examine on instruction for the start push button (PB) directs the processor to see if the reference address I:010/00 is ON. In the same way, the examine on instruction for the "tank level not high" input instructs the processor to see if the reference address I:010/01 is OFF. If there is logic continuity through both instructions, the output coil instruction at address O: 000/00 is turned ON. Logic continuity from the left side to the right end of a rung is called "Logical Power Flow" in PLC programming.

This same output bit is then used to "seal in" the start push button instruction. It also turns on the energized instruction bit O:000/01 so as to turn on the pump run light.

In the IEC LD programming language standard, the input address I:010/00 would have the form %IX010.00 and the output address O:000/01 would be listed as %QX000.0. Now not all PLC manufacturers fully comply with the IEC programming standard, so many programming examples used in this book will not comply with the IEC standard. Before using any programmable example from this book, a PLC manufacturer's instruction manual must be consulted for the proper programming language format.

Instruction List

Instruction List (IL) is a textual programming language that you can use to create the code for a PLC control program. Its syntax for statements is similar to microprocessor assembly language and consists of instructions followed by addresses on which the instructions act. The IL language contains a comprehensive range of instructions for creating a complete user program. For example, in the Siemens S7 programming software package, over 130 different basic IL instructions and a wide range of addresses are available, depending on the model PLC used.

IL instruction statements have two basic structures. The first is a statement consisting of an instruction by itself (for example, NOT) and another where the statement consists of an instruction and an address. The most common structure is for the statement to have an instruction and an address. The address of an instruction statement indicates a constant or the location where the instruction finds a value on which to perform an operation.

The most basic type of IL instructions are Boolean Bit Logic Instructions. These instructions perform logic operations on single bits in PLC memory. The basic Bit Logic Instructions are (1) "And" (A) and its negated form, "And Not" (AN); (2) "Or" (O); and (3) "exclusive Or " (Or) and its negated form, "Exclusive Or Not" (XN). These instructions check the signal state of a bit address to establish whether the bit is activated "1" or not activated "0."

Bit logic instructions are also called *relay logic instructions* since they can execute commands that can replace a relay logic circuit. Figure 11-18 is an example of "And" logic operation. The IL program is listed on the left side, and the relay logic circuit is shown on the right side for comparison. In this example, the statement list program uses an AND instruction (A) to program two normally open (NO) contacts in series. Only when the signal state of both the normally open contacts is changed to "1" can the state of output Q4.0 be changed to "1" and the coil be energized.

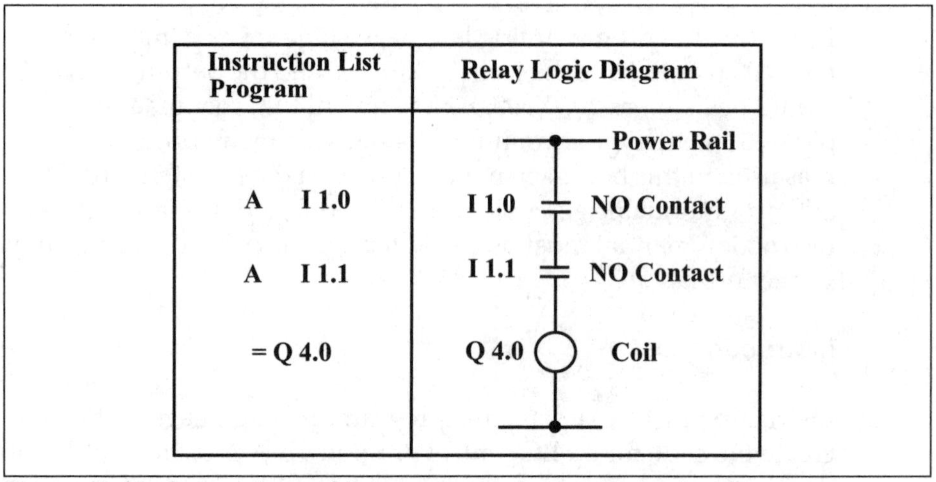

Instruction List Program	Relay Logic Diagram
	Power Rail
A I 1.0	I 1.0 NO Contact
A I 1.1	I 1.1 NO Contact
= Q 4.0	Q 4.0 Coil

Figure 11-18. Comparison of IL program and relay logic circuit

Function Block Diagram (FBD)

The Function Block Diagram (FBD) is a graphical programming language. It allows the programmer to build complex control procedures by taking existing functions from the FBD library and wiring them in a graphic diagram area. An FBD describes a relationship or function between input and output variables. A function is a set of elementary function blocks, as shown in Figure 11-19. Input and output variables are connected to blocks by connection lines.

You can build an entire function operated by a FBD program by using standard elementary function blocks from the FBD library. Each elementary function block has a fixed number of input connection points and output connection points. For example, the Boolean AND function block shown in Figure 11-19 has two inputs and only one output. The inputs are connected on its left border; the outputs are connected on its right border. An elementary function block performs a single function between its inputs and its outputs. For example, the elementary function block shown in Figure 11-19 performs the Boolean AND operation on its two inputs and produces a result at the output. The name of the function to be per-

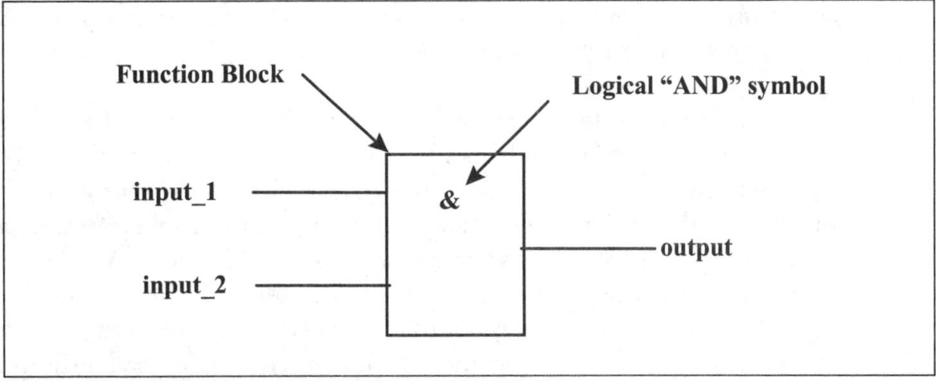

Figure 11-19. Typical elementary function block

formed by the block is written in its symbol. In the case of the "AND" function the symbol is "&."

Programming Devices

The programming device is used to enter, store, and monitor the programmable controller software. Programming devices can be a dedicated portable unit or a personal computer-based system. The personal computer-based systems normally have basic components: a keyboard, visual display or CRT, personal computer, printer, and communications interface card and cable (see Figure 11-20).

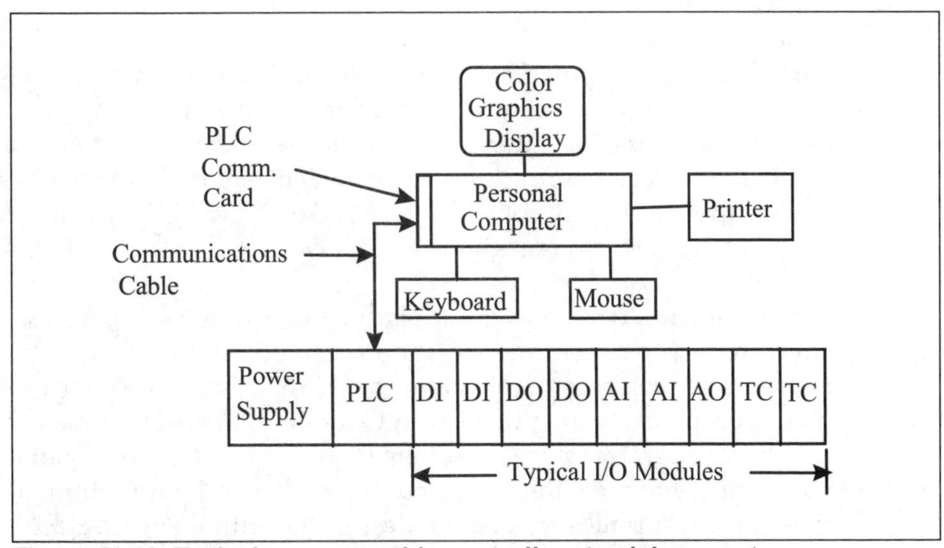

Figure 11-20. Typical programmable controller standalone system

The programming devices are normally only connected to the programmable controller system during the programming, startup, or troubleshooting of the control system. Otherwise, the programming device is disconnected from the system.

The programming terminals are normally either a handheld programmer or a personal computer-based system. The handheld programmers are inexpensive portable units, which are normally used to program small programmable controllers. Most of these units resemble portable calculators but with larger displays and a somewhat different keyboard. The displays are generally LED (light-emitting diode) or dot matrix LCD (liquid crystal display), and the keyboard consists of alphanumeric keys, programming instruction keys, and special-function keys. Even though they are mainly used to write and edit the control program, the portable programmers are also used to test, change, and monitor the program.

The standard programming terminal is a personal computer, such as that shown in Figure 11-20, in which the programming software is loaded on the hard drive. These units can perform program editing and storage. They also have added features such as the capacity to print out programs and to be connected to local area networks (LAN). LANs give the programmer or engineer access to any programmable controller in the communications network, so he or she can monitor or control any device in the network. Normally, laptop PCs are used as the programming terminal because they are light and portable and can be easily used in the field when the control program is tested, started up, and modified.

Power Supply

The power supply converts AC line voltages to DC voltages in order to power the electronic circuits in a programmable controller system. These power supplies rectify, filter, and regulate voltages and currents so as to supply the correct amounts of voltage and current to the system. The power supply normally converts 120 VAC or 240 VAC line voltage into direct current (DC) voltages such as +5 VDC, –15 VDC, or +15 VDC.

You may integrate the power supply for a programmable controller system with the processor, memory, and I/O modules into a single housing. It may also be a separate unit connected to the system through a cable. As a system expands to include more I/O modules or special-function modules, most programmable controllers require an additional or auxiliary power supply to meet the increased power demand. Programmable controller power supplies are usually designed to eliminate the electrical noise present on the AC power and signal lines of industrial plants so it does not introduce errors into the control system. The power supplies are also designed to operate properly in the higher- temperature and humidity environments present in most industrial applications.

Communication Devices

The main function communication device or port is to communicate with the programming to enter, modify, and monitor the PLC control plan. In a small standalone PLC, there may be a single serial port to connect with the programming device. However, most PLC systems have more than one communication port or device. In a typical PLC system, there is the standard serial RS-232C port for programming and a vendor proprietary communication network that is used to transfer information between the remote I/O equipment racks and the other PLCs in the system. In larger PLC systems, there is normally an Ethernet link to communicate with PCs and other networks connected to the system.

Process Visualization

There are two methods commonly used to provide operators with real-time visual interface to programmable controller-based control systems. The first method is to wire directly from the programmable controller I/O modules to hard-wired lights and digital indicators on a display or control panel. This method is cost effective in small systems that will not be changed, but it is generally not recommended or used in larger control systems that might be expanded in the future.

The more common method is to use PCs with HMI software that can generate a visual representation of the process or machine being controlled. The process or machine display screens are usually based on the process and instrument drawings or mechanical drawings for the system being controlled and provide the operators with an overall view of the process or machine under control.

The software-based graphical display method has advantages in that the process display screens can be easily modified for process changes, and the software can perform other functions, such as control, alarms, messaging, report generation, historical trending, integrated real-time statistical process control (SPC), and batch process recipe control.

The process overview pictures are generally interesting to visitors but historical trending and alarm message display are more widely viewed and used by the operators in daily operations.

Alarms must be quickly analyzed and responded to by the operators to prevent product or process damage or injury to personnel. Most HMI software packages allow the operator to configure or tailor the messages to the process or plant being controlled. The messages can be derived from individual I/O bits or analog signals that are out of range or process limits. Message systems minimize process or machine downtime by alerting the operators to out-of-tolerance conditions or events.

The historical trending of key process or machine parameters allows the operator to detect and then prevent potential production or machine problems. The SPC software continuously records process or machine sequences and events that are related to product quality. This makes it possible to verify product quality on a continuous basis.

HMI software provides standard screen objects, such as pushbuttons, selector switches, and panel indicators to create user interfaces that are appropriate to the process being controlled. HMI systems not only acquire process data, messages, and events, the software may also store the information in archives and then the information can be made available on a filtered or sorted basis. The configuration engineer can design a message or alarm so that an operator must acknowledge the condition.

In a typical system, a Microsoft SQL Server is used to archive messages. The system archives information when a message or an alarm occurs or when there is a change in status. Message and alarm sequence logs are used to document events on a chronological basis. The system can generally print out all changes in status (new, departed or acknowledged) of the alarms or messages. Process values are generally archived cyclically on an event or limits violation basis. The analog process values can also be archived on a condensed basis using signal averaging.

The Microsoft SQL Server uses effective loss-free data compression functions so that memory resources are conserved. However, a large process can generate an archive of considerable size so the user must set a maximum archiving period, such as one week or month, to limit the data volume. Each data archive can be segmented (e.g., per week, per month) and then the completed individual archives can be exported to a long-term archive server.

HMI software packages have reporting systems that allow printing of process data in various formats, such as message sequence logs, system message logs, or user production reports. Before the reports are printed out, they can be reviewed on a monitor and saved to a report file. Most HMI packages make it possible to start a report on a time- or event-driven basis or on command from the operator. The user can generally assign separate printers to each print job.

The HMI graphics design software is generally simple and easy to use. Graphic objects can be easily linked to PLC internal tags for animation or control purposes. When a new object has been placed on a graphics screen, an easy-to-edit dialog box appears on the screen. Graphics designer software allows the operator to specify and animate virtually all object properties. Using OPC, the system operator can connect the HMI software to any communication device for which an OPC-compliant server or driver is available. HMI systems can function as both a native OPC client and OPC server, and most HMI software packages support browsing for OPC server addresses.

The HMI software packages provide wizards (software assistants) to aid the configuration engineer in the system development. For example, a message wizard might offer default settings that the developers can confirm, modify, or reject. Preview windows are used to display the effects of any chosen parameters. If the configuration designer confirms the setting, the solution is implemented by the system.

HMI systems provide a library of configured display and control modules like valves, process tanks, pumps, pipes, instruments, and switches. The developer can also create custom objects and store them in the library for future use. Objects are generally stored by topics in a library. When needed, objects are simply brought into the process screen using the standard computer drag and drop technique.

A completed control and graphics configuration can be tested without having to connect to a PLC by using simulation software that is part of the HMI software system. To simulate PLC inputs or outputs, each point can be assigned a characteristic value (analog or discrete). When a picture appears on the PC screen during a test, it becomes clear, via a color or value change whether or not the control, display, or alarm objects are functioning properly.

Plantwide Computer-Based System

The modern plantwide data collection and control system combines features of DCS and programmable controllers. It also uses high-speed local area networks (LANs) to integrate all hardware and software capabilities into a complete production and business operations system.

A local area network (LAN) is a user-owned and -operated data transmission system that operates within a building or set of buildings. LANs allow many different machines and processes to exchange large amounts of information at high speed over a limited distance. LANs link together communicating devices—such as computers, programmable controllers, process controllers, terminals, printers, and mass storage units—within a single process or manufacturing building or plant.

Communication networks enable computers to share data resources, hardware resources, and software. For example, a typical manufacturing facility might tie together the process control system computers and a central computer used by purchasing to accelerate the order process for the plant's raw materials.

Figure 11-21 shows a typical plantwide computer-based control and data collection system for an industrial facility. It consists of three levels of control and data collection with three associated communication networks. The highest level (Level III) is the information network used by groups at the plant like accounting and purchasing. This network has the highest-speed communication (100 Mbps and higher) network because it must handle large amounts of data and information. The middle level (Level II) is used to perform the automation and control of the industrial plant and normally has a communications network with speeds from 1 Mbps to 100 Mbps. This network is slightly slower than the Level III network because process and machine control require somewhat lower data rates. The lowest level in the plant (Level I) is used to connect programmable control and other controllers directly to the field devices, such as weight, flow, pressure, level transmitters. The Level I switches and communication speed are generally less than 1 Mbps.

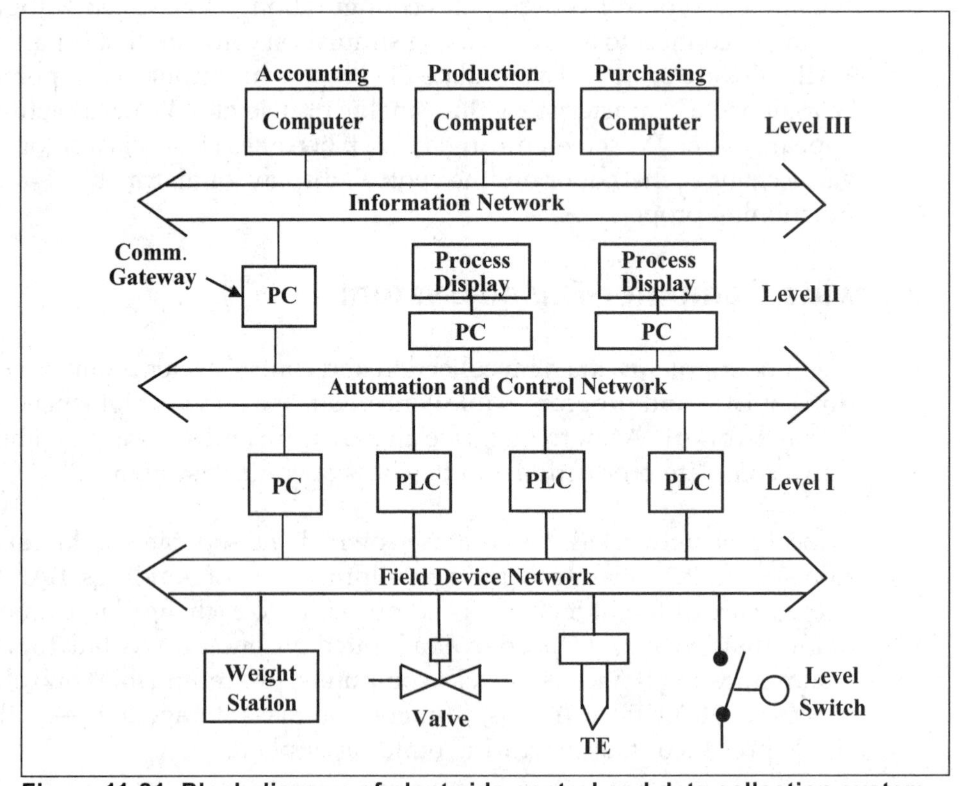

Figure 11-21. Block diagram of plantwide control and data collection system

Personal Computer-Based Control Systems

The use of personal computers (PCs) to directly replace PLCs in control applications is becoming more widespread. One disadvantage of programmable controllers is that they use proprietary hardware and software, which require additional engineering to adapt to plant or technological changes.

After decades of exponential growth in capabilities, PCs now come in rugged packages, with ever-faster processors, larger memory, and a multitude of Windows-based software products to support process control. All these improvements in the performance of PCs have come with lower cost due to higher competition and high-volume manufacturing.

There are two types of PC-based controllers used to replace proprietary PLCs in control applications. The first is a system that consists of an industrial grade PC running under Windows operating software and connected to standard PLC input and output modules. A typical example is the Allen-Bradley SoftLogix 5 Controller. This is a software package that runs on an industrial grade PC using Microsoft Windows operating engine.

Using a SoftLogix 5 controller, the user can integrate programming, control, and HMI functions on a single PC. The Windows NT/Windows 2000 soft control engine provides the open architecture and power of PCs, combined with the functionality of the Allen-Bradley PLC-5 processor. The system can monitor and control across ControlNet, DeviceNet, Universal Remote I/O, and some third-party networks. The SoftLogix 5 system is programmed using Allen-Bradley software running on a PC.

The second type is a software system called SoftPLC® that converts an industrial PC (x86 based) into a full-function PLC-like process controller. The SoftPLC system combines the standard PID; discrete and analog control functions typical of PLCs with the data handling, computing and network capabilities of PCs.

The SoftPLC system provides a complete instruction set and almost unlimited user program and data table memory space. SoftPLC is normally configured as an embedded system and runs on PC hardware much like a PLC but with no hard drive, monitor, keyboard, or mouse attached to the system. It runs as an embedded 32-bit real-time multitasking kernel on 386, 486, Pentium, and other x86-compatible processor platforms. Since a SoftPLC controller does not have a Windows operating system installed, there are no concerns about the system becoming non-deterministic or failing suddenly.

SoftPLC's ladder logic instruction set is a superset of the Allen-Bradley PLC-5 ladder logic set. The standard ladder instructions include contacts, coils and branching; timers and counters; data comparisons and moves; math and logical operation; shift registers and sequencers; jumps and subroutines, and special functions (such as PID control, messaging, and diagnostics).

The ladder logic program and data table application can be very large. The program size is only limited by the amount of RAM in the processor. The ladder logic programs can contain up to 998 program files that are organized as a main program with ladder subroutine areas. The data table can be configured with up to 1000 files of 1000 elements each.

The main advantage of the SoftPLC system is that it is an open architecture, non-proprietary system, so the user can easily interface to other vendor control devices and can easily connect to PC networks and other industrial networks.

EXERCISES

11.1 List some of the disadvantages of the early centralized control computers.

11.2 Discuss the purpose and function of the four levels in a typical distributed control system.

11.3 What are the functions and features of a typical DCS?

11.4 Explain the operation and purpose of the processor in a typical programmable controller system.

11.5 What is the main purpose of the input and output system in a PLC system?

11.6 List some discrete input devices typically found in process industries.

11.7 List some typical discrete output devices encountered in industrial applications.

11.8 List some typical analog signal values found in process applications.

11.9 Explain the difference between volatile and nonvolatile memory.

11.10 List some common applications for personal computers in programmable controller systems.

11.11 Which device is most commonly used to program PLCs?

11.12 Discuss the three basic instructions used in ladder logic programs.

11.13 What is power flow in a relay ladder diagram?

11.14 Explain the concept of logical power flow in a ladder diagram program.

11.15 What are the two basic structures used in IL instruction statements?

Appendix A

Standard Graphics Symbols for Process Control and Instrumentation

ISA-5.1-1984 (R1992), Instrument Symbols and Identification, is generally used in the instrumentation and control field to document projects.

This standard provides sufficient information to allow a person with limited knowledge of process control to understand the means of measurement and control of a process when reviewing process documents and drawings. The information provided in this appendix is very limited and intended only to introduce the subject of graphics symbols. It is recommended that the standard be obtained from ISA.

One important function the standard performs is to provide a standard list of line symbols to be used in documentation, as shown in Table A-1.

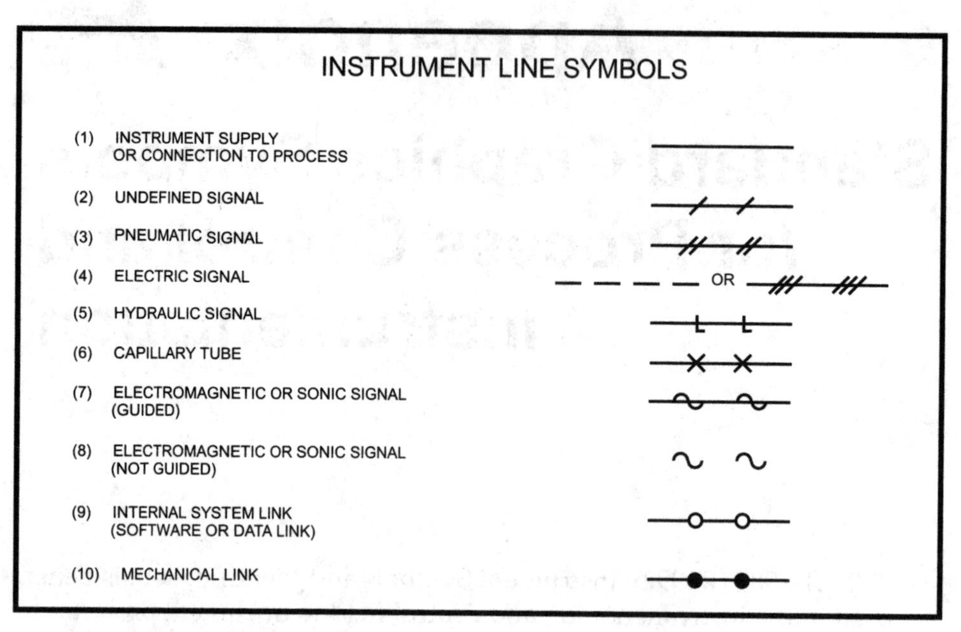

*The following abbreviations can be used to denote the types of supply: AS—Air Supply, IA—Instrument Air, PA—Plant Air, ES—Electric Supply, GS—Gas Supply, HS—Hydraulic Supply, NS—Nitrogen Supply, SS—Steam Supply, WS—Water Supply.

Table A-1. Instrument Line Symbols (1)

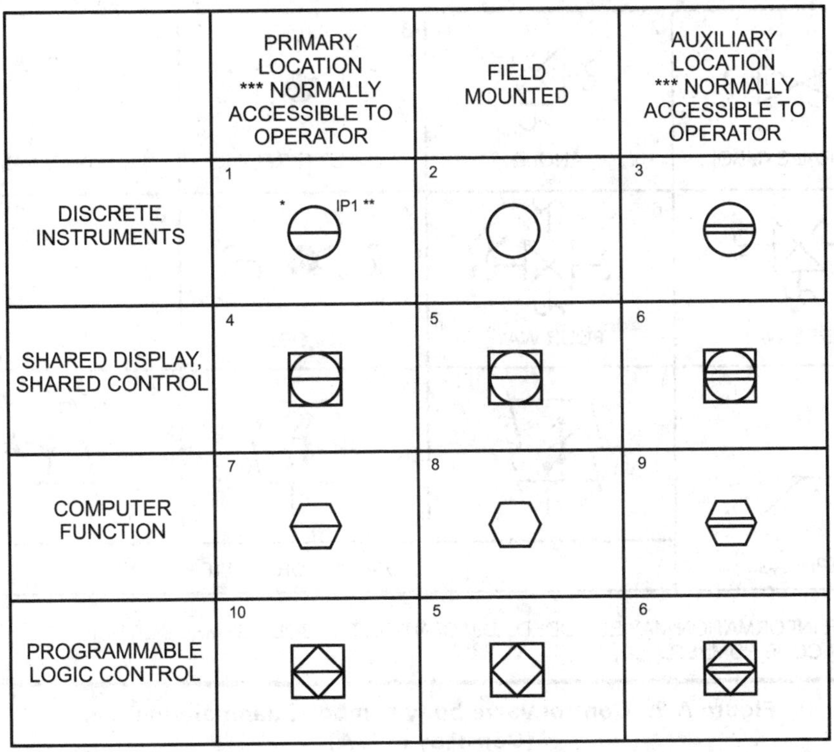

	PRIMARY LOCATION *** NORMALLY ACCESSIBLE TO OPERATOR	FIELD MOUNTED	AUXILIARY LOCATION *** NORMALLY ACCESSIBLE TO OPERATOR
DISCRETE INSTRUMENTS	1 * IP1 **	2	3
SHARED DISPLAY, SHARED CONTROL	4	5	6
COMPUTER FUNCTION	7	8	9
PROGRAMMABLE LOGIC CONTROL	10	5	6

* SYMBOL SIZE MAY VARY ACCORDING TO THE USER'S NEEDS AND THE TYPE OF DOCUMENT. A SUGGESTED SQUARE AND CIRCLE SIZE FOR LARGE DIAGRAMS IS SHOWN ABOVE. CONSISTENCY IS RECOMMENDED

** ABBREVIATIONS OF THE USER'S CHOICE SUCH AS IP1 (INSTRUMENT PANEL #1), IC2 (INSTRUMENT CONSOLE #2), CC3 (COMPUTER CONSOLE #3), ETC., MAY BE USED WHEN IT IS NECESSARY TO SPECIFY INSTRUMENT OR FUNCTION LOCATION.

*** NORMALLY INACESSIBLE OR BEHIND-THE-PANEL DEVICES OR FUNCTIONS MAY BE DEPICTED BY USING THE SAME SYMBOLS BUT WITH DASHED HORIZONTAL BARS, I.E.

Figure A-1. General instrument or function symbols (Courtesy of ISA)

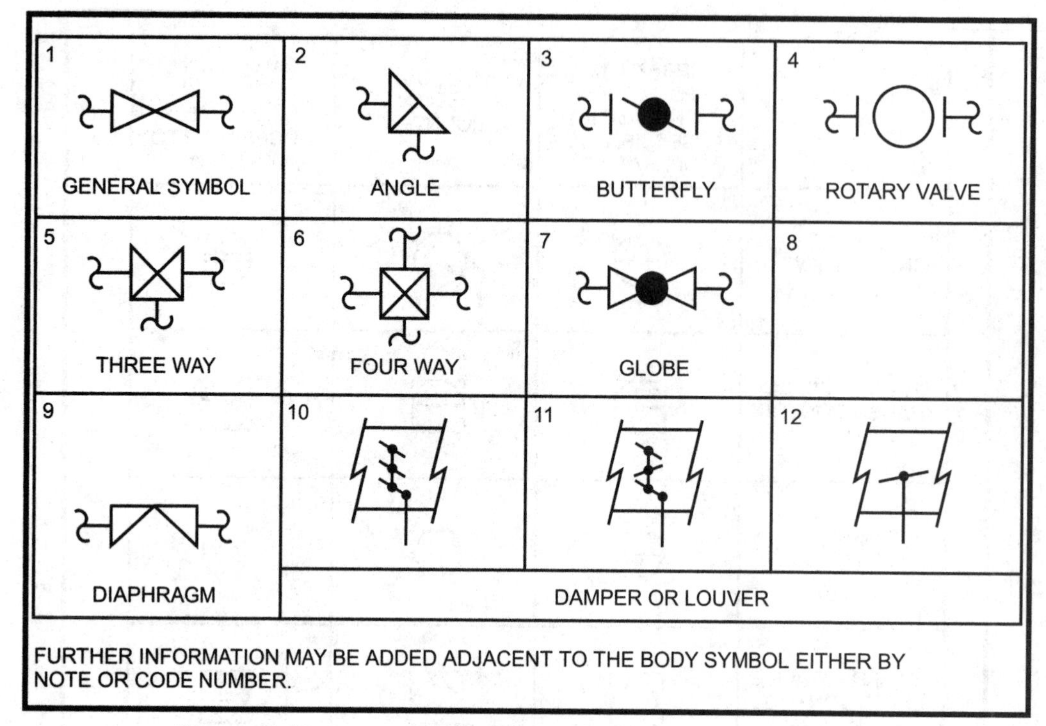

Figure A-2. Control valve body symbols, damper symbols (Courtesy of ISA)

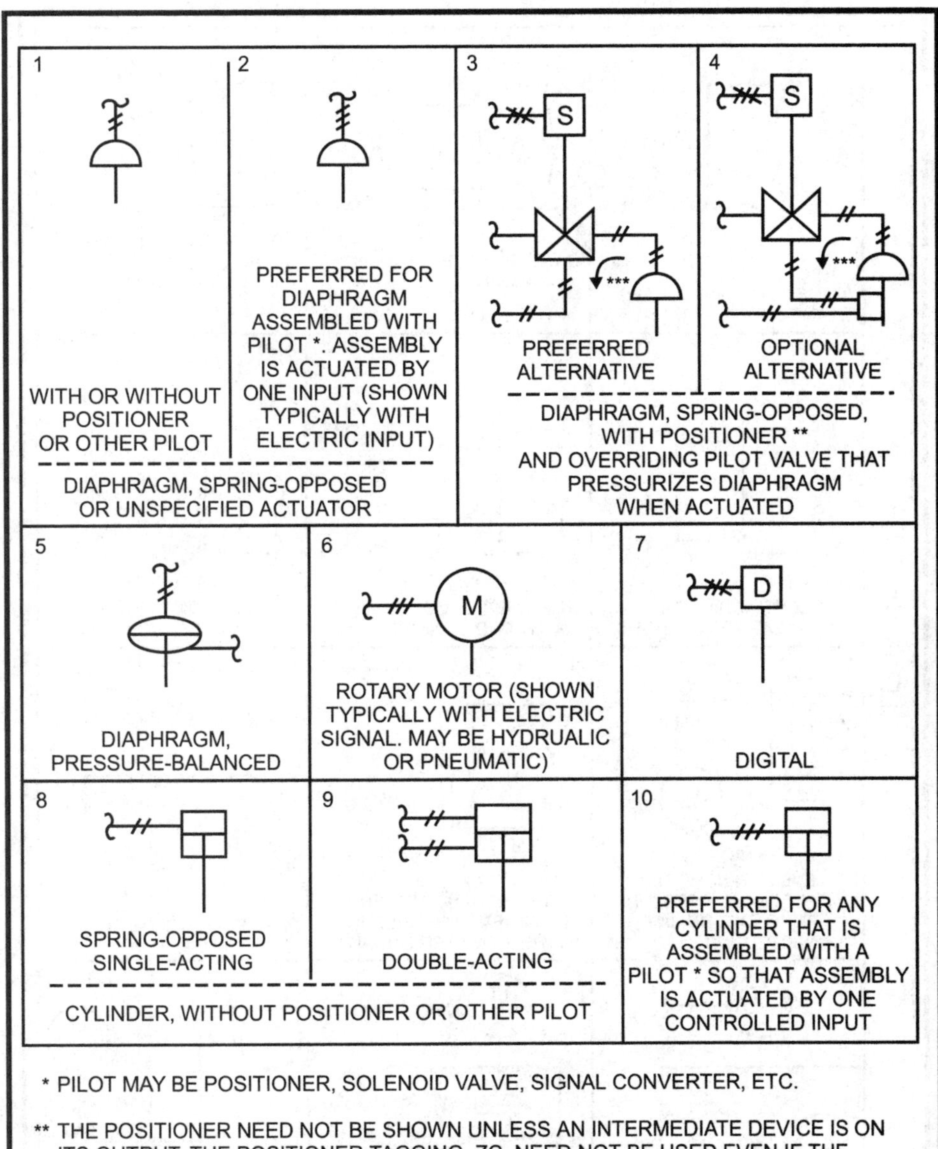

* PILOT MAY BE POSITIONER, SOLENOID VALVE, SIGNAL CONVERTER, ETC.

** THE POSITIONER NEED NOT BE SHOWN UNLESS AN INTERMEDIATE DEVICE IS ON ITS OUTPUT. THE POSITIONER TAGGING, ZC, NEED NOT BE USED EVEN IF THE POSITIONER IS SHOWN. THE POSITIONER SYMBOL, A BOX DRAWN ON THE ACTUATOR SHAFT, IS THE SAME FOR ALL TYPES OF ACTUATORS. WHEN THE SYMBOL IS USED, THE TYPE OF INSTRUMENT SIGNAL, I.E., PNEUMATIC, ELECTRIC, ETC., IS DRAWN AS APPROPRIATE. IF THE POSITIONER SYMBOL IS USED AND THERE IS NO INTERMEDIATE DEVICE ON ITS OUTPUT, THEN THE POSITIONER OUTPUT SIGNAL NEED NOT BE SHOWN.

*** THE ARROW REPRESENTS THE PATH FROM A COMMON TO A FAIL OPEN PART. IT DOES NOT CORRESPOND NECESSARILY TO THE DIRECTION OF FLUID FLOW.

Figure A-3. Actuator function symbols (Courtesy of ISA)

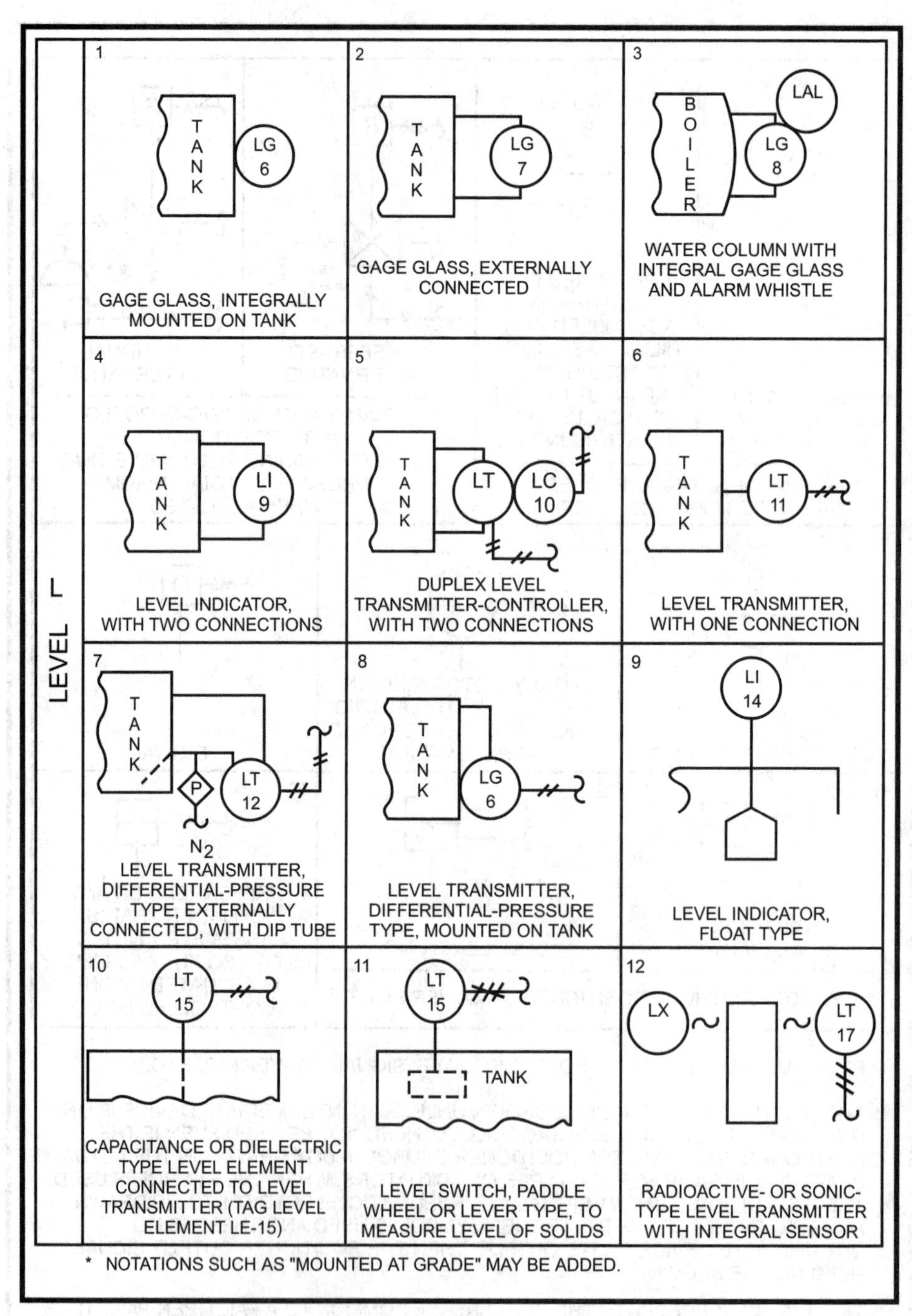

Figure A-4. Primary element symbols (Courtesy of ISA)

Table A-2. Typical tag numbers

TYPICAL TAG NUMBER	
TIC 103	- Instrument Identification or Tag Number
T 103	- Loop Identification
103	- Loop Number
TIC	- Functional Identification
T	- First-letter
IC	- Succeeding-Letters
EXPANDED TAG NUMBER	
10-PAH-5A	- Tag Number
10	- Optional Prefix
A	- Optional Suffix
Note: Hyphens are optional as separators	

Another important function performed by the standard is to identify each instrument by an alphanumeric code or tag number.

The functional identification of an instrument consists of letters from Table A-3 and includes one first letter designating the measured or initiating variable and one or more succeeding letters denoting the functions performed.

Instrument tag numbering is according to the functions performed and not according to the construction. For example, a differential pressure element measuring flow is identified by FE and not by DPE.

The following notes for Table A-3 were taken directly from ISA-5.1-1984 (R1992), paragraph 5.1; used with permission.

Notes for Table A-3

1. A "user's choice" letter is intended to cover unlisted meanings that will be used repetitively in a particular project. If used, the letter may have one meaning as a first-letter and another meaning as a succeeding-letter. The meanings need to be defined only once in a legend, or other place, for that project. For example, the letter *N* may be defined as "modulus of elasticity" as a first-letter and "oscilloscope" as a succeeding-letter.

2. The unclassified letter *X* is intended to cover unlisted meanings that will be used only once or used to a limited extent. If used, the letter may have any number of meanings as a first-letter and any number of meanings as a succeeding-letter. Except for its use with distinctive symbols, it is expected that the meanings will be

defined outside a tagging bubble on a flow diagram. For example, *XR-2* may be a stress recorder and *XX-4* may be a stress oscilloscope.

3. The grammatical form of the succeeding-letter meanings may be modified as required. For example, "indicate" may be applied as "indicator" or "indicating," "transmit" as "transmitter" or "transmitting," etc.

4. Any first-letter, if used in combination with modifying letters *D* (differential), *F* (ratio), *M* (momentary), *K* (time rate of change), *Q* (integrate or totalize), or any combination of these is intended to represent a new and separate measured variable, and the combination is treated as a first-letter entity. Thus, instruments *TDI* and *TI* indicate two different variables, namely, differential-temperature and temperature. Modifying letters are used when applicable.

5. First-letter *A* (analysis) covers all analyses not described by a "user's choice" letter. It is expected that the type of analysis will be defined outside a tagging bubble.

6. Use of first-letter *U* for "multivariable" in lieu of a combination of first-letters is optional. It is recommended that nonspecific variable designators such as *U* be used sparingly.

7. The use of modifying terms "high," "low," "middle" or "intermediate," and "scan" is optional.

8. The term "safety" applies to emergency protective primary elements and emergency protective final control elements only. Thus, a self-actuated valve that prevents operation of a fluid system at a higher-than-desired pressure by bleeding fluid from the system is a back-pressure-type *PCV*, even if the valve is not intended to be used normally. However, this valve is designated as a *PSV* if it is intended to protect against emergency conditions, *i.e.*, conditions that are hazardous to personnel and/or equipment and that are not expected to arise normally.

   The designation *PSV* applies to all valves intended to protect against emergency pressure conditions regardless of whether the valve construction and mode of operation place them in the category of the safety valve, relief valve, or safety relief valve. A rupture disc is designated *PSE*.

9. The passive function *G* applies to instruments or devices that provide an uncalibrated view, such as sight glasses and television monitors.

10. "Indicate" normally applies to the readout—analog or digital—of an actual measurement. In the case of a manual loader, it may be used for the dial or setting indication, *i.e.*, for the value of the initiating variable.

11. A pilot light that is part of an instrument loop should be designated by a first-letter followed by the succeeding-letter *L*. For example, a pilot light that indicates an expired time period should be tagged *KQL*. If it is desired to tag a pilot light that is not part of an instrument loop, the light is designated in the same way. For example, a running light for an electric motor may be tagged *EL*, assuming voltage to be the appropriate measured variable, or *YL*, assuming the operating status is being monitored. The unclassified variable X should be used only for applications which are limited in extent. The designation *XL* should not be used for motor running lights, as these are commonly numerous. It is permissible to use the user's choice letters *M, N* or *O* for a motor running light when the meaning is previously defined. If *M* is used, it must be clear that the letter does not stand for the word "motor," but for a monitored state.

12. Use of a succeeding-letter *U* for "multifunction" instead of a combination of other functional letters is optional. This nonspecific function designator should be used sparingly.

13. A device that connects, disconnects, or transfers one or more circuits may be either a switch, a relay, an ON-OFF controller, or a control valve, depending on the application.

    If the device manipulates a fluid process stream and is not a hand-actuated ON-OFF block valve, it is designated as a control valve. It is incorrect to use the succeeding-letters *CV* for anything other than a self-actuated control valve. For all applications other than fluid process streams, the device is designated as follows:

    A switch, if it is actuated by hand.

    A switch or an ON-OFF controller, if it is automatic and is the first such device in a loop. The term "switch" is generally used if the device is used for alarm, pilot light, selection, interlock, or safety.

    The term "controller" is generally used if the device is used for normal operating control.

    A relay, if it is automatic and is not the first such device in a loop, *i.e.*, it is actuated by a switch or an ON-OFF controller.

14. It is expected that the functions associated with the use of succeeding-letter *Y* will be defined outside a bubble on a diagram when further definition is considered necessary. This definition need not be made when the function is self-evident, as for a solenoid valve in a fluid signal line.

15. The modifying terms "high," and "low," and "middle" or "intermediate" correspond to values of the measured variable, not to values of the signal, unless otherwise noted. For example, a high-level alarm derived from a reverse-acting level transmitter signal should be an *LAH*, even though the alarm is actuated when the signal falls to a low value. The terms may be used in combinations as appropriate. (*See* Section 6.9A.)

16. The terms "high" and "low," when applied to positions of valves and other open-close devices, are defined as follows: "high" denotes that the valve is in or approaching the fully open position, and "low" denotes that it is in or approaching the fully closed position.

17. The word "record" applies to any form of permanent storage of information that permits retrieval by any means.

18. For use of the term "transmitter" versus "converter," see the definitions in Section 3.

19. First-letter *V*, "vibration or mechanical analysis," is intended to perform the duties in machinery monitoring that the letter *A* performs in more general analyses. Except for vibration, it is expected that the variable of interest will be defined outside the tagging bubble.

20. First-letter *Y* is intended for use when control or monitoring responses are event-driven as opposed to time- or time schedule-driven. The letter *Y*, in this position, can also signify presence or state.

21. Modifying-letter *K*, in combination with a first-letter such as *L*, *T*, or *W*, signifies a time rate of change of the measured or initiating variable. The variable *WKIC*, for instance, may represent a rate-of-weight-loss controller.

22. Succeeding-letter *K* is a user's option for designating a control station, while the succeeding-letter *C* is used for describing automatic or manual controllers. (*See* Section 3, Definitions.)

Table A-3. Instrument Identification Letters (1)

	FIRST-LETTER (4)		**SUCCEEDING-LETTERS** (3)		
	MEASURED OR INITIATING VARIABLE	**MODIFIER**	**READOUT OR PASSIVE FUNCTION**	**OUTPUT FUNCTION**	**MODIFIER**
A	Analysis (5,19)		Alarm		
B	Burner, Combustion		User's Choice (1)	User's Choice (1)	User's Choice(1)
C	User's Choice (1)			Control (13)	
D	User's Choice (1)	Differential (4)			
E	Voltage		Sensor (Primary Element)		
F	Flow Rate	Ratio (Fraction) (4)			
G	User's Choice (1)		Glass, Viewing Device (9)		
H	Hand				High (7, 15, 16)
I	Current (Electrical)		Indicate (10)		
J	Power	Scan (7)			
K	Time, Time Schedule	Time Rate of Change 4, 21)		Control Station (22)	
L	Level		Light (11)		Low (7, 15, 16)
M	User's Choice (1)	Momentary(4)			Middle, Intermediate(7,15)
N	User's Choice (1)		User's Choice (1)	User's Choice (1)	User's Choice (1)
O	User's Choice(1)		Orifice, Restriction		
P	Pressure, Vacuum		Point (Test) Connection		
Q	Quantity	Integrate, Totalize(4)			
R	Radiation		Record (17)		
S	Speed, Frequency	Safety (8)		Switch (13)	
T	Temperature			Transmit (18)	
U	Multivariable(6)		Multifunction(12)	Multifunction (12)	Multifunction (12)
V	Vibration, Mechanical Analysis(19)			Valve, Damper, Louver (13)	
W	Weight, Force		Well		
X	Unclassified (2)	X Axis	Unclassified (2)	Unclassified (2)	Unclassified (2)
Y	Event, State or Presence (20)	Y Axis		Relay, Compute, Convert (13, 14, 18)	
Z	Position, Dimension	Z Axis		Driver, Actuator, Unclassified Final Control Element	

NOTE: Numbers in parentheses refer to specific explanatory notes in Section 5.1.

BIBLIOGRAPHY

1. ISA-5.1-1984 (R1992), *Instrument Symbols and Identification*, ISA, Research Triangle Park, NC, 1984.

Appendix B Thermocouple Tables

The thermocouple tables presented here give the output voltage in millivolts (mV) over a range of temperatures in degrees centigrade (°C) for several different types of TCs. The first material listed is the positive terminal when the temperature measured is higher than the reference temperature.

Table B-1. Type J: Iron-Constantan

°C	0	5	10	15	20	25	30	35	40	45
–150	–6.50	–6.66	–6.82	–6.97	–7.12	–7.27	–7.40	–7.54	–7.66	–7.78
–100	–4.63	–4.83	–5.03	–5.23	–5.42	–5.61	–5.80	–5.98	–6.16	–6.33
–50	–2.43	–2.66	–2.89	–3.12	–3.34	–3.56	–3.78	–4.00	–4.21	–4.42
–0	0.00	–0.25	–0.50	–0.75	–1.00	–1.24	–1.48	–1.72	–1.96	–2.20
+0	0.00	0.25	0.50	0.76	1.02	1.28	1.54	1.80	2.06	2.32
50	2.58	2.85	3.11	3.38	3.65	3.92	4.19	4.46	4.73	5.00
100	5.27	5.54	5.81	6.08	6.36	6.63	6.90	7.18	7.45	7.73
150	8.00	8.28	8.56	8.84	9.11	9.39	9.67	9.95	10.22	10.50
200	10.78	11.06	11.34	11.62	11.89	12.17	12.45	12.73	13.01	13.28
250	13.56	13.84	14.12	14.39	14.67	14.94	15.22	15.50	15.77	16.05
300	16.33	16.60	16.88	17.15	17.43	17.71	17.98	18.26	18.54	18.81
350	19.09	19.37	19.64	19.92	20.20	20.47	20.75	21.02	21.30	21.57
400	21.85	22.13	22.40	22.68	22.95	23.23	23.50	23.78	24.06	24.33
450	24.61	24.88	25.16	25.44	25.72	25.99	26.27	26.55	26.83	27.11
500	27.39	27.67	27.95	28.23	28.52	28.80	29.08	29.37	29.65	29.94
550	30.22	30.51	30.80	31.08	31.37	31.66	31.95	32.24	32.53	32.82
600	33.11	33.41	33.70	33.99	34.29	34.58	34.88	35.18	35.48	35.78
650	36.08	36.38	36.69	36.99	37.30	37.60	37.91	38.22	38.53	38.84
700	39.15	39.47	39.78	40.10	40.41	40.73	41.05	41.36	41.68	42.00

Source: Ref 1

Table B-2. Type T: Copper-Constantan

°C	0	5	10	15	20	25	30	35	40	45
–150	–4.60	–4.71	–4.82	–4.92	–5.02	–5.11	–5.21	–5.29	–5.38	
–100	–3.35	–3.49	–3.62	–3.76	–3.89	–4.01	–4.14	–4.26	–4.38	–4.49
–50	–1.80	–1.97	–2.14	–2.30	–2.46	–2.61	–2.76	–2.91	–3.06	–3.21
–0	0.00	–0.19	–0.38	–0.57	–0.75	–0.93	–1.11	–1.29	–1.46	–1.64
+0	0.000	0.193	0.389	0.587	0.787	0.990	1.194	1.401	1.610	1.821
50	2.035	2.250	2.467	2.687	2.908	3.132	3.357	3.584	3.813	4.044
100	4.277	4.512	4.749	4.987	5.227	5.469	5.712	5.957	6.204	6.453
150	6.703	6.954	7.208	7.462	7.719	7.987	8.236	8.497	8.759	9.023
200	9.288	9.555	9.823	10.09	10.36	10.64	10.91	11.18	11.46	11.74
250	12.02	12.29	12.58	12.86	13.14	13.43	13.71	14.00	14.29	14.57
300	14.86	15.16	15.45	15.74	16.04	16.33	16.63	16.93	17.22	17.52
395	17.82	18.12	18.43	18.73	19.03	19.34	19.64	19.95	20.26	20.57

Table B-3. Type S: Platinum-Platinum/10% Rhodium

°C	0	5	10	15	20	25	30	35	40	45
+0	0.000	0.028	0.056	0.084	0.113	0.143	0.173	0.204	0.235	0.266
50	0.299	0.331	0.364	0.397	0.431	0.466	0.500	0.535	0.571	0.607
100	0.643	0.680	0.717	0.754	0.792	0.830	0.869	0.907	0.946	0.986
150	1.025	1.065	1.166	1.146	1.187	1.228	1.269	1.311	1.352	1.394
200	1.436	1.479	1.521	1.564	1.607	1.650	1.693	1.736	1.780	1.824
250	1.868	1.912	1.956	2.001	2.045	2.090	2.135	2.180	2.225	2.271
300	2.316	2.362	2.408	2.453	2.499	2.546	2.592	2.638	2.685	2.731
350	2.778	2.825	2.872	2.919	2.966	3.014	3.061	3.108	3.156	3.203
400	3.251	3.299	3.347	3.394	3.442	3.490	3.539	3.587	3.635	3.683
450	3.732	3.780	3.829	3.878	3.926	3.975	4.024	4.073	4.122	4.171
500	4.221	4.270	4.319	4.369	4.419	4.468	4.518	4.568	4.618	4.668
550	4.718	4.768	4.818	4.869	4.919	4.970	5.020	5.071	5.122	5.173
600	5.224	5.275	5.326	5.377	5.429	5.480	5.532	5.583	5.635	5.686
650	5.738	5.790	5.842	5.894	5.946	5.998	6.050	6.102	6.155	6.207
700	6.260	6.312	6.365	6.418	6.471	6.524	6.577	6.630	6.683	6.737
750	6.790	6.844	6.897	6.951	7.005	7.058	7.112	7.166	7.220	7.275
800	7.329	7.383	7.438	7.492	7.547	7.602	7.7656	7.711	7.766	7.821
850	7.876	7.932	7.987	8.042	8.098	8.153	8.209	8.265	8.320	8.376
900	8.432	8.488	8.545	8.601	8.657	8.714	8.770	8.827	8.883	8.940
950	8.997	9.054	9.111	9.168	9.225	9.282	9.340	9.397	9.455	9.512
1000	9.570	9.628	9.686	9.744	9.802	9.860	9.918	9.976	10.04	10.09
1050	10.15	10.21	10.27	10.33	10.39	10.45	10.57	10.56	10.62	10.68
1100	10.74	10.80	10.86	10.92	10.98	11.04	11.10	11.16	11.22	11.28
1150	11.34	11.40	11.46	11.52	11.58	11.64	11.70	11.76	11.82	11.88
1200	11.94	12.00	12.06	12.12	12.18	12.24	12.30	12.36	12.42	12.48
1250	12.54	12.60	12.66	12.72	12.78	12.84	12.90	12.96	13.02	13.08
1300	13.14	13.20	13.26	13.32	13.38	13.44	13.50	13.59	13.62	13.68
1350	13.74	13.80	13.86	13.92	13.98	14.04	14.10	14.16	14.22	14.28
1400	14.34	14.40	14.46	14.52	14.58	14.64	14.70	14.76	14.82	14.86
1450	14.94	15.00	15.05	15.11	15.17	15.23	15.92	15.35	15.41	15.47
1500	15.53	15.59	15.65	15.71	15.77	15.83	15.89	15.95	16.01	16.07
1550	16.12	16.18	16.24	16.30	16.36	16.42	16.48	16.54	16.60	16.66
1600	16.72	16.78	16.83	16.89	16.95	17.10	17.07	17.13	17.19	17.25
1650	17.31	17.36	17.42	17.48	17.54	17.60	17.66	17.72	17.77	17.83
1700	17.89	17.95	18.01	18.07	18.12	18.18	18.24	18.30	18.36	18.42

Source: Ref 1

Table B-4. Type K

°C	0	5	10	15	20	25	30	35	40	45
–150	–4.81	–4.92	–5.03	–5.14	–5.24	–5.34	–5.43	–5.52	–5.60	–5.68
–100	–3.49	–3.64	–3.78	–3.92	–4.06	–4.19	–4.32	–4.45	–4.58	–4.70
–50	–1.89	–2.03	–2.20	–2.37	–2.54	–2.71	–2.87	–3.03	–3.19	–3.34
–0	0.00	–0.19	–0.39	–0.58	–0.77	–0.95	–1.14	–1.32	–1.50	–1.68
+0	0.00	0.20	0.40	0.60	0.80	1.00	1.20	1.40	1.61	1.80
50	2.02	2.23	2.43	2.64	2.85	3.05	3.26	3.47	3.68	3.89
100	4.10	4.31	4.51	4.72	4.92	5.13	5.33	5.53	5.73	5.93
150	6.13	6.33	6.53	6.73	6.93	7.13	7.33	7.53	7.73	7.93
200	8.13	8.33	8.54	8.74	8.94	9.14	9.34	9.54	9.75	9.95
250	10.16	10.36	10.57	10.77	10.98	11.18	11.39	11.59	11.80	12.01
300	12.21	12.42	12.63	12.83	13.04	13.25	13.46	13.67	13.88	14.09
350	14.29	14.50	14.71	14.92	15.13	15.34	15.55	15.76	15.98	16.19
400	16.40	16.61	16.82	17.03	17.24	17.46	17.67	17.88	18.09	18.30
450	18.51	18.73	18.94	19.15	19.36	19.58	19.79	20.01	20.22	20.43
500	20.65	20.86	21.07	21.28	21.50	21.71	21.92	22.14	22.35	22.56
550	22.78	22.99	23.20	23.42	23.36	23.84	24.06	24.24	24.49	24.70
600	24.91	25.21	25.34	25.55	25.76	25.98	26.19	26.40	26.61	26.82
650	27.03	27.24	27.45	27.66	27.87	28.08	28.29	28.50	28.72	28.93
700	29.14	29.35	29.56	29.77	29.97	30.18	30.39	30.60	30.81	31.04
750	31.23	31.44	31.65	31.85	32.06	32.27	32.48	32.68	32.89	33.09
800	33.30	33.50	33.71	33.91	34.12	34.32	34.53	34.73	34.93	35.14
850	35.34	35.54	35.75	35.95	36.15	36.35	36.55	36.76	36.96	37.16
900	37.36	37.56	37.76	37.97	38.16	38.36	38.56	38.76	38.95	39.15
950	39.35	39.55	39.75	39.94	40.14	40.34	40.53	40.73	40.92	41.12
1000	41.34	41.51	41.70	41.90	42.09	42.29	42.48	42.67	42.87	43.06
1050	43.25	43.44	43.63	43.83	44.02	44.29	44.48	42.67	42.87	44.97
1100	45.16	45.35	45.54	45.73	45.92	46.11	46.29	46.48	46.48	46.85

Source: Ref 1

BIBLIOGRAPHY

1. *Thermocouple Reference Tables*. 1979. NBS Monograph 125, Washington, DC: National Bureau of Standards.

Appendix C Answers to Exercises

Chapter 1

1.1 The three main factors or terms are manipulated variables, disturbances, and controlled variables. Manipulated variables include valve position, motor speed, damper position, and blade pitch. Examples of controlled variables are temperature, level, position, pressure, pH, density, moisture content, weight, and speed. Some examples of disturbances to a process include: variations in ambient temperature, changes in demand for product, or changes in supply of feed material.

1.2 The four essential elements of a process control system are: *process*, *measurement*, *evaluation*, and *control*.

1.3 The primary function performed by a process controller is evaluation.

1.4 a) PIC-200, Pressure Indicating Controller

b) FV-250, Flow Valve

c) LC-500, Level Controller

d) HS-100, Hand Switch.

1.5 The primary requirement of any process control system is that it is reasonably stable.

1.6 The system error is the difference between the value of the control variable set point and the value of the measured variable.

1.7 The four common design criteria used are settling time, maximum error, offset error, and error area.

1.8 a) $K_C = 0.67$ and b) PB = 500%.

1.9 The main reason for using integral action with proportional control is to automatically correct for "off-set" present in proportional only controllers.

1.10 Reset windup exists in control loops using proportional plus integral control because under certain conditions the controller will continue to integrate and attempt to change the output even outside the operating range of the controller.

1.11 The controller used on the heat exchanger shown in Figure 1-17 is a temperature indicating controller and it is located on the front of a main control panel.

1.12 Processes that are slow to respond to disturbances and have long response time benefit the most from the use of PID control.

Chapter 2

2.1 In the feedback control loop shown in Figure 2-1, a measurement is made of the variable to be manipulated. This measured process value (PV) is then compared with a set point (SP) to generate an error signal (e = PV-SP). If a difference or error exists between the actual value and the desired value of the process, the process controller takes the necessary corrective action to return the process to the desired value.

2.2 The term *lag* in process control means any relationship in which some result happens after some cause.

2.3 The sensor output one second after the input changes rapidly from 20°C to 22°C is 21.77°C.

2.4 $$[R \times C] = \left[\frac{\text{volts}}{\text{coulombs/seconds}}\right]\left[\frac{\text{coulombs}}{\text{volts}}\right] = \text{seconds}$$

2.5 $$[R \times C] = \left[\frac{°\text{C}}{\text{calories/seconds}}\right]\left[\frac{\text{calories}}{°\text{C}}\right] = \text{seconds}$$

2.6 The system time constant (τ) is 32 seconds.

2.7 Dead time is the time period during which a system does not respond to a change to the input of the system.

2.8 The dead time is one second.

2.9 Ratio Control is typically used on processes where two or more process streams are mixed continuously to maintain a steady composition in the resulting mixture.

2.10 To tune a Cascade control loop, first place the master controller into the manual mode and tune the inner controller to obtain stable control. Then place the master controller in the auto mode and tune the entire loop by adjusting the tuning parameters of the master controller only.

2.11 PI control is effective in most flow control applications because the integral action of the PI controller provides a dampening action on the noise inherent in most flow control processes.

2.12 $K_C = 0.36$ psi/ft, $T_i = 1$ min, and $T_d = 0.25$ min

2.13 $K_C = 0.08$ psi/°C, $t_i = 2$ min, and $t_d = 0.5$ min.

Chapter 3

3.1 The current flow is 15 amps.

3.2 The current is 4.8 amps.

3.3 The area is 9 cmil.

3.4 The resistance of 500 feet of copper wire having a cross-sectional area of 4110 cmil is 1.29Ω.

3.5 The resistance of 500 feet of silver wire with a diameter of 0.001 inch is 4900Ω.

3.6 The distance to the point where the wire is shorted to ground is 1238.2 ft.

3.7 The total current (I_t) in a series circuit with two 250 Ω resistors and an applied voltage is 24 VDC is 48 mA. Also, $V_1 = 12$ volts and $V_2 = 12$ volts.

3.8 $I_1 = 1$ A, $I_2 = 0.5$ A, and $I_t = 1.5$ A

3.9 $I_1 = 1$ mA, $I_2 = 5$ mA, $R_x = 400$ W, $V_1 = 2$ volts, and $V_2 = 10$ volts.

3.10 The three methods used to increase the magnetic effect of a coil of wire are: 1) increase the amount current by increasing the applied voltage; 2) increase the turns of wire in the coil; or 3) insert an iron core through the center of the coil.

3.11 One purpose of a transformer is to transfer power from the primary to the secondary. Another function is to isolate the primary circuit from the secondary load. A final purpose is to step up or step down the value of the input AC voltage to provide a different value of AC voltage to the output load.

3.12 A full-wave bridge rectifier circuit has a much smoother direct current output signal than a half-wave bridge circuit.

3.13 Electric relays operate as follows: when a changing electric current is applied, a strong magnetic field is produced and the resulting magnetic force moves the iron core that is connected to a set of contacts. These contacts in the relay are used to make or break electrical connections in control circuits.

3.14 The three common types of solenoid valves are direct acting, internal pilot-operated, and manual reset.

Chapter 4

4.1 $1010_2 = 10_{10}$.

4.2 $10011_2 = 19_{10}$.

4.3 $10101011_2 = 253_8$.

4.4 $101111101000_2 = 5750_8$.

4.5 $33_{10} = 41_8$.

4.6 $451_{10} = 703_8$.

4.7 $47_{16} = 71_{10}$.

4.8 $157_{16} = 343_{10}$.

4.9 $56_{10} = 38_{16}$.

4.10 The BCD code of the decimal number 37 is 0011 0111.

4.11 The BCD code of the decimal number 270 is 0010 0111 0000.

4.12 The ASCII code for *Level Low* is 4C 65 76 65 6C 20 4C 6F 77 in hexadecimal format.

4.13 To verify the logic identity $A + 1 = 1$, we let $B = 1$ in the two input OR truth table that follow

Inputs	Output
A B	Z
0 1	1
1 1	1

Since $B = Z = 1$ in the truth table, then $A + 1 = 1$

4.14 Let the Start function = A, the Stop function = B and the Run Request = C, so the logic equation is $(A + B)\bar{B} = C$.

4.15 Let the tank high level = A, the tank low level = B and the Pump Running = C, so the logic equation is $(\bar{A}C)(B + C) = C$.

4.16 The logic gate circuit for a standard Start/Stop circuit is shown in Figure 4-16e.

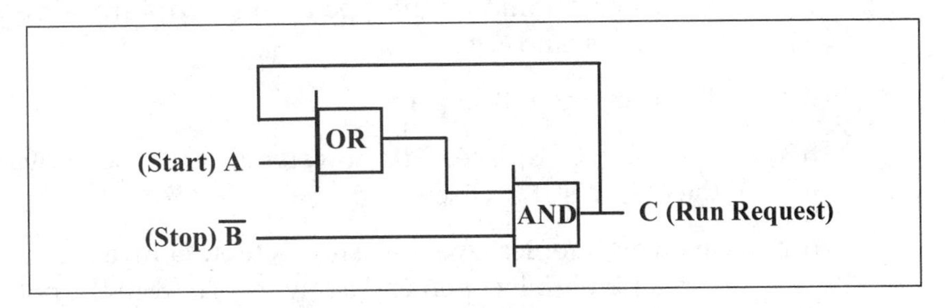

Figure 4-16e. Answer to exercise 4.16.

Chapter 5

5.1 The force F_2 is 250 lbs.

5.2 The new volume is 3.33 ft^3.

5.3 The new pressure is 15.5 psi.

5.4 $P_g = 35.5$ psi.

5.5 $P_a = 35.3$ psi.

5.6 The displacement is 2.048 inches.

5.7 The pressure detected is 2.13 psi.

5.8 The pressure is 1.95 psi.

5.9 The common sensing elements used in pressure gages are: Bourdon tubes, diaphragms, and bellows.

5.10 In the first method, the capacitance change is detected by measuring the magnitude of an AC voltage across the plates when excited. In the second method, the sensing capacitor forms part of an oscillator, and the electronic circuit changes the frequency to tune the oscillator.

5.11 (a) Error = (±0.02)(100 inH_2O) = ±2.0 inH_2O. So, the actual pressure reading is in the range of 48 to 52 inH_2O.

(b) Error = (±0.01)(0-100) inH_2O = ±1 inH_2O. Thus, the actual pressure range is 49-51 inH_2O.

(c) Error = (±0.005)(50 inH_2O) = ±0.25 inH_2O. Thus, the pressure is in the range of 49.75 to 50.25 inH_2O.

5.12 $\Delta R = 0.24\Omega$.

5.13 The instrument span is 11 to 66 inH_2O.

Chapter 6

6.1 The three common sight-type level sensors are: glass gauges, displacers, and tape.

6.2 The pressure is 3.43 psi.

6.3 If an object displaces 3 ft^3 of water at 20°C, the buoyancy force on the object is 186.9 lbs.

6.4 In an air bubbler level-measurement system, a dip tube is installed in a tank with its open end a few inches from the bottom. Air is forced through the bubbler tube; when the fluid bubbles just escape from the open end, the pressure in the tube equals the hydrostatic head of the liquid. As liquid level varies, the pressure in the dip tube changes correspondingly. A pressure-sensitive device measures the change in pressure.

6.5 The first disadvantage of bubbler systems is limited accuracy. Another disadvantage is that bubblers might introduce foreign matter into the process. Also, liquid purges can upset the material balance of the process, and gas purges can overload the vent system on vacuum processes. If the purge medium fails, not only is the level indication on the tank lost, but the system also is exposed to process material that can cause plugging, corrosion, freezing, or safety hazards.

6.6 In a capacitance probe, variations in fluid level cause changes in capacitance that can be measured with an electronic circuit in the level instrument. In a typical capacitance instrument, a metal probe that is installed in a process tank is insulated from the vessel and forms one plate of the capacitor; the metal vessel forms the other plate. The material between the two plates forms the dielectric of a capacitor. As the liquid level rises, vapors in the tank with a low dielectric constant are displaced by liquid with a high dielectric value. Capacitance changes are detected with an electronic instrument calibrated in units of level.

6.7 An ultrasonic level measurement system measures the time required for sound waves to travel through material. The velocity of a sound wave is a function of the type of wave being transmitted and the density of the medium in which it travels. When a sound wave strikes a solid medium such as a liquid surface, only a small amount of the sound energy penetrates the surface; a large percentage of the sound wave is reflected. The reflected sound wave is called an echo.

6.8 A nuclear level-detection system uses a low-level gamma-ray source or sources on one side of a process vessel and a radiation detector on the other side of the tank. The material in the tank has transmissibility different from that of air, so that the instrument determines the level of the material in the container based on the amount of gamma rays reaching the radiation detector.

6.9 The basic principle of a guided wave level probe is Time Domain Reflectometer (TDR) technology. TDR uses pulses of electromagnetic energy, which are transmitted down the probe tube. When a radar pulse reaches a liquid surface that has a higher dielectric constant than the air in the process tank, the pulse is reflected back to the electronic unit. A high-speed electronic timing circuit precisely measures the transit time and calculates an accurate measurement of the liquid level in the tank.

6.10 The common level switches used are: inductive, thermal, float, rotating paddle, and ultrasonic.

6.11 In a magnetic float-type level switch, a reed switch is positioned inside a sealed and nonmagnetic guide tube at a point where rising or falling liquid level should activate the switch. The float, which contains an annular magnet, rises or falls with liquid level and is guided by the tube.

6.12 The basic principle behind operation of a thermal level switch is the difference in thermal conductivity of air and liquid. This difference is sensed by the thermistor in the thermal level switch. The internal temperature of any device depends upon the heat dispersion of its surrounding environment. Since heat dispersion is greater in a liquid than in a gas or air, the resistance of a thermistor level switch will change sharply whenever the probe enters or leaves a liquid.

6.13 In the instrument, a low-power synchronous motor keeps the paddle in motion at a very low speed when no solids are present. Under this condition, there is very low torque on the motor drive. When the level in the storage tank or bin rises to the paddle switch, a higher torque is applied to the motor drive and the paddle stops. The level switch detects the increased torque and actuates a switch or set of switches in the instrument.

Chapter 7

7.1 $T = 271.4$°F and $T = 406.15$ Kelvin.

7.2 $T = 204.44$°C and $T = 477.59$ Kelvin.

7.3 $T = 39.2$°C.

7.4 The expansion in the rod is 0.0166 meters.

7.5 The normal operating ranges are: 1) Type J range: –190 to 760°C, 2) Type K range: –190 to 1260°C, and 3) Type S range: 0 to 1482°C.

7.6 The Seebeck voltage is 1.8 mV.

7.7 The temperature is 252°C.

7.8 The temperature is 357.4°C.

7.9 The isothermal block is a good heat conductor, and this holds measurement junctions at the same temperature. The absolute block temperature is unimportant because the two junctions act in opposition and cancel out the voltage produced.

7.10 The resistance ratio is 1.315.

7.11 The temperature is 56.75°C.

7.12 A typical application for a radiation pyrometer is the measurement of the temperature of molten metal or molten glass during a smelting and forming operation.

Chapter 8

8.1 The conductance of each solution is:

a) C = 20μ mhos; b) C = 5μ mhos; and c) C = 4μ mhos

8.2 $[H^+] = 10^{-3}$, pH = 3, and the solution is acidic.

8.3 The no-load voltage of the battery is 12.48 volts.

8.4 The span is 12.5 inches of pressure.

8.5 The maximum moisture content of the air is 150 grain/lb.

8.6 The frequency of an electromagnetic radiation source that has a wavelength of 100 meters is 3×10^6 Hz.

8.7 The photon energy is 6.63×10^{-25} joules and the number of photons is 1.5×10^{24} photons.

8.8 The intensity of a 1000-w point light source at 10 meters is 0.796 w/m^2 and at 20 meters is 0.199 w/m^2.

8.9 The maximum wavelength for a resistance change by photon absorption for a CdS semiconductor is $\lambda_{max} = 0.514 \mu m$.

8.10 A photovoltaic cell generates 0.3 volts open-circuit when exposed to 10 w/m^2 of radiation intensity. The open-circuit voltage of the cell at 20 w/m^2 is 0.389 volts.

8.11 In a photomultiplier tube, the cathode is maintained at a high negative voltage and is coated with a photoemissive material.

Numerous dynodes maintained at successively more positive voltages follow the cathode. The final electrode is the *anode,* which is grounded through a resistor R. When a light photon strikes the photoemissive cathode with sufficient energy, several electrons are ejected from the surface, and the voltage potential difference accelerates them to the first dynode. Each electron from the cathode that strikes the first dynode ejects several electrons. All of these electrons are accelerated to the second dynode where each one strikes the surface with sufficient energy to again eject several electrons. This process is repeated for each dynode until the electrons that reach the anode have greatly multiplied and they produce a large current flow through the output resistor. The voltage produced across the output resistor R is directly proportional to the light striking the cathode.

8.12 In the typical application shown in Figure 8-17, a turbidity value is developed from a test sample under controlled conditions. In this application, a laser beam is split and passed through two mediums to matched photodetectors. One medium is a carefully selected standard sample of fixed turbidity. The other medium is an in-line process liquid. If the in-line process liquid attenuates the laser beam more than the standard or reference sample, the electronic circuit triggers an alarm or takes some appropriate control action to reduce turbidity.

8.13 The most popular type of measurement cell used in oxygen analyzers is the electrochemical zirconium cell.

Chapter 9

9.1 The fluid velocity is 25.3 ft/sec.

9.2 The volumetric flow is 5.23 cfm and the mass flow is 325.3lb/min.

9.3 The fluid velocity is 10.34 ft/s and volumetric flow rate is 0.9 cfs.

9.4 The Reynolds number is 82,036 and the flow is turbulent.

9.5 The Reynolds number is 3667.15 and the flow is transitional.

9.6 The volumetric flow, if the pressure drop across the restriction increases to 5-inH_2O, is 10 cfs.

9.7 The fluid velocity is 6.6 cfs.

9.8 Eccentric and segmental orifices are preferable to concentric orifices for measurement of slurries or dirty liquids as well as for measurement of gas or vapor where liquids may be present, especially large slugs of liquid. Where the stream contains particulate matter, the segmental orifice, which provides an open path at the bottom of the pipe, may be preferable. However, when conditions permit, the eccentric orifice is preferred because of its greater ease of manufacture to precise tolerances and generally more accurate and repeatable performance.

9.9 Differential pressure.

9.10 The air velocity is 8.02 ft/s.

9.11 The meter coefficient is 2 pulses/gallon and the scaling factor is 20.

9.12 The fluid velocity is 17.6 ft/s.

Chapter 10

10.1 The rangeability is 16.

10.2 The diaphragm area required to fully open the control valve is $2x10^{-3}$ m^2.

10.3 The valve-sizing coefficient C_v is experimentally determined for each different size and style of valve using water in a test line under carefully controlled standard conditions. The standard test piping arrangement established by the Fluid Controls Institute (FCI) to provide uniform measurement of C_v data is shown in Figure 10-5. Using this test setup, control valve manufacturers determine and publish C_v valves for their valves.

10.4 The volumetric flow rate is 1060.2 cfm.

10.5 The valve-sizing coefficient is 56.57 and a 3 in. valve should be selected.

10.6 The valve will not cavitate and the valve-sizing coefficient is 253.

10.7 If the pressure at the outlet of the valve is still below the vapor pressure of the liquid, bubbles will remain in the downstream system and result in *flashing*. If the downstream pressure recovery is sufficient to raise the outlet pressure above the liquid vapor pressure, the bubbles will collapse, or implode, producing *cavitation*.

10.8 The torque constant is 0.143oz-in./V and damping constant is 4.8×0^{-3} oz-in./rad/s.

10.9 The torque (K_t) constant is 0.06 oz-in./V and damping constant (D_m) is 7.16 oz-in./rad/s.

10.10 The parallel capacitance required to make the winding appear purely resistive is 16.6 microfarads.

10.11 The total head developed is 19.03 ft.

10.12 Rotary pumps function by continuously producing reduced-pressure cavities on the suction side, which fills with fluid. The fluid is moved to the discharge side of the pump, where it is compressed and then discharged from the pump.

Chapter 11

11.1 Some of the disadvantages of early centralized control computers were as follows: 1) very slow operating speed, 2) very small computer memory size (4K to 8K), 3) all programming had to be done in machine language, and 4) lack of programming experience by control vendors and users.

11.2 The process data collected at level 1 is transferred to level 2. At this level, process operators and engineers use consoles with a keyboard, mouse, and video display to view and adjust the various processes being controlled and monitored by the system. At level 2, process and control engineers implement advanced control functions and strategies, and operations management members perform advanced data collection and analysis on process information. The various plant management systems—such as inventory management and control, billing and invoicing, and statistical quality control—exist at level 3. The highest level is used to provide extensive process and operations information to corporate management in large industrial plants.

11.3 The functions and features of a typical DCS are control and monitoring, operations interface, and advanced applications software.

11.4 The processor reads the inputs, executes logic as determined by the application program, performs calculations, and controls the outputs accordingly.

11.5 The main purpose of the I/O system is to provide the physical connection between the process equipment and the PLC system.

11.6 Some typical discrete input devices found in process industries are process switches for temperature, flow, and pressure.

11.7 Some typical discrete output signals encountered in industrial applications are control relays, solenoid valves, and motor starters.

11.8 Some typical analog signal values found in process control applications are 4-to-20 maDC current, 0-to-10 VDC, and 1-to-5 VDC.

11.9 The difference between volatile and nonvolatile memory is that volatile memory will lose its programmed contents if all operating power is lost or removed. Nonvolatile memory will retain its data and program even if there is a complete loss of operating power.

11.10 The common applications for personal computers in programmable controller systems are programming, graphical unit interfaces, and data collection.

11.11 The personal computer is the most common device used to program PLCs.

11.12 The three basic instructions used in ladder logic programs are the 1) normally open, 2) normally closed, and 3) coil output instructions.

11.13 Power flow in a relay ladder diagram is the flow of electric current from the left side to the right side of a ladder rung.

11.14 Logical power flow is obtained in a LD program if there is logical continuity from the left side to the right side of a ladder logic rung.

11.15 IL instruction statements have two basic structures. One, a statement made up of an instruction alone (for example, NOT), and another where the statement is made up of an instruction and an address.

Index